NEUROMETHODS

Series Editor
Wolfgang Walz
University of Saskatchewan
Saskatoon, SK, Canada

For further volumes:
http://www.springer.com/series/7657

Neuromethods publishes cutting-edge methods and protocols in all areas of neuroscience as well as translational neurological and mental research. Each volume in the series offers tested laboratory protocols, step-by-step methods for reproducible lab experiments and addresses methodological controversies and pitfalls in order to aid neuroscientists in experimentation. *Neuromethods* focuses on traditional and emerging topics with wide-ranging implications to brain function, such as electrophysiology, neuroimaging, behavioral analysis, genomics, neurodegeneration, translational research and clinical trials. *Neuromethods* provides investigators and trainees with highly useful compendiums of key strategies and approaches for successful research in animal and human brain function including translational "bench to bedside" approaches to mental and neurological diseases.

Single Cell 'Omics of Neuronal Cells

Edited by

Jonathan V. Sweedler

The Beckman Institute, University of Illinois Urbana-Champaign, Urbana, IL, USA

James Eberwine

Dept Pharmacology, 38 John Morgan Bldg, University of Pennsylvania, Philadelphia, PA, USA

Scott E. Fraser

Bridge Institute, University of Southern California, Los Angeles, CA, USA

Humana Press

Editors
Jonathan V. Sweedler
The Beckman Institute
University of Illinois Urbana-Champaign
Urbana, IL, USA

James Eberwine
Dept Pharmacology, 38 John Morgan Bldg
University of Pennsylvania
Philadelphia, PA, USA

Scott E. Fraser
Bridge Institute
University of Southern California
Los Angeles, CA, USA

ISSN 0893-2336 ISSN 1940-6045 (electronic)
Neuromethods
ISBN 978-1-0716-2524-8 ISBN 978-1-0716-2525-5 (eBook)
https://doi.org/10.1007/978-1-0716-2525-5

Cover Caption: Image courtesy of Dr. Scott Fraser.

This Humana imprint is published by the registered company Springer Science+Business Media, LLC, part of Springer Nature.
The registered company address is: 1 New York Plaza, New York, NY 10004, U.S.A.

Preface to the Series

Experimental life sciences have two basic foundations: concepts and tools. The Neuromethods series focuses on the tools and techniques unique to the investigation of the nervous system and excitable cells. It will not, however, shortchange the concept side of things as care has been taken to integrate these tools within the context of the concepts and questions under investigation. In this way, the series is unique in that it not only collects protocols but also includes theoretical background information and critiques which led to the methods and their development. Thus, it gives the reader a better understanding of the origin of the techniques and their potential future development. The Neuromethods publishing program strikes a balance between recent and exciting developments like those concerning new animal models of disease, imaging, *in vivo* methods, and more established techniques, including, for example, immunocytochemistry and electrophysiological technologies. New trainees in neurosciences still need a sound footing in these older methods in order to apply a critical approach to their results.

Under the guidance of its founders, Alan Boulton and Glen Baker, the *Neuromethods* series has been a success since its first volume published through Humana Press in 1985. The series continues to flourish through many changes over the years. It is now published under the umbrella of Springer Protocols. While methods involving brain research have changed a lot since the series started, the publishing environment and technology have changed even more radically. Neuromethods has the distinct layout and style of the Springer Protocols program, designed specifically for readability and ease of reference in a laboratory setting.

The careful application of methods is potentially the most important step in the process of scientific inquiry. In the past, new methodologies led the way in developing new disciplines in the biological and medical sciences. For example, physiology emerged out of anatomy in the nineteenth century by harnessing new methods based on the newly discovered phenomenon of electricity. Nowadays, the relationships between disciplines and methods are more complex. Methods are now widely shared between disciplines and research areas. New developments in electronic publishing make it possible for scientists that encounter new methods to quickly find sources of information electronically. The design of individual volumes and chapters in this series takes this new access technology into account. Springer Protocols makes it possible to download single protocols separately. In addition, Springer makes its print-on-demand technology available globally. A print copy can therefore be acquired quickly and for a competitive price anywhere in the world.

Saskatoon, SK, Canada ***Wolfgang Walz***

Preface

As editors of this book, it gives us great pleasure to ruminate about the rapid, ongoing development of the field of single cell omics. It is almost cliché to initiate a discussion of single cell analyses by recounting that Robert Hooke coined the word "cell" in his 1655 Royal Society monograph, *Micrographia*. It seemed an apt descriptor for the "minute bodies" that he observed in a thin slice of cork using his homemade microscope; this was followed by Leeuwenhoek's description of live single cells – "animicules." Together, they provided the foundation for the evolution of Modern Cell Theory over 150 years later. Both of these origin stories were enabled and fueled by technological advancement – the development of the microscope. Technology has always been a driver for scientific progress, and indeed, each step in the advancement of single cell analyses has been driven by new tools and technologies. For example, single cell transcriptomics was initiated in the early 1990s with the advent of RNA amplification methodologies, and then became more mainstream in the mid-2010s with the development of NextGen Sequencing and high-throughput single cell isolation techniques.

Time and time again, technological innovation has not only made scientific data collection faster, more sensitive, and more complete, it has also redefined the impossible. In the mid-twentieth century, the idea of sequencing an entire human genome would have been considered insane. Even in 1990, many viewed it as folly, yet a draft genome was made available in just a decade, and today, the cost and time involved in sequencing a genome have improved by orders of magnitude. More recently, with the success of the Human Genome Project and the development of the associated field of genomics (the collective quantification and characterization of genes), which was fueled by high throughput DNA sequencing, it became clear that applying a genome scale – or omics perspective – for other aspects of cell biology would provide similarly unique and novel insights into cell function.

Single cell omics often entails far more than sequencing a cell's genome; it can include the characterization of a cell's RNA (transcriptome), proteins (proteome), and metabolites (metabolome) as well as sub-areas such as the epigenome and lipidome, to name just two. *This change from analysis of single genes to multiple genes, single RNAs to multiple RNAs, single proteins to multiple proteins, etc., all in the same cell, is the revolution that single cell biology has spawned.* The single cell is the correct granularity for studies of regulatory biology, as it is the ligands impinging on a cell's receptors that carry information through secondary messengers to alter the metabolism or gene expression in the cell. This clear causal relationship would become confused if studied in a homogenate of many cells. Understanding the regulatory impact of multiple genes, proteins, or metabolites is made more tenable when studying an individual cell, where the chemical constituents and responses can be ascribed to that cell. Single cell analysis provides hope for understanding how multiple genes or proteins work in concert to orchestrate cell physiology and function.

It is with these factors in mind that we compiled this Springer Nature volume, with contributions offering insights into a diverse set of single cell analysis tools. Li and Eberwine illustrate how the single cell transcriptome can be quantified at high depth, Misal and Kelly show how cells can be isolated in nanoliter volumes and analyzed by nano-LC-MS/MS for single cell proteomics measurement, and Hansen et al. demonstrate how single cell proteins can be characterized by immunoblotting. Though there are other methods available to study these cellular chemicals, these particular approaches highlight a wide range of technologies, from high-yield chemical amplification to mass spectrometry and nanotechnology, for analysis of the chemical constituents of cells.

Most of the single cell datasets in hand come from a single analysis modality, such as probing the transcriptome or the morphology of a cell. While these datasets can be rich in information, for example, in helping to define cell type, the marriage of multiple analysis procedures on the same cell (multi-modal analyses) provides even more informative data. In addition to providing greater information about a cell, the conjoined data obtained by multi-modal, or multi-omic, analyses foster the analysis of correlations and interactions between the different data types. Collecting the combination of data types in multi-modal analyses can be challenging because of incompatibilities between collection or detection methodologies, for example, transcriptome detection may not be compatible with methods that quantify the proteins within that same cell. The harmonization of methods to deeply phenotype the diverse chemical inventories of a cell requires innovation and troubleshooting. In their chapter, Geurs et al. highlight how the genome and transcriptome can be assessed along with the epigenomic state of the genomic DNA, information they use to assess single nucleotide variants in single cells. Aaron Streets presents his effort to develop a computational framework for analyzing multi-omics datasets – an exceedingly difficult task for which his methodology shows significant promise. Portero and colleagues use the power of electrospray mass spectrometry coupled with capillary electrophoresis to assess proteins and metabolites in the same single cells. Antilla et al. combine capillary electrophoresis-mediated chemical cytometry with scRNA seq to assess both enzyme kinetics and the same cell's transcriptome. The chapter by Rubakhin, Romanova, and Sweedler describes how to measure the lipids, metabolites, and peptides in thousands of cells per experiment using mass spectrometry; intriguingly, the method leaves much of the cell behind, allowing follow-up characterization efforts.

By combining information on cellular physiology and function with constituent chemical quantification, we can provide more key opportunities for single cell analysis, as manipulation of cellular function and dysfunction is the ultimate goal for single cell biologists. To illustrate how such functional/chemical couplings can be accomplished, a few chapters were developed around particular technologies to assess single cell function. The saying "seeing is believing" is illustrated by Nguyen et al. as they highlight the power of combining various imaging modalities from nano secondary ion mass spectrometry (NanoSIMS) with single molecule fluorescence and other imaging methods. Neurons and other cells function via transfer of electrical signals between cells and Juhasz has shown how this critical cellular physiology can be quantitated in the same cell as its cognate transcriptome. While electrophysiology can be relatively easily assessed, Keyes et al. capture some of the other physiological parameters that are not so easily measured in single cells using biosensors.

The impetus for this book derives from our belief that the democratization of technology requires that it be disseminated broadly through publication, lectures, and courses. Technologies are adopted and utilized most readily when scientists learn their "ins and outs" and adapt them to their experimental needs. The three of us, along with some of the chapter authors, have taught and/or organized the Cold Spring Harbor Laboratory Course on Single Cell Analyses or the Single Cell Analysis Conference (Dr. David Stewart, Executive Director of Meetings and Courses) as well as the Neuroscience School of Advanced Studies (NSAS) Course on Single Cell Omics (Dr. Luigi Pulvirenti, Director of NSAS). These courses, meetings, and workshops present the single cell analysis toolkit to relatively small sets of participants, but with the advantages of associated benchwork and extensive didactic discussion. These courses are taught with the goals of presenting the details of the techniques and also showing how technological innovation is needed to derive the data which, in turn, will yield the knowledge that will push the field forward and foster the next technological advances.

The overarching goal of this book is to share lessons about the development and fruits of cutting-edge technologies by reaching a wider audience than any set of courses might. While no set of chapters can present an exhaustive set of every multi-modal approach for single cell analysis, this compilation highlights selected approaches that are at the forefront of combining different detection modalities to investigate single cells, enabling us to deepen our understanding of a cell's biology. As we move from analysis of single cells both to higher resolution subcellular omics analysis and to multi-cellular circuits analyzed at single cell omics levels, further technological innovation will be needed. We hope this volume, and the courses that inspired it, motivates the next generation of innovators to embrace these challenges.

We conclude by thanking the many members of the field, whose activities, presentations, and discussions helped to shape the work contained here. More importantly, we thank the numerous students whose questions, puzzlement, and problems shaped this work more than they will ever know. Finally, we owe a huge debt to Stephanie Baker of the University of Illinois; without her diligent efforts to coordinate chapter solicitation, "nag" authors (including us), and format their contributions, there would be no book.

Urbana, IL, USA — *Jonathan V. Sweedler*
Philadelphia, PA, USA — *James Eberwine*
Los Angeles, CA, USA — *Scott E. Fraser*

Contents

Contents

Contributors

NANCY L. ALLBRITTON • *University of Washington, Seattle, WA, USA*
MATTHEW M. ANTTILA • *University of North Carolina at Chapel Hill, Chapel Hill, NC, USA; University of Washington, Seattle, WA, USA*
SAM B. CHOI • *Department of Chemistry & Biochemistry, University of Maryland, MD, USA*
JAMES EBERWINE • *Department of Systems Pharmacology and Translational Therapeutics Perelman School of Medicine, University of Pennsylvania, Philadelphia, PA, USA*
ANDREW G. EWING • *Department of Chemistry and Molecular Biology, University of Gothenburg, Gothenburg, Sweden*
LUKE A. GALLION • *University of North Carolina at Chapel Hill, Chapel Hill, NC, USA; University of Washington, Seattle, WA, USA*
SARAH GEURS • *Department of Human Genetics, University of Leuven, KU Leuven, Leuven, Belgium*
LOUISE L. HANSEN • *Department of Bioengineering, University of California, Berkeley, Berkeley, CA, USA; The UC Berkeley-UCSF Graduate Program in Bioengineering, Berkeley, CA, USA*
AMY E. HERR • *Department of Bioengineering, University of California, Berkeley, Berkeley, CA, USA; The UC Berkeley-UCSF Graduate Program in Bioengineering, Berkeley, CA, USA; Chan Zuckerberg Biohub, San Francisco, CA, USA*
GÁBOR JUHÁSZ • *ELTE NAP Neuroimmunology Research Group, Department of Biochemistry, Institute of Biology, ELTE Eötvös Loránd University, Budapest, Hungary; Laboratory of Proteomics, Institute of Biology, ELTE Eötvös Loránd University, Budapest, Hungary*
KATALIN ADRIENNA KÉKESI • *Laboratory of Proteomics, Institute of Biology, ELTE Eötvös Loránd University, Budapest, Hungary; Department of Physiology and Neurobiology, Institute of Biology, ELTE Eötvös Loránd University, Budapest, Hungary*
RYAN T. KELLY • *Department of Chemistry and Biochemistry, Brigham Young University, Provo, UT, USA*
JEREMIAH KEYES • *Department of Biology, Behrend College, Pennsylvania State University, Erie, PA, USA*
THOMAS LEFEVRE • *Department of Human Genetics, University of Leuven, KU Leuven, Leuven, Belgium*
JIE LI • *Department of Chemistry & Biochemistry, University of Maryland, MD, USA*
JIFEN LI • *Department of Systems Pharmacology and Translational Therapeutics Perelman School of Medicine, University of Pennsylvania, Philadelphia, PA, USA*
GABRIELA LOMELI • *Department of Bioengineering, University of California, Berkeley, Berkeley, CA, USA; The UC Berkeley-UCSF Graduate Program in Bioengineering, Berkeley, CA, USA*
ALICIA A. LORK • *Department of Chemistry and Molecular Biology, University of Gothenburg, Gothenburg, Sweden*
ANNIE MASLA • *Department of Bioengineering, University of California – Berkeley, Berkeley, CA, USA; UC Berkeley – UCSF Graduate Program in Bioengineering, Berkeley, CA,*

USA; Center for Computational Biology, University of California, Berkeley, Berkeley, CA, USA

SOHUM MEHTA • *Department of Pharmacology, University of California, San Diego, La Jolla, CA, USA*

SANTOSH A. MISAL • *Department of Chemistry and Biochemistry, Brigham Young University, Provo, UT, USA*

DÁNIEL MITTLI • *Laboratory of Proteomics, Institute of Biology, ELTE Eötvös Loránd University, Budapest, Hungary; Department of Physiology and Neurobiology, Institute of Biology, ELTE Eötvös Loránd University, Budapest, Hungary*

PETER NEMES • *Department of Chemistry & Biochemistry, University of Maryland, MD, USA*

THO D. K. NGUYEN • *Department of Chemistry and Molecular Biology, University of Gothenburg, Gothenburg, Sweden*

LEENA R. PADE • *Department of Chemistry & Biochemistry, University of Maryland, College Park, MD, USA*

BRAE V. PETERSEN • *University of Washington, Seattle, WA, USA*

NHU T. N. PHAN • *Department of Chemistry and Molecular Biology, University of Gothenburg, Gothenburg, Sweden*

ERIKA P. PORTERO • *Department of Chemistry & Biochemistry, University of Maryland, MD, USA*

ELENA V. ROMANOVA • *Department of Chemistry, Beckman Institute for Advanced Science and Technology, University of Illinois at Urbana–Champaign, Urbana, IL, USA*

STANISLAV S. RUBAKHIN • *Department of Chemistry, Beckman Institute for Advanced Science and Technology, University of Illinois at Urbana–Champaign, Urbana, IL, USA*

ZOË STEIER • *Department of Bioengineering, University of California – Berkeley, Berkeley, CA, USA; UC Berkeley – UCSF Graduate Program in Bioengineering, Berkeley, CA, USA; Center for Computational Biology, University of California, Berkeley, Berkeley, CA, USA*

AARON STREETS • *Department of Bioengineering, University of California – Berkeley, Berkeley, CA, USA; UC Berkeley – UCSF Graduate Program in Bioengineering, Berkeley, CA, USA; Center for Computational Biology, University of California, Berkeley, Berkeley, CA, USA; Chan Zuckerberg Biohub, San Francisco, CA, USA*

JONATHAN V. SWEEDLER • *Department of Chemistry, Beckman Institute for Advanced Science and Technology, University of Illinois at Urbana–Champaign, Urbana, IL, USA*

VANDA TUKACS • *Laboratory of Proteomics, Institute of Biology, ELTE Eötvös Loránd University, Budapest, Hungary*

MANIBARATHI VAITHIYANATHAN • *University of Washington, Seattle, WA, USA*

JULEA VLASSAKIS • *Department of Bioengineering, University of California, Berkeley, Berkeley, CA, USA*

THIERRY VOET • *Department of Human Genetics, University of Leuven, KU Leuven, Leuven, Belgium; KU Leuven Institute for Single Cell Omics (LISCO), KU Leuven, Leuven, Belgium; CASM, Wellcome Sanger Institute, UK*

JIN ZHANG • *Department of Pharmacology, University of California, La Jolla, CA, USA; Department of Bioengineering, University of California, La Jolla, CA, USA; Department of Chemistry and Biochemistry, University of California, La Jolla, CA, USA*

Chapter 1

Single Cell aRNA-Seq for Transcriptomics and Cellular Phenotyping

Jifen Li and James Eberwine

Abstract

In this chapter, we describe a single cell transcriptome analysis protocol – linear antisense RNA amplification and sequencing (aRNA-seq). A simple and rapid molecular procedure that can amplify femtogram amounts of nucleic acid to microgram amounts and can be used to obtain RNA expression profiles from each individual cell. aRNA has been adapted for use in CEL-seq, InDROP-seq, and LIANTI. aRNA-seq is a useful tool to assess biological function of various cellular and tissue systems in single cell level. Because tissues are heterogeneous, and cells exist in different states and microenvironments, pooling of all cells in one reaction will average out signals across all cell types and physiological stages. As a result, rare or underrepresented cells will be gone unnoticed. Single cell analysis by RNA transcriptomics enables the measurement of cell variability at high resolution regardless of biochemical state. These data are useful in understanding cell responses to normal physiology and disease etiology. They are provided a map for manipulating the dynamics of signal transduction, thereby opening the door to discover novel druggable target that may be useful in developing clinical therapeutics.

Key words Single cell, Linear amplification, RNA sequencing, Transcriptome

1 Tips for Working with Single Cell

In this chapter, the application of aRNA analysis [1–4] to single primary neuron cell analysis [5, 6] will be used to illustrate the standard operating procedure that enables efficient and robust amplification of the mRNA transcriptome from single cells. Working with single cell RNA is similar to working with RNA in other contexts, except that there is much smaller initial quantities. Therefore, it is necessary to be more precise and careful than usual to avoid loss of material and contamination. Cautions that should always be invoked include the following:

1. Wear gloves and don't touch things on the bench without wearing gloves.
2. Change gloves often.

Jonathan V. Sweedler et al. (eds.), *Single Cell 'Omics of Neuronal Cells*,
Neuromethods, vol. 184, https://doi.org/10.1007/978-1-0716-2525-5_1,

3. Be careful in how tubes are handled and be careful to not touch the rim or the inside of the lid.
4. Don't talk or breathe over tubes when they are open.
5. Always use filter tips and change pipette tips between each sample to prevent cross-contamination.
6. Pay attention to precise pipetting. With such small amounts of starting material, it is important to not lose material during investigator-initiated manipulations, especially during the steps leading to the first round of amplification where there is a limited amount of material. Before discarding tip, it is important to note that no liquid is stuck on or inside it.
7. Make working aliquots of reagents such as nuclease-free water and ethanol each day to insure that they are RNAse-free.
8. Make fresh dilutions of your amplification primers for each experiment.
9. Work carefully to avoid contamination.
10. Use products like RNaseZAP to clean bench, but wash it down afterward with 70% ethanol, because such products contain detergents that may inhibit subsequent enzymatic reactions.

2 Materials

2.1 Reagents

ERCC RNA Spike-In Mix (Thermo Fisher Scientific 4456740).

SuperScript III (supplied with first-strand buffer and DTT) (Thermo Fisher Scientific 18080044).

Nuclease-free water (Thermo Fisher Scientific AM9937).

dNTP mix (10 mM each) (Thermo Fisher Scientific 18427088).

T7-oligo(dT): GGAGGCCGGAGAATTGTAATACGACTCAC TATAGGGA GACGCGTGTTTTTTTTTTTTTTTTTTTTTTTV(V=any nucleotide except T), RP-LC purified.

RNasin (Promega N2511).

Second-strand buffer (Thermo Fisher Scientific 10812014).

DNA polymerase I (Thermo Fisher Scientific 18010025).

RNase H (Thermo Fisher Scientific 18021014).

T4 DNA polymerase (Thermo Fisher Scientific 18005025).

T7 Megascript kit (Thermo Fisher Scientific AMB1334-5).

SUPERaseIn (Thermo Fisher Scientific AM2696).

Random primers (Promega C1181).

MinElute kit (Qiagen 28206).

TruSeq Stranded mRNA Library Preparation (Illumina 20020594).

TruSeq RNA Single Indexes (12 indexes, 24 samples) Set A (Illumina 20020492).

TruSeq RNA Single Indexes (12 indexes, 24 samples) Set B (Illumina 20020493).

Agencourt RNAClean XP beads (Beckman Coulter A63987).

Agencourt AMPure XP beads (Beckman Coulter A63881).

High Sensitivity D5000 reagent (Agilent Technologies 5067-5593).

High Sensitivity D5000 ScreenTape (Agilent Technologies 5067-5592).

D5000 ScreenTape (Agilent Technologies 5067-558).

D5000 Reagents (Agilent Technologies 5067-5589).

High Sensitivity RNA ScreenTape (Agilent Technologies 5067-5579).

High Sensitivity RNA ScreenTape Sample Buffer (Agilent Technologies 5067- 5580).

High Sensitivity RNA ScreenTape Ladder (Agilent Technologies 5067-5581).

RNA ScreenTape (Agilent Technologies 5067-5576).

RNA ScreenTape Sample Buffer (Agilent Technologies 5067-5577).

RNA ScreenTape Ladder (Agilent Technologies 5067-5578).

2.2 Equipment

Agilent 2200 TapeStation (Agilent Technologies G2964-90000).

DynaMag-2 magnet (Thermo Fisher Scientific 12321D).

Heat blocks.

Thermocycler.

2.3 Materials for Single Cell Pipette Isolation

1 × HBSS (Gibco 14175-079).

Capillaries for micropipettes (Kimble Chase 34500-99).

MP Series pressure type electrode holder (Warner Instruments 64-1263 custom).

1/16-inch inner diameter PVC tubing (Thermo Fisher Scientific 8000-0004).

Nylon 3-way stopcock, 2 female to male luers (Kimble Chase 420163-4503).

Luer-lock tip 3-ml syringes (BD 309657).

Nunc 35-mm tissue culture dishes (Thermo Fisher Scientific 12-565-90).

A PC computer installed with live imaging system.

3 Harvesting Single Cells Using a Patch Pipette (Figs. 1 and 2)

1. Plate primary mouse neurons on 12-mm coverslip coated with laminin (1 ug/ml)/poly-D-lysine (80 μg/ml) in 35-mm dish at a concentration of 1.0×10^5/ml for 7–10 days (note: cells should be plated at a low enough density that there is space around each cell).
2. At day of cell isolation, warm up 1 × HBSS at 37 °C water bath for 10 min, prepare 1.5 ml Eppendorf tubes containing 1× first-strand cDNA buffer (cell harvesting buffer) on ice, and turn on microscope and living imaging system.
3. Add 1.5 ml 1 × HBSS to a lid of 35-mm dish, and transfer a coverslip with mouse neurons to it (note: 1 × HBSS provides physiological ions, low phosphate concentration, and no calcium or magnesium).
4. Place the cells in the dish lid on the microscope stage, and find a region of cells suitable for harvest (note: cells should be relatively isolated from its neighbors to prevent harvest of surrounding cell (Fig. 1)). Position the cells to be harvested in the center of the field of view.
5. Turn to 10× objective lens. Secure a glass pipette (custom made) in the pipette holder (note: for more information about pipettes, refer to the Sutter Pipette Cookbook).
6. Use the coarse micromanipulators to position the pipette in the field of view.
7. Lower the pipette into the solution. Adjust the focus so you can see the pipette tip on computer screen.
8. Turn to 40× objective lens. Use the coarse micromanipulators to carefully lower the pipette; meanwhile, adjust the focus to the cells. Before the tip comes fully into focus, switch to using the fine micromanipulators. Stop when the tip touches the cell body (you will see the cell deform slightly).
9. Release the positive pressure on the pipette and use the syringe to aspirate the cell into the pipette (Fig. 2). The shear forces on the cell will cause it to lyse in the process.
10. Drag and simultaneously rotate the pipette around the inner wall of harvesting tube in order to break off the tip. Push any remaining liquid out of the pipette using syringe with 30-g needle. Spin the tube briefly in a microcentrifuge to collect the tip and its contents in the bottom of the tube. Keep on ice until ready to start first-strand synthesis.

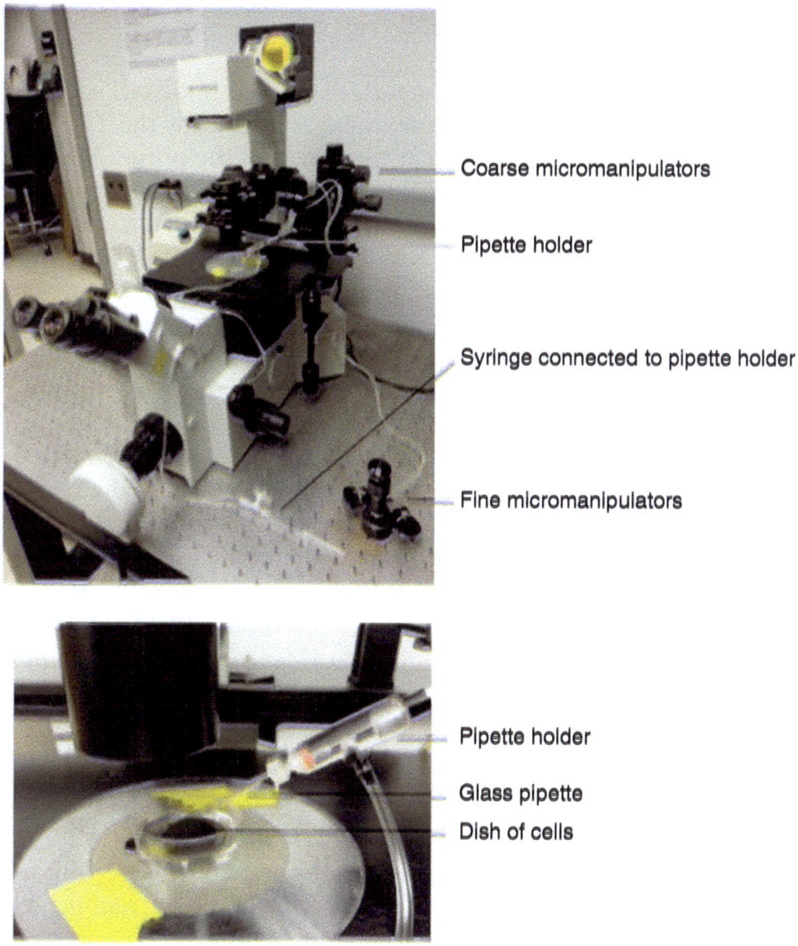

Fig. 1 Pipette harvesting single cell setup. Olympus converted microscope equipped with micromanipulators (top panel). Enlarged image showing dish of cells with a harvesting pipette on the microscope stage (bottom panel)

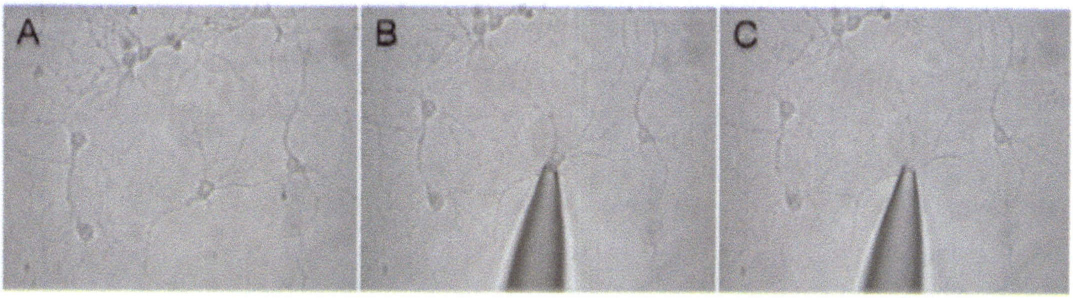

Fig. 2 Example of a successful harvest of a single neuron on the coverslip. We selected a cell in the area that cells well isolated (A) and advanced the pipette tip toward the desired cell (B). Panel (C) shows the field of view after the cell had been harvested. Note there was no cell left under the pipette tip, but the surrounding processes of the neuron remain intact. (Reproduced from Ref. [11] with permission from JoVE)

4 Linear aRNA Amplification First Round (Fig. 3)

4.1 First-Strand cDNA Synthesis (Fig. 3 Round 1)

In this stage, the total RNA from a single cell is used as a template. First-strand cDNA synthesis is primed by the T7-oligo(dT) primer. The poly(dT) tract at the 3′ end of the primer binds and selects the poly(A) tail of mRNAs; the 5′ end of the primer contains the T7 RNA polymerase promoter sequence, which is used for in vitro transcription of RNA. The primer is extended by SuperScript III reverse transcriptase using dNTPs, to create a complementary DNA (cDNA) strand bound to the RNA template. This is the most critical step in single cell transcriptome analysis. If the first-strand cDNA synthesis isn't efficient, then only a portion of the transcriptome will be amplified. In other nucleic acid amplification protocols, only ~10% of the transcriptome is converted into cDNA during first-strand cDNA synthesis. Using the protocol described herein, ~60% of the transcriptome is converted to cDNA providing more complexity to the final amplified product. In this version of the protocol, the aRNA primer has an oligo-dT that will bind to the poly-A+ tail of mRNA; therefore, it is this fraction of cellular RNA that will be amplified. If rather one would like to amplify non-poly-A containing RNA, then a aRNA primer where the oligo-dT is replaced with random priming sequence can be used to prime the single cell RNA. To avoid RNA degradation by RNases, the RNase inhibitor RNasin is added. DTT is added to stabilize RNasin, and to prevent RNase release from RNasin.

1. Thaw all reagents to room temperature, mix well, and spin down. Never vortex enzymes and remove enzymes from −20 °C before use. Keep on ice.
2. Make ERCC RNA spike-in mix 1:4,000,000 dilution with nuclease-free water by three serial dilutions, 1:100 × 1: 100 × 1:400.
3. Make T7-oligo(dT) primer 1:100 dilution (10 ng/μl) with nuclease-free water from stock solution 1 μg/μl.
4. Prepare first-strand cDNA primer-dNTP mix:

First-strand buffer (5×)	2.4 μl
ERCC RNA spike-in mix (1:4 × 10^6)	0.9 μl
dNTPs (2.5 mM each)	1.2 μl
T7-oligo(dT) (10 ng/μl)	0.3 μl
DTT (100 mM)	0.45 μl
Total	5.25 μl

5. Add 5.25-μl primer mix to each tube containing single cells, and mix by pipetting.

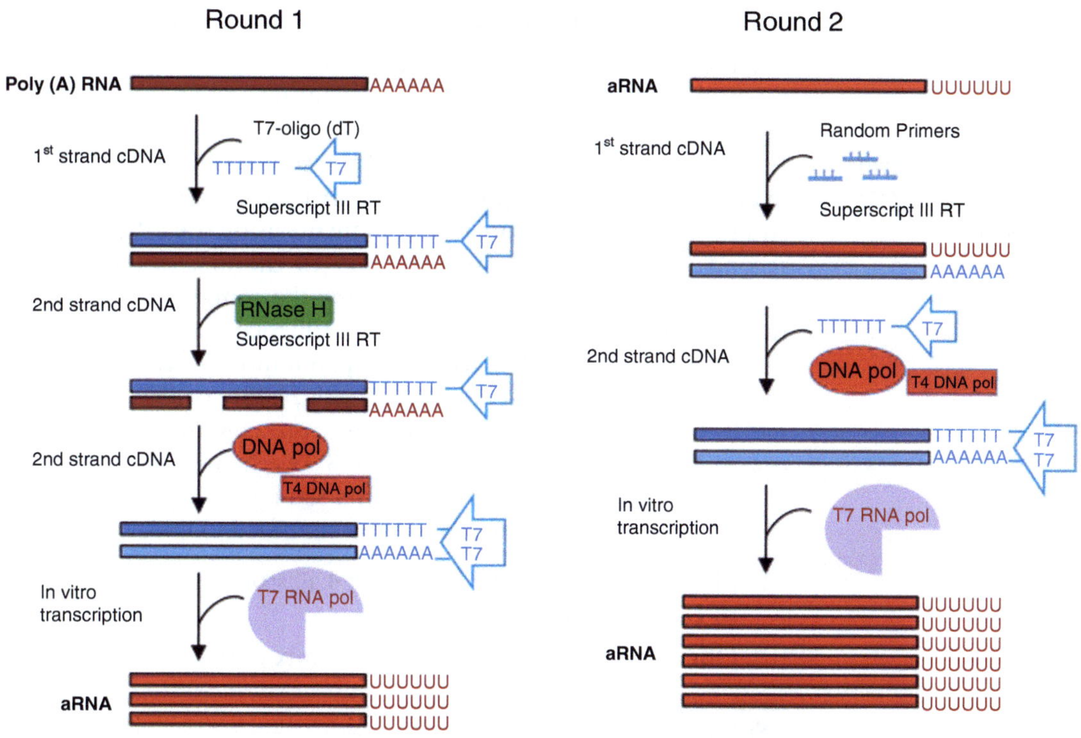

Fig. 3 Schematic of two rounds of aRNA amplification procedure. In the first round (Round 1), first-strand cDNA is synthesized by reverse transcription (RT) after oligo(dT)-T7 primer anneals to the poly(A) tail of mRNA and primed from mRNA. RNase H is used to create RNA fragments by digest portions of the bound mRNA from cDNA/mRNA complex. The small RNA fragments serve to prime second-strand cDNA by DNA polymerase (pol). Finally, aRNA is amplified via linear in vitro transcription by T7 RNA polymerase, using the T7 RNA polymerase promoter incorporated in the double-stranded cDNA and generating antisense RNA (aRNA) with polyU at 5′ end. In the second round (Round 2), first-strand synthesis is primed by random primers instead of the oligo (dT)-T7 primer by reverse transcriptase using the aRNA as a template instead of mRNA. After RNA denaturation, second-strand synthesis is primed with the oligo(dT)-T7 primer, which binds to the poly(A) tail of the cDNA created during first-strand synthesis by DNA polymerase. Finally, RNA is linearly in vitro transcribed through the enzymatic activity of T7 RNA polymerase acting on T7 promoter site that is incorporated into the double-stranded cDNA

6. Incubate for 5 min at 70 °C, and immediately place on ice. Spin briefly, and return to ice.
7. On ice, make the following enzyme master mix in the order listed:

Nuclease-free water	1.0 μl
RNasin (40 U/μl)	0.3 μl
SuperScript III RT (200 U/μl)	0.45 μl
Total	1.75 μl

8. Add 1.75-μl master mix to each tube and mix by pipetting.
9. Incubate 30 min at 42 °C, and 15 min at 70 °C, spin briefly, and place on ice. Store at −20 or continue on to the next stage.

4.2 Second-Strand cDNA Synthesis (Fig. 3 Round 1)

In this stage, RNase H nicks the RNA strand of the RNA-cDNA hybrid, creating priming sites for DNA polymerase I, which upon addition extends the short RNA fragments using dNTPs, creating a second cDNA strand complementary to the first (second-strand cDNA synthesis). This includes creating the complementary strand to the T7 RNA polymerase promoter sequence that is incorporated during first-strand cDNA synthesis. All but most 5' RNA fragments are displaced by the 5′ → 3′ exonuclease activity of DNA polymerase I. T4 DNA polymerase fills in 5′ overhangs with its 5′ → 3′ polymerase activity, and removes the 3′ overhangs (resulting from the loss of the RNA primers) with its strong 3′ → 5′ exonuclease activity, creating blunt-ended, double-stranded cDNA.

1. Make 0.69-mM dNTP by adding 4-μl stock solution (2.5 mM) to 10.47-μl nuclease-free water
2. On ice, make the following master mix in the order listed:

Nuclease-free water	8.0 μl
Second-strand buffer (5×)	5.6 μl
dNTPs (0.69 mM each)	0.75 μl
DNA polymerase I (10 U/μl)	1.0 μl
RNase H (2 U/μl)	0.25 μl
Total	15.6 μl

3. Add 15.6-μl master mix to each tube and mix by pipetting up and down ten times.
4. Incubate for 2 h at 16 °C.
5. Add 1-μl T4 DNA polymerase (5 U/μl). Mix by pipetting. Incubate for 10 min at 16 °C.
6. Incubate for 10 min at 75 °C.

4.3 In Vitro Transcription (IVT, Megascript T7 Kit) (Fig. 3 Round 1)

The double-stranded cDNA made in the previous two steps can be used as a template for IVT. Using the T7 RNA polymerase promoter that was "built" into the cDNA, T7 RNA polymerase synthesizes thousands of copies of antisense RNA from each cDNA template. The RNA is antisense in orientation as it is a copy of the initial poly-A+ RNA that was initially isolated from the single cell and used as a template for first-strand cDNA synthesis. RNase inhibitor SUPERaseIn is added to protect the newly synthesized RNA from degradation.

1. At room temperature, make the master mix below in the order listed:

ATP, CTP, GTP, UTP	6 µl each
Reaction buffer (10×)	6 µl
Enzyme mix (10×)	6 µl
SUPERaseIn (20 U/µl)	3 µl
Total	39 µl

Note: make sure the reaction buffer does not contain any precipitate.

2. For each sample, add 39-µl master mix to a PCR tube.
3. Add each of your samples (30 µl) to a PCR tube containing master mix.
4. Incubate in a thermocycler with heated lid using program 37 °C for 14 h, 4 °C hold.

4.4 IVT Cleanup (Agencourt RNAClean XP Beads)

In this stage, Agencourt RNAClean XP beads (RNA beads) are used to purify the antisense RNA produced during IVT by removing unused nucleotides, primers, enzymes and other proteins, and salts.

1. Get RNA beads out of the fridge to equilibrate to room temperature (RT).
2. Label 1.5-ml microcentrifuge tubes (one for each sample).
3. Transfer your first-round aRNA from the PCR tubes to the 1.5-ml tubes.
4. Vortex RNA beads to resuspend magnetic particles that may have settled.
5. Add 1.8× volume of Agencourt RNAClean beads (124.2 µl) to reaction. Mix by pipetting up and down ten times.
6. Let stand at RT for 5 min. Meanwhile, prepare 5-ml 70% ethanol.
7. Open lids and place on magnetic stand for 5 min.
8. Slowly remove and discard 173-µl supernatant without disturbing beads.
9. Add 200-µl freshly prepared 70% ethanol. Incubate for 10 s, and remove ethanol.
10. Repeat 70% ethanol wash.
11. Remove last traces of ethanol from the bottom of the tube with P10 pipette.
12. Let stand at RT for 5 min to dry, then close lids, and remove from magnetic stand.

13. Add 32.5-μl elution buffer (add 5 × FS buffer 4–150-μl nuclease-free water).
14. Incubate at RT for 2 min. Place on magnetic stand and wait for 5 min.
15. Label 1.5-ml microcentrifuge tubes (one for each sample), and transfer 30-μl supernatant to new tubes.
16. Concentrate supernatant to 4 μl by SpeedVac (Eppendorf Vacufuge) at 45 °C for 60 min.

5 Linear aRNA Amplification Second/Third Round (Fig. 3)

5.1 First-Strand cDNA Synthesis (Fig. 3 Round 2)

To further amplify the cellular RNA, the aRNA produced during IVT can be used as a template for a new round of first-strand cDNA synthesis by SuperScript III reverse transcriptase. cDNA synthesis requires priming with random hexamers as the aRNA is antisense in orientation and does not have a poly-A tail on its 3′ end. Note rather that the aRNA has a poly(U) tract at its 5′ end.

1. Dilute 500 ng/μl stock of random primers 1:10 in nuclease-free water, and add 1 μl to each tube.
2. Incubate for 10 min at 70 °C. Immediately place on ice for at least 2 min, and spin briefly.
3. On ice, make the following master mix in the order listed:

First-strand buffer (5×)	2 μl
dNTPs (2.5 mM each)	0.5 μl
RNasin (40 U/μl)	0.5 μl
SuperScript III (200 U/μl)	1 μl
DTT (100 mM)	1 μl
Total	5 μl

4. Add 5-μl master mix to each tube.
5. Incubate for 10 min at 25 °C, 30 min at 42 °C, and 5 min at 95 °C.
6. Place on ice for at least 2 min, and spin briefly. Store at −20 or continue on to the next stage.

5.2 Second-Strand cDNA Synthesis (Fig. 3 Round 2)

To perform second-strand cDNA synthesis, the RNA-cDNA hybrid is denatured, and the first-strand cDNA is used as a template for second-strand synthesis by DNA polymerase I. Because aRNA has a 5′ poly(U) tract, the first-strand cDNA made in the previous step now has a 3′ poly(A) tail, which can be primed using the aRNA primer, T7-oligo(dT).

1. Make a fresh 1:100 dilution of T7-oligo(dT) (1 μg/μl). Add 2 μl of diluted (10 ng/μl) T7-oligo(dT) to each tube and incubate for 6 min at 70 °C.
2. Immediately place on ice for at least 2 min. Spin briefly, and return to ice.
3. On ice, make the following master mix in the order listed:

Nuclease-free water	43.5 μl
Second-strand buffer (5×)	15 μl
dNTPs (2.5 mM)	1.5 μl
DNA polymerase I (10 U/μl)	2 μl
Total	62 μl

4. Add 62-μl master mix to each tube. Mix thoroughly by pipetting.
5. Incubate for 2 h at 16 °C.
6. Add 2-μl T4 DNA polymerase (5 U/μl). Mix thoroughly by pipetting.
7. Incubate for 10 min at 16 °C.

5.3 cDNA Cleanup (MinElute Kit) (Fig. 3 Round 2)

At this stage in the amplification procedure, double-stranded cDNA is purified using the Qiagen MinElute kit to remove enzymes and other proteins, dNTPs, primers, and salts. This is necessary to increase amplification yield by removing the dNTPs which are competitive inhibitors of the enzymatic activity of T7 RNA polymerase. Due to potential loss of material during this purification procedure, please note that this step is not performed during the first round of aRNA amplification.

1. Add 300-μl Buffer ERC (stored at RT) to each tube, mix, and apply the sample to a MinElute column (stored at 4 °C).
2. Centrifuge at 12,000 × g for 1 min.
3. Use a pipette to remove and discard the flow-through.
4. To wash, add 750-μl Buffer PE (stored at room temp) to the MinElute column.
5. Centrifuge at 12,000 × g for 1 min.
6. Use a pipette to remove and discard the flow-through.
7. Centrifuge at 12,000 × g for a further 1 min.
8. Place the MinElute column in a clean 1.5-ml microcentrifuge tube.
9. To elute cDNA, add 13-μl nuclease-free water to the middle of MinElute column, let stand at room temperature for 1 min, and then centrifuge at 12,000 × g for 1 min.

5.4 IVT (Megascript T7 Kit) (Fig. 3 Round 2)

This step is identical to first-round IVT, except that the volumes are smaller because it was possible to concentrate the material after purification.

1. At room temp, make the following master mix, in the order listed:

ATP, CTP, GTP, UTP	2 μl each
Reaction buffer (10×)	2 μl
Enzyme mix (10×)	2 μl
SUPERaseIn (20 U/μl)	1 μl
Total	13 μl

2. For each sample, add 13-μl master mix to a PCR tube.
3. Add 10 μl of your eluted cDNA from the MinElute cleanup to the PCR tube containing IVT master mix. Save the remaining 3 μl, labeled as second- or third-round cDNA, and store at −80 °C for any future analyses needed.
4. Incubate in a thermocycler with heated lid, using the program 37 °C for 14 h, 4 °C hold.

5.5 IVT Cleanup (Agencourt RNA Clean XP Beads)

To clean up the aRNA for sequencing library preparation, Agencourt RNAClean XP beads (RNA beads) can be used to purify the aRNA produced after IVT to remove unused nucleotides, primers, enzymes and other proteins, and salts.

1. Get RNA beads out of the fridge to equilibrate to room temperature.
2. Label 1.5-ml microcentrifuge tubes (one for each sample).
3. Transfer your second-round/third-round aRNA from the PCR tubes to the 1.5-ml tubes.
4. Vortex RNA beads to resuspend magnetic particles that may have settled.
5. Add 1.8× volume of RNA beads 41.4 μl to reaction. Mix by pipetting up and down ten times.
6. Let stand at RT for 5 min. Meanwhile, prepare 5-ml 70% ethanol.
7. Open lids and place on magnetic stand for 5 min.
8. Slowly remove and discard 54-μl supernatant without disturbing beads.
9. Add 200-μl freshly prepared 70% ethanol. Incubate for 10 s, and remove ethanol.
10. Repeat 70% ethanol wash.

11. Remove last traces of ethanol from the bottom of the tube with P10 pipette.
12. Let stand at RT for 5 min to dry, and remove from magnetic stand.
13. Add 32.5-μl elution buffer (add 5 × FS buffer 4–150-μl nuclease-free water; for third-round, nuclease-free water only).
14. Incubate at RT for 2 min. Place on magnetic stand and wait for 5 min.
15. Transfer 30-μl supernatant to a new tube.
16. Save 5 μl to a fresh tube for TapeStation analysis with Agilent High Sensitivity RNA Kit.
17. Concentrate supernatant to 4 μl by SpeedVac at 45 °C for 60 min.
18. At the third round, skip **steps 16** and **17**, and analyze the final product using Nanodrop and Agilent RNA Kit. Proceed to library preparation (below).

6 Library Preparation (Illumina TruSeq Stranded mRNA Kit) (Fig. 4)

The kit reagents are stored at −20 °C unless otherwise noted in the manufacturer's instructions. Thaw all reagents to room temperature, mix well, and spin down. Never vortex enzymes and remove enzymes from −20 °C before use. Keep on ice.

6.1 First-Strand cDNA Synthesis (Fig. 4)

In this stage, amplified antisense RNA is used as a template for first-strand synthesis by SuperScript III reverse transcriptase. The synthesis is primed by random primers contained in the fragment, prime, finish mix. Note that the aRNA is not fragmented, because the lengths are short and appropriate for sequencing library generation. If the aRNA products are long, then fragmentation as per the manufacturer's suggested protocol works well. At the end of this step, the 70 °C denaturation step separates the first-strand cDNA from the RNA template.

1. Start with 200 ng aRNA in 5-μl nuclease-free water. If your samples are not concentrated enough, you can reduce the amount of RNA to as little as 50 ng, but ideally you should use the same amount for each sample.
2. Add 13-μl fragment, prime, finish mix (FPF).
3. Remove and discard 1 μl.
4. Make the following master mix (volume for one sample):

5× first-strand synthesis actinomycin D mix (FSA)	7.2 μl
SuperScript III (200 U/μl)	0.8 μl
Total	8.0 μl

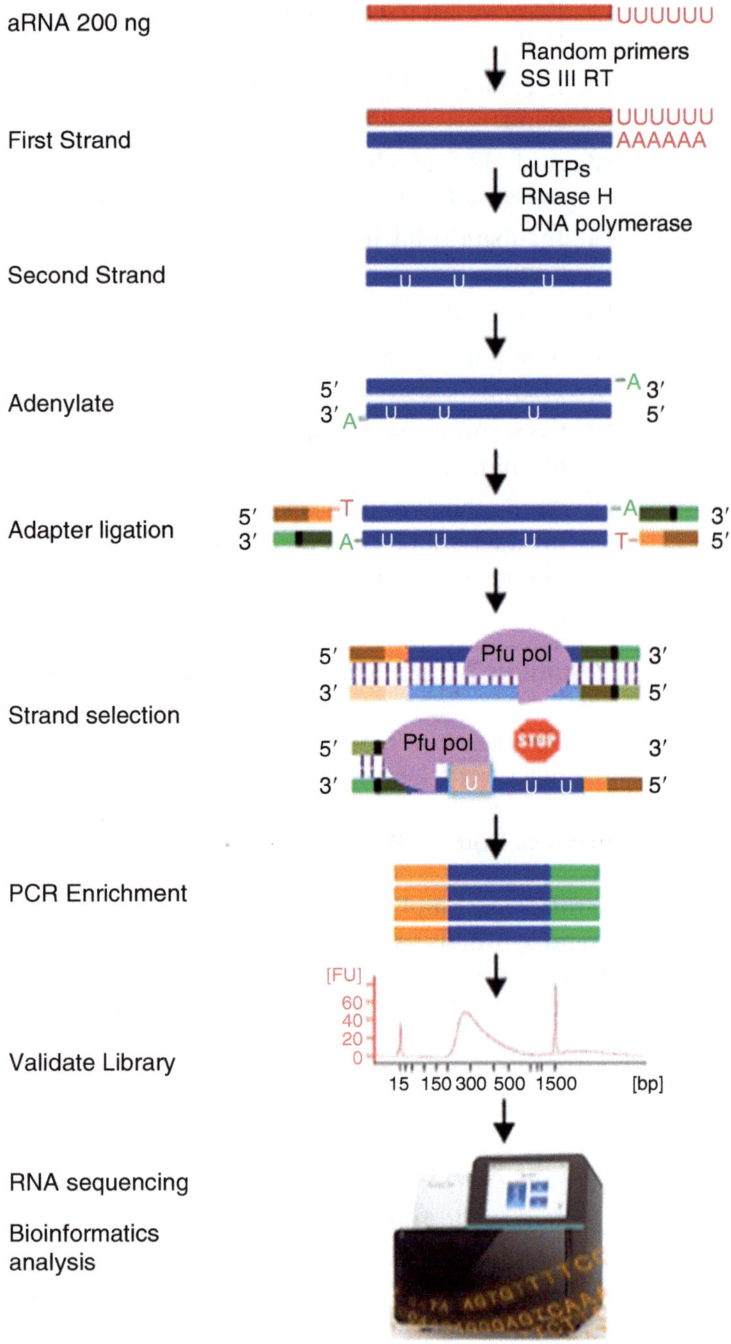

Fig. 4 Schematic of stranded library preparation protocol using aRNA. First-strand cDNA synthesis is primed by random primer using 200 ng aRNA as a template. Second-strand cDNA is synthesized using short RNA fragment created by RNase H hydrolysis of cDNA/RNA complex. Note dUTP is added to the reaction to selectively label the second strand preventing its being amplified during PCR amplification. Next, "A" base is added to 3′ ends of both cDNA to allow ligation of the sequencing adaptor by pairing its 3′ "T" base. Note a sequencing adapter contains index or "barcode" for multiplexing. Finally, PCR is used to enrich forward strand cDNA fragment that contains adapters at both ends. Meanwhile, reverse strand is selectively destructed by Pfu DNA polymerase during PCR

5. Add 8-μl master mix to each tube, pipette up and down six times to mix, and incubate at:

   25 °C for10 min

   42 °C for 15 min

   70 °C for 5 min

6. Spin briefly and place on ice.

6.2 Second-Strand Synthesis and End Repair (Fig. 4)

Next, the first-strand cDNA is used as a template for second-strand cDNA synthesis. Importantly, dUTP is incorporated into the second strand, which prevents it from being amplified during PCR enrichment. This allows the sequencing data to be "stranded" enabling identification of the sense and antisense orientation of the final sequenced product. This is useful in identifying endogenous antisense RNA transcripts.

1. Make a master mix of the following (volume for one sample):

Resuspension buffer (RSB)	5 μl
5× second-strand mix (SSM)	20 μl
Total	25 μl

2. Add 25 μl to each tube and pipette six times to mix. Incubate for 1 h at 16 °C.
3. Get Agencourt AMPure XP beads (DNA beads) out of the fridge to equilibrate to RT.
4. Cleanup:
   (i) Vortex DNA beads, add 90 -μl to each tube, and pipette up and down ten times.
   (ii) Incubate at RT for 5 min. Meanwhile, prepare 5-ml 80% ethanol.
   (iii) Open lids and place on magnetic stand, and wait for 5 min.
   (iv) Remove and discard 135-μl supernatant without disturbing beads.
   (v) Add 200-μl freshly prepared 80% ethanol. Incubate at RT for 10 s, and remove.
   (vi) Repeat 80% ethanol wash. Remove last traces of ethanol from the bottom of the tube with P10 pipette.
   (vii) Let stand at RT for 5 min to dry.
   (viii) Remove from the magnetic stand and resuspend in 17.5-μ l RSB.

(ix) Incubate for 2 min at RT, place on magnetic stand, and wait for 5 min. Transfer 15-μl supernatant to new tubes.

(x) Samples may be stored at −80 °C for a few days.

6.3 3′ Adenylation (Fig. 4)

An A base is added to the 3′ ends of both cDNA strands in this step to allow the adapter complex to transiently anneal via its unpaired 3′ T bases.

1. Make the following master mix (volume for one sample):

A-tailing mix (ATM)	12.5 μl
Resuspension buffer (RSB)	2.5 μl
Total	15 μl

2. Add 15-μl master mix to each tube.
3. Incubate for 30 min at 37 °C.
4. Incubate for 5 min at 70 °C. Then on ice 2 min. Spin briefly and place on ice.

6.4 Ligate Adapters (Fig. 4)

After adenylation, the sequencing adaptors are ligated onto the cDNA to enable subsequent PCR amplification of the library and to enable Illumina sequencing.

1. Add 2.5-μl adapter index to each tube and mix. Refer to the index guide to determine which adapter indexes to use. Please make note for which sample has which index.
2. Make the following master mix (volume for one sample):

Ligation mix (LGM)	2.5 μl
Resuspension buffer (RSB)	2.5 μl
Total	5.0 μl

3. Add 5-μl master mix to each tube.
4. Set a pipette to 40 μl and pipette up and down ten times to mix.
5. Incubate for 10 min at 30 °C.
6. Add 5-μl Stop ligation buffer (SLB) and pipette up and down ten times to mix.
7. Clean up.

   First wash:

1. Vortex DNA beads, add 42 μl to each tube, and pipette up and down ten times.
2. Incubate at RT for 5 min. Meanwhile, prepare 10-ml 80% ethanol.
3. Place on magnetic stand, and wait for 5 min.
4. Remove and discard 79.5-μl supernatant without disturbing beads.
5. Add 200-μl freshly prepared 80% ethanol. Incubate at RT for 10 s, and remove ethanol.
6. Repeat 80% ethanol wash. Remove last traces of ethanol from the bottom of the tube with P10 pipette.
7. Let stand at RT for 5 min to dry.
8. Remove from the magnetic stand and resuspend in 52.5-μl RSB.
9. Incubate for 2 min at RT, place on magnetic stand, and wait for 5 min.
10. Transfer 50-μl supernatant to new tubes.

Second wash:

1. Vortex DNA beads, add 50 μl to each tube, and pipette up and down ten times.
2. Incubate at RT for 5 min.
3. Place on magnetic stand, and wait for 5 min.
4. Remove and discard 95-μl supernatant without disturbing beads.
5. Add 200-μl freshly prepared 80% ethanol. Incubate at RT for 10 s, and remove ethanol.
6. Repeat 80% ethanol wash. Remove last traces of ethanol from the bottom of the tube with P10 pipette.
7. Let stand at RT for 5 min to dry.
8. Remove from the magnetic stand and resuspend in 22.5-μl RSB.
9. Incubate for 2 min at RT, place on magnetic stand, and wait for 5 min. Transfer 20-μl supernatant to PCR tubes.
10. Samples may be stored at −80 °C for several days.

6.5 Enrich DNA Fragments (Fig. 4)

PCR is used to selectively amplify those cDNA fragments that have adapters ligated to them, by using PCR primers that bind to the adapters. In the first step of the PCR reaction, this selectively amplifies the forward strand, since the dUTP-containing reverse strand is not amplified. The enzyme used for this PCR amplification is Pfu DNA polymerase which stalls, and then falls off, when it encounters dUTP in the template.

1. Make the following master mix (volume for one sample):

PCR primer cocktail (PPC)	5.0 μl
PCR master mix (PMM)	25 μl
Total	30 μl

2. Add 30-μl master mix to each tube and mix by pipetting.
3. Run the following PCR program. Your total volume will be 50 μl:

   98 °C for 30 s

   13 cycles of:

      98 °C for 10 s

      60 °C for 30 s

      72 °C for 30 s.

   72 °C for 5 min

   4 °C hold
4. Clean up.

6.6 First Wash

1. Transfer PCR product to labeled 1.5-ml microcentrifuge tube (one per sample).
2. Vortex DNA beads and add 50 μl to each tube. Pipette up and down ten times.
3. Incubate at RT for 5 min.
4. Place on magnetic stand, and wait for 5 min.
5. Remove and discard 95-μl supernatant without disturbing beads.
6. Add 200-μl freshly prepared 80% ethanol. Incubate at RT for 10 s, and remove ethanol.
7. Repeat 80% ethanol wash. Remove last traces of ethanol from the bottom of the tube with P10 pipette.
8. Let stand at RT for 5 min to dry.
9. Remove from the magnetic stand and resuspend in 52.5-μ l RSB.
10. Incubate for 2 min at RT, place on magnetic stand, and wait for 5 min.
11. Transfer 50-μl supernatant to new tubes.

6.7 Second Wash

1. Vortex DNA beads and add 50 μl to each tube. Pipette up and down ten times.
2. Incubate at RT for 5 min.
3. Place on magnetic stand, and wait for 5 min.

4. Remove and discard 95-μl supernatant without disturbing beads.
5. Add 200-μl freshly prepared 80% ethanol. Incubate at RT for 10 s, and remove ethanol.
6. Repeat 80% ethanol wash. Remove last traces of ethanol from the bottom of the tube with P10 pipette.
7. Let stand at RT for 5 min to dry.
8. Remove from the magnetic stand and resuspend in 22.5-μ l RSB.
9. Incubate for 2 min at RT, place on magnetic stand, and wait for 5 min.
10. Transfer 20-μl supernatant to new tubes.

6.8 Analyze Library with Agilent TapeStation Using DNA 5000 Kit (Fig. 5)

For efficient sequencing on Illumina sequencers, the sequencing libraries should have an average size of 400 bp. Before submitting for sequencing, libraries are diluted to 10 nM in RSB and pooled. The concentration is checked with TapeStation High Sensitivity D5000 Kit. The amount of library loaded onto the sequencer is critical as too little will result in limited number of identified reads and too much will result in over-clustering of the library on the sequencer which will reduce the amount of reliable sequence information that can be generated.

6.9 Index Guide (Single Index)

To allow for multiplexing of different libraries in a single flow cell lane, each library has a 6 bp index or "barcode" incorporated with the unique index adapter that was ligated onto the PCR product. During sequencing of the index, A and C bases are read by excitation with a red laser, while G and T bases are read by excitation with a green laser. Within a set of multiplexed libraries, there should be roughly equal numbers of bases of each color at each base position (1–6); otherwise, the run will fail. This is referred to as "balancing" the indexes. The index adapters you will use can be found in Illumina document "Adapter sequences." Illumina document "index adapters pooling guide" provides recommendations for optimizing color balance when you pool the indexed libraries. We generally make 12 libraries in one pool and run on NextSeq 500 flow cell lanes.

6.10 Agilent TapeStation Analysis (Fig. 5)

Agilent TapeStation analysis gives information about nucleic acid concentration, length, and the presence of contaminants, such as proteins, ethanol, or glycogen.

1. After two rounds of aRNA amplification, RNA can be detected with Agilent High Sensitivity RNA Kit.
2. After three rounds of aRNA amplification, RNA can be detected with Nanodrop, as well as Agilent RNA Kit.

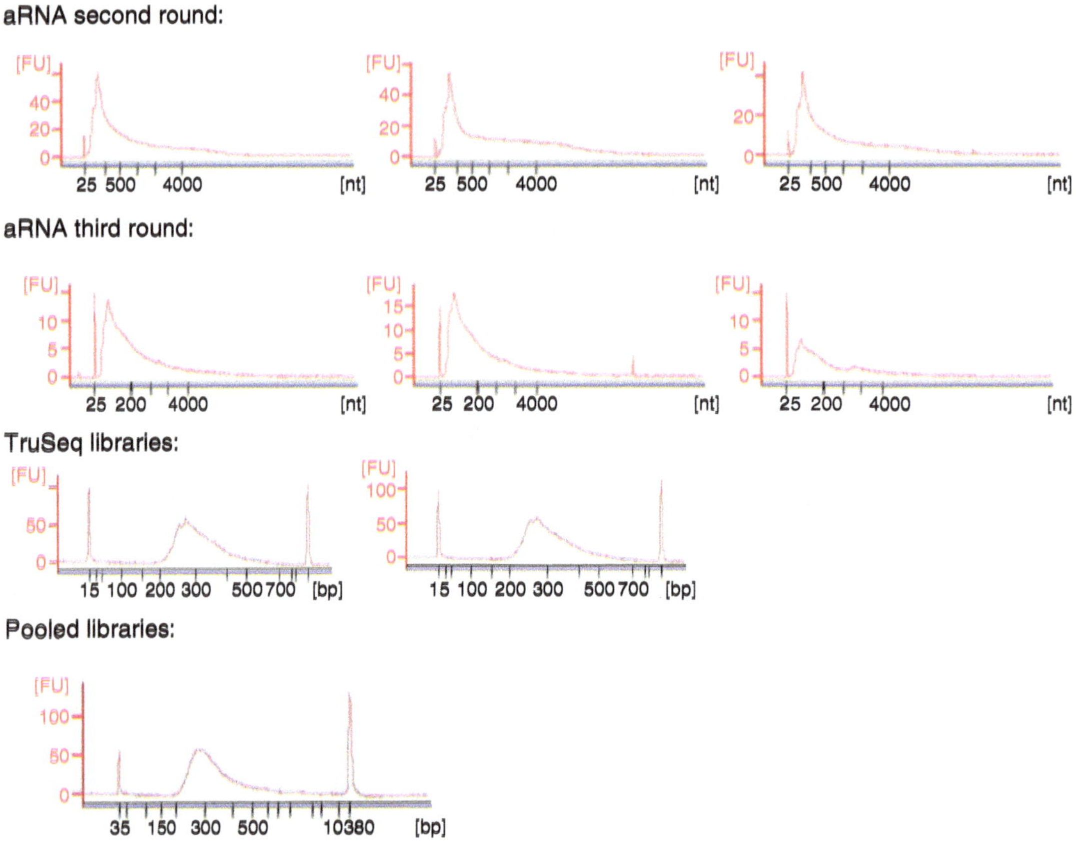

Fig. 5 Representative Agilent TapeStation quantification after each aRNA amplification and library preparation. Following amplification, analysis using Agilent TapeStation is recommended to examine the distribution and quantity of aRNA and libraries. A successful amplification from single cell material shows on the electropherogram as a high-amplitude large, wide peak. The size of the products is indicated in nucleotides (nt) or base pairs (bp) on the *X* axis. The amount of the product is indicated in the arbitrary fluorescence units (FU) on the *Y* axis

3. After library construction, library DNAs can be analyzed with Agilent DNA 5000 Kit.
4. After pooling libraries, library DNAs can be analyzed with Agilent High Sensitivity D5000 Kit.

7 Conclusions

As an example of how the aRNA approach can be applied to single cells, when using this protocol, we have observed approximately 10,000 expressed genes at varying abundances per cell with 50% of reads in a cell covering 430 highly expressed genes [7]. The most abundant expressed cell-type-specific genes comprise 2% of reads on average, over 1000 times more than the median gene expression [7]. This highlights the deep read-depth and high sensitivity that can be achieved with the aRNA-seq technique.

The high yield and unbiased amplification of antisense RNA-based single cell sequencing has been used for detection of G-protein-coupled receptor mRNA in serotonergic neurons [8], cell-type-specific transcriptional genes from live neurosurgically resected adult human brains [9]. It also formed the basis for development of multiplexing of samples for high-throughput RNA sequencing (CEL-seq) [1, 2] as well as whole-genome sequencing using linear amplification via T7 polymerase promoter site-bearing transposon insertion (LIANTI) [4].

A future direction that is in the offing that is doable because of the high sensitivity of the aRNA procedure is to assess the genomics of subcellular compartments. Applying the aRNA procedure to the analysis of different subcellular chemicals within single cells will enable the measurement of multiple subcellular chemicals simultaneously within a cell [10]. This harkens the possibility of understanding the systems chemistry and systems biology of individual cells. In so doing, it promises to provide the data for the discovery of novel chemical/signal pathways that will more fully illustrate the role of cell-to-cell variation in the functioning and interplay of single cells among a cohort of cells.

Acknowledgments

We appreciate and acknowledge the work of students, post-docs, and research assistants from the Eberwine lab over the years, who have contributed to the development of this optimized aRNA protocol and the various chemical detection methods based upon linear nucleic acid amplification.

References

1. Hashimshony T et al (2012) CEL-seq: single-cell RNA-seq by multiplexed linear amplification. Cell Rep 2:666–673
2. Hashimshony T et al (2016) CEL-Seq2: sensitive highly-multiplexed single-cell RNA-seq. Genome Biol 17:77
3. Klein AM et al (2015) Droplet barcoding for single-cell transcriptomics applied to embryonic stem cells. Cell 161:1187–1201
4. Chen C et al (2017) Single-cell whole-genome analyses by linear amplification via transposon insertion (LIANTI). Science 356:189–194
5. Van Gelder RN et al (1990) Amplified RNA synthesized from limited quantities of heterogeneous cDNA. Proc Natl Acad Sci U S A 87: 1663–1667
6. Eberwine J et al (1992) Analysis of gene expression in single live neurons. Proc Natl Acad Sci U S A 89:3010–3014
7. Dueck H et al (2015) Deep sequencing reveals cell-type-specific patterns of single-cell transcriptome variation. Genome Biol 16:122
8. Spaethling JM et al (2014) Serotonergic neuron regulation informed by in vivo single-cell transcriptomics. FASEB J 28:771–780
9. Spaethling JM et al (2017) Primary cell culture of live neurosurgically resected aged adult human brain cells and single cell transcriptomics. Cell Rep 18:791–803
10. Li J, Eberwine J (2018) The successes and future prospects of the linear antisense RNA amplification methodology. Nat Protoc 13: 811–818
11. Morris J et al (2011) Transcriptome analysis of single cells. J Vis Exp 25:2634

Chapter 2

Single-Cell Genomic and Epigenomic DNA Analyses

Sarah Geurs, Thomas Lefevre, and Thierry Voet

Abstract

This chapter will be dedicated to methods for single-cell genome and epigenome analysis of the human brain. Examining the (epi)genome at a high resolution and sensitivity, allowing for the detection of heterogeneity among individual neurons within the brain, is paramount to investigate and understand genome instability as well as epigenomic variation and transcriptional regulation within cells. Firstly, we discuss tissue disaggregation methods to collect individual neurons or neuronal nuclei from brain tissue for genome analysis methods. Subsequently, an overview of whole-genome amplification (WGA) strategies and possible WGA artifacts is provided, and we conclude on which WGA methods are best suited to study different types of genomic aberrations. Additionally, direct library preparation methods or alternative WGA methods are contrasted to the classic WGA methods. Secondly, modification to the epigenome, such as DNA methylation, posttranslational histone modifications, and alterations to chromosome conformation and chromatin accessibility play a crucial regulatory role in gene expression and are involved in many biological processes. We list the available approaches to study these epigenomic features at a single-cell level in human neurons. Lastly, multi-omics techniques that scrutinize multiple layers of information of the same single cell in parallel are discussed. We provide an overview of methods for simultaneous genome-and-transcriptome analysis, analysis of multiple (epi)-genomic layers and techniques that analyze the (epi)-genome and transcriptome in parallel.

Key words Single-cell/nucleus sequencing, Genomics, Epigenomics, Brain, Neuron, Genome instability

1 Single-Cell Genomics

1.1 Somatic Mosaic CNVs and SNVs in Brain

1.1.1 The Origin of CNVs and SNVs in Cells and Neurons

Mutations, such as single-nucleotide variants (SNVs), copy number variants (CNVs), or structural variants (SVs), can arise with each cell cycle and division. Estimated per-locus per generation mutation rates for genomic rearrangements range between 2×10^{-5} and 1.25×10^{-4}, which is about three to four orders of magnitude greater than for nucleotide substitution rates [1]. It is implied that in dividing tissues and during development, genetic heterogeneity is likely to arise. Specifically, somatic CNVs can arise in G1-phase through DNA breakage and repair by nonhomologous end-joining [2]. Using DNA replication processes, CNVs can also result from

Jonathan V. Sweedler et al. (eds.), *Single Cell 'Omics of Neuronal Cells*,
Neuromethods, vol. 184, https://doi.org/10.1007/978-1-0716-2525-5_2,

fork stalling and template switching (FoSTeS) or microhomology-mediated break-induced replication (MMBIR), inducing complex rearrangements [3, 4]. Furthermore, DNA damage induced by replication stress will trigger DNA damage repair mechanisms such as (nonallelic) homologous recombination ((NA)HR) in the G2-phase of the cell cycle to correct the DNA damage, with CNVs as possible product [3, 5]. Finally, during M-phase, chromosome nondisjunction or lagging leads to segregation errors causing whole-chromosome aneuploidies. Also, a lagging chromosome may be incorporated into a micronucleus, which may spark complex rearrangements such as chromothripsis and chromoanasynthesis, where tens/hundreds of CNVs and other rearrangements simultaneously originate from a single event. Chromothripsis is characterized by alternating copy number profiles with loss and preservation of heterozygosity, while in chromoanasynthesis duplications or triplications on one parental allele are frequent [2, 6–9].

However, also postmitotically, somatic CNVs can emerge, meaning that somatic mutations in human neurons can still arise after corticogenesis is completed. Errors induced during transcription-related DNA damage and retrotransposon activity along with epigenetic modification, cellular respiration, environmental stressors, or undesired effects of internal cell chemistry in combination with erroneous double-strand break (DSB) repair, such as homologous recombination (HR) and nonhomologous end-joining (NHEJ), can introduce CNVs, ranging from megabase-scale CNVs to smaller indels, or SVs in the genome of terminally differentiated neurons [10–13].

SNVs in the neuronal genome often result from replication errors during the development of the brain. Nevertheless, the majority of SNVs in the brain arise in postmitotic neurons because of endogenous mutagens generated as side effects of the cellular metabolism [14, 15]. Each day, on average 20,000 to 100,000 errors arise in the genome of each nucleated cell of the human body. DNA damage repair mechanisms, such as base excision repair (BER), single-strand break repair, nucleotide excision repair (NER), and base mismatch repair, attempt to correct for the mutation; however, also DNA repair mechanisms inevitably introduce mistakes in the neuronal genome [15].

1.1.2 Burden of CNVs and SNVs Detected in Neuronal Tissue

Using cytogenetic techniques, Rehen et al. detected aneuploidies in 33% of mouse neuroblasts, while Yurov et al. reported 10% and 30–35% of neurons carrying aneuploidies in human adult and fetal brain, respectively [16–18]. Since cytogenetic approaches are prone to false-positive calls and only target a few loci with low resolution, more accurate and higher-resolution analyses enabled through next-generation sequencing of individual neurons are required to reveal the levels of somatically acquired genetic variation in brain tissue. Through single-cell genome analysis, one has shown that in

2.7% of the neurons of a postmortem healthy human brain, full-chromosome aneuploidies can be observed and 5–41% of the neurons carry at least one mega-base-scale CNV [19], with the fraction of CNV-burdened neurons seemingly decreasing with age [20, 21]. In contrast to CNVs, SNVs appear to accumulate with age. In young individuals, approximately 1000 SNVs per neuron were observed, which increased to up to 3000–6000 SNVs per neuron at ~80 years of age [14, 22].

1.1.3 Consequences of Mosaic CNVs and SNVs in Human Brain

The high burden of somatic genetic variations illustrates the need for the analysis of the human brain at single-cell resolution. Indeed, since the majority of these somatic mutations are not clonal or at maximum appear in a small subset of the neurons, they would often not be detected by conventional bulk DNA sequencing analysis. Furthermore, it is not understood what the cellular phenotypic consequences of this genetic variation among neurons are, stressing the need for single-cell multi-omics analyses which enable coupling genetic variation to function. On the one hand, these observed mosaic mutations could contribute to normal phenotypic variation and functional decline with aging [23, 24]; however, increasing evidence indicates that these somatic genetic variations play a role in neurodevelopmental and neuropsychiatric brain disorders [25, 26] and are even suggested to contribute to several neurodegenerative disorders [10].

For example, somatic SNVs, especially loss-of-function mutations, were detected in autism spectrum disorder (ASD) candidate genes in the neurons of postmortem ASD donor brains with targeted deep sequencing [27]. Also low-level mosaic aneuploidy and structural variants are reported to play a role in autism pathology, illustrated by an ASD case with mosaic tetrasomy of chromosome 3q and a ring chromosome 17 [28–30].

That CNVs in a minority of the neurons can cause widespread brain dysfunction is demonstrated by a mosaic tetrasomy of chromosome 1q in only 20% of the neurons which caused hemimegalencephaly in a patient reported by Cai et al. [24]. Furthermore, hemimegalencephaly and focal cortical dysplasia, usually have pediatric epilepsy as a consequence and are commonly caused by mosaic gain-of-function mutations in the PI3K-AKT3-mTOR pathway only present in the genome of neurons located within the brain lesion [31].

Only aneuploidies of chromosome 13, 18, 21, and the sex chromosomes are constitutionally compatible with life birth; aneuploidies of other chromosomes may only be tolerated as somatic mosaicism during development and can have intellectual disability as a consequence [26, 32]. Furthermore, an increase in mosaic aneuploidy of chromosomes 18 and X is reported in postmortem schizophrenia donor brains [33]. Additionally, L1 retrotransposition in or near psychiatric disease-associated genes are shown to be increased in schizophrenia donor brains in comparison to healthy controls [33, 34].

Somatic mutations are also hypothesized to have pathological roles in neurodegenerative disorders as illustrated by CNVs and potentially chromoanasynthesis that were recently detected in neurons from synucleinopathy patients [35, 36]. Moreover, several studies indicate a contribution of somatic mutations to the pathogenesis of Alzheimer's disease (AD). For example, mutant forms of amyloid beta (APP) and presenilin-1 (PSEN-1) were shown to effectuate chromosome mis-segregation and aneuploidy in Alzheimer's disease (AD) patients [37, 38]. Also, a somatic *PSEN1* mutation in only 14% of cerebral cortex neurons induced a fully penetrant AD case [39]. Moreover, Bushman et al. described an 8% increase of total DNA content and a fourfold increase of nuclei with more than two *APP* copies in cortical nuclei derived from AD donors over non-diseased controls [40]. And as final example, a two-to-threefold increase in DNA content and a higher vulnerability to cell death of these hyperploid neurons in neurodegeneration-prone regions of AD donor brains was unveiled with slide-based cytometry [41, 42].

1.2 Methods for Genomic DNA Analysis in Brain

1.2.1 Cytogenetic and Cytometric Analysis

Historically, the detection of gains and losses of genetic material was performed through cytogenetic analysis using karyotyping, fluorescence in situ hybridization (FISH), and (array) comparative genomic hybridization (aCGH). Karyotyping requires the chromosomes at their most condensed conformation during (pro)-metaphase [43]. By adding AT-specific DNA ligands (e.g., quinacrine, Giemsa), the characteristic banding patterns for each chromosome are visualized [44, 45]. The resolution of karyotyping is limited to microscopic ~5 Mb changes [46]. FISH, on the other hand, enables genomic analysis in nondividing cells and can therefore be executed on adult postmitotic neurons. Probes targeting specific regions of the genome are labeled with a fluorophore allowing the identification of chromosomal aneuploidy, often assuming that the signal from, e.g., centromeric probes would translate to the entire chromosome [47, 48]. Through the combination of several locus-specific probes with different fluorophores targeting adjacent regions of a chromosome, multicolor banding (MCB) can be achieved, rendering information on the copy number of labeled chromosomal regions and chromosomal rearrangements within targeted chromosomes [17, 49]. FISH on metaphase chromosomes provides a resolution of roughly 1 Mb. However, resolution can be increased to 50–100 kb or 1–400 kb when applying FISH on interphase cell nuclei (interphase FISH) or naked DNA fibers (fiber FISH) respectively [50, 51].

Array CGH allows for the detection of duplications and deletions of specific regions of interest by differentially labeling a reference and test DNA sample with a fluorescent dye and hybridizing both on the same microarray to bind to specific probes. The fluorescence intensity of the test sample versus the reference sample indicates the presence of CNVs [52]. aCGH does not require cells to undergo cell division and can therefore be executed on neurons, such that aberrations can be detected to a resolution of 5–10 Mb [53]. More recently, high-resolution CGH using bulk DNA as input material is made possible through oligoarrays [54], and improved array-based approaches have been implemented to allow screening of copy number changes in single cells after whole-genome amplification [55–57].

In the past, also cytometric analysis was used to estimate the level of DNA content variation in neurons. In slide-based cytometry, the tissue is fixed on a microscopy slide and stained with a DNA-intercalating dye, DNA content is computationally determined based on fluorescence intensity, and neuronal density is determined to allow downscaling to the resolution of a single neuron [21, 41]. During flow cytometry, neurons with increased DNA content can be detected as a population of neuronal nuclei with higher intensities of DNA-intercalating dye signals [40]. Nevertheless, cytometric analysis can merely indicate the presence of DCV, no information on the type of DCV or which chromosomes are affected can be obtained.

1.2.2 Single-Cell Genomics Methods

The cytogenetics field has undoubtedly rendered numerous breakthroughs and particularly aided in developing understanding of the etiopathogenesis of diseases as well as diagnosing those. Furthermore, sequencing has revolutionized the genetics field by allowing the read-out of the base pairs in the genome. Sequencing of the first complete human genome with Sanger sequencing, a first-generation sequencing method, took 13 years [58], while nowadays, up to 48 complete human genomes can be sequenced in only 44 hours on next-generation sequencing (NGS) platforms from, e.g., Illumina®. The strength and innovation of NGS lies in the possibility to massively parallel sequence billions of DNA molecules per assay, rather than one by one as with Sanger dideoxy-based sequencing, at dramatically reduced cost per base as well as to multiplex samples. However, bulk sequencing analysis provides only an average profile of the genome from a DNA sample obtained from thousands or millions of cells. As a consequence, the cellular heterogeneity that occurs within the population of cells is often concealed. Even after elaborate statistical and bioinformatics analysis, merely the most prevalent heterogeneities present in at least 10% of the cells can be detected [59]. However, over the past 10–15 years, technical improvements and discoveries made sequencing the DNA contained within an individual single cell possible, enabling the study of the genome at a much greater resolution and with much higher sensitivity, allowing for the detection of heterogeneity among individual cells within a tissue.

Single-Cell/Single Nuclei Isolation

Single-cell sequencing technologies require the isolation of individual cells from tissues or cell cultures. A first technique to isolate single neurons is to dissociate the tissue into a single-cell suspension by means of enzymatic or mechanical tissue disaggregation. However, the ability to generate a cell suspension from brain tissue depends on the possibility to collect fresh neuronal tissue, which is only possible on rare occasions, for example, when resections of parts of the brain are necessary when treating epilepsy. Furthermore, even if one accomplishes to collect fresh brain tissue, the isolation of whole neurons remains a highly complex and erroneous process. Since neuronal tissue consists of a highly interconnected network of neurons, single-cell isolation through enzymatic or mechanical treatment will likely result in a suspension of easy-to-dissociate cell types, while morphologically more complex or lower abundant neurons will be damaged or lacking in the cell suspension (Fig. 1). Nevertheless, several commercially available kits are on the market to isolate single neurons from mouse and rat primary brain tissue, and an enzymatic cell-isolation procedure for human fetal brain is described by Zhong et al. [60]. Furthermore, for *Drosophila* brain, which is less complex and contains an approximate 90 to 10 ratio of neurons to glia and lacks the presence of myelin, cell isolation is feasible through enzymatic treatment with shaking and subsequent washes [61]. Once one obtains a single-neuron suspension, cell isolation can be executed following manual micro-pipetting, fluorescence-activated cell sorting (FACS), or with microfluidic devices. Micro-pipetting is recommended when only few cells are available and offers visual confirmation, although this method introduces operator bias. FACS on the other hand allows for high-throughput cell isolation when a large number of cells are available with an additional possibility for selection based on cellular phenotype. A drawback of FACS is that the neurons might get damaged in the process. Microfluidic devices allow for automated high-throughput cell isolation with visual confirmation in standardized nanoliter reactions. It is less prone to operator bias and can be executed at low cost per cell, but the likelihood of losing cells is higher with this method.

Secondly, single cells can be isolated with laser capture dissection (Fig. 1). However, for extracting neurons, this is a highly complicated process as neurites may get damaged and lost, and a high probability for contamination with other cells emerges.

The third and most commonly used method to prepare brain tissue for single-cell genome sequencing is through single-nucleus isolation (Fig. 1). Most frequently, brain tissue for research is obtained via biobanks, where donated brains are stored fixated or frozen. Since it is not feasible to extract whole neurons from frozen brain tissue, single-nucleus isolation provides a way forward for single-cell genomics. However, the major drawback of single-

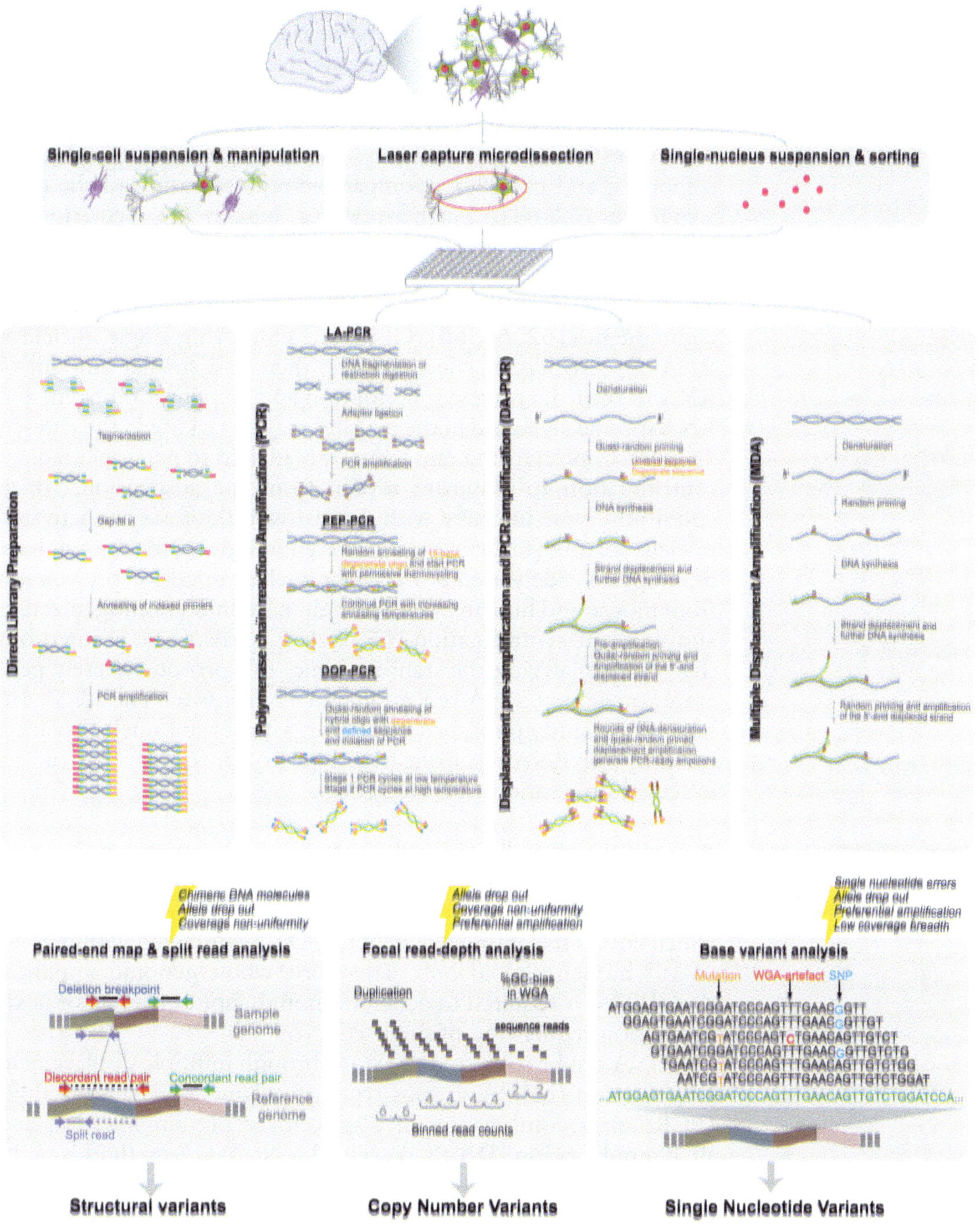

Fig. 1 Overview of single-cell genomics of brain tissue workflow. This figure depicts the different steps of the single-cell genomics workflow. The workflow starts with the extraction of single neurons from brain tissue. Single-cell isolation can be executed through single-cell suspension and manipulation, laser capture micro-dissection, or single-nucleus suspension and sorting. After deposition of individual cells or nuclei into micro-well plates, direct library preparation directly prepares the genomic DNA for next-generation sequencing, while whole-genome amplification (WGA) methods multiply the genomic DNA for library preparation. Three WGA methods are illustrated, polymerase chain reaction amplification through linker adaptor-PCR (LA-PCR),

nucleus isolation for genome analysis is that genomic material contained within micronuclei as well as mitochondria will most likely be excluded from the analysis. Nonetheless, transcriptome analysis of single neuronal nuclei is often favored over single cells. In comparison to single-cell RNA-seq, single-nucleus RNA-seq introduces less dissociated bias, eliminates transcriptional stress responses, and provides a comparable representation of the transcriptome [62–64]. Furthermore, a higher representation of nascent transcripts mapping to introns, noncoding RNA genes, and pseudogenes is obtained with single-nucleus RNA-seq, which can serve as useful cell-type markers in addition to the protein coding genes [62, 65, 66]. Of course, employing single nuclei for RNA-seq also has its consequences; there is a smaller amount of RNA to start the reverse transcription and amplification from, and RNA specifically found in the neurites may be lacking in the dataset. To extract nuclei from brain tissue, it is advised to perform gradient centrifugation to eliminate myelin from the suspension, which could otherwise interfere with downstream fluorescence-activated nucleus sorting (FANS) or droplet technologies. Swiech et al. have developed a widely used protocol for nuclei suspension preparation from mouse and human brain through mechanical homogenization and gradient centrifugation [65, 67]. It needs to be pointed out that when analyzing the transcriptome, this protocol merely generates qualitative RNA-seq results for brain tissue. The Regev lab has subsequently published a toolbox with several nuclei isolation methods for RNA-seq from fresh and frozen tumors that can be executed on a variety of tissues [68].

Whole-Genome Amplification

A single human diploid cell contains approximately 6–7 pg of genomic material. Microarrays and next-generation sequencing require at least 1 ng up to 1 μg of genomic DNA as start material, which equals the amount of genomic DNA contained within $>10^2$ to 10^5 human diploid cells. Therefore, whole-genome amplification (WGA) is required to generate enough input material for next-generation sequencing of single cells [69–72].

DNA can be subjected to WGA through multiple displacement amplification (MDA), various types of polymerase chain reactions (PCR), or a combination of displacement pre-amplification and PCR amplification (DA-PCR) (Fig. 1). No WGA method is faultless and all introduce some sort of amplification bias. However, each strategy has its own specific advantages and limitations making

Fig. 1 (continued) primer extension pre-amplification-PCR (PEP-PCR), or degenerate oligonucleotide primed-PCR (DOP-PCR), displacement pre-amplification and PCR amplification (DA-PCR), and multiple displacement amplification (MDA). WGA artifacts affecting the different types of analyses for SVs, CNVs, and SNVs are indicated with a lightning bolt. (Figure adapted from [69] and [147] with images from smart.servier.com)

them more or less suitable for certain downstream applications and the detection of specific types of genomic mutations. Therefore, a well-considered choice of WGA method is of utmost importance for the success of a single-cell sequencing experiment. In Table 1 and the following section, we list the possible WGA artifacts and subsequently provide an overview of the different WGA techniques and when to use them.

Types of WGA Artifacts

Locus Dropout (LDO) (Coverage Breadth)

The breadth of genomic coverage is the percentage of the genome that is represented in the sequencing data of a cell's WGA product. None of the WGA methods will amplify each and every base of the genome, and additionally nonuniformity among amplified genomic regions also leads to locus dropout (LDO) in the sequencing data [71–73]. For single-nucleotide variant detection, a broad genomic coverage is desired, as every SNV located in a dropped-out locus will not be detected. However, for DNA copy number analysis, a less broad representation of the genome is acceptable as these aberrations generally span more than one or a few loci.

Allele Dropout (ADO)

The phenomenon of allele dropout occurs when not all alleles that are present in the cell's genome at a certain locus are represented in the sequencing data of the WGA product [71–74]. This may result from preferential amplification (PA) due to technical factors such as nonrandom priming during the first rounds of amplification, and is particularly affecting single-nucleotide variant analysis, where ADO leads to local loss of heterozygosity (LOH) of a SNP or failure to detect a true SNV.

Coverage Nonuniformity (Coverage Depth)

The average number of times each locus of the genome is represented in the sequencing data of the WGA product represents the depth of coverage. Ideally, every base of the genome is amplified equally; however, in practice, WGA suffers from preferential amplification, and additionally some loci, e.g., genomic regions that are rich in guanine and cytosine bases, may amplify less efficiently [71–74]. Coverage nonuniformity, where certain loci across the genome have a falsely higher or lower representation in the sequencing data, is detrimental for accurate copy number calling as it leads to false-positive calling of duplications or deletions in the affected regions. Coverage nonuniformity may also affect single-nucleotide variant detection when coverage and representation of a variant allele is perturbed and low.

Single-Nucleotide Errors

With every round of amplification, single-nucleotide errors are introduced in the WGA product [71]. However, depending on the fidelity of the DNA polymerase used in each WGA method (Table 1), one method may be more suited than another for SNV calling.

Table 1
Characteristics of the three types of whole-genome amplification [71, 73]

	MDA (REPLI-g®)	PCR (Ampli1® and GenomePlex®)	DA-PCR (PicoPLEX®)
Breadth of coverage (…%) (LDO)	72–97%	74% Ampli1; 39% GenomePlex	52%
Yield	1–2 μg/cell	2 μg/cell	2–5 μg/cell
Amplicon length	2000–70,000 bp > =12 kb	50–2000 bp	200–1000 bp
Proofreading activity	YES = > phi29 enzyme	NO	NO
False-positive discovery rate	2.6–6.7*10^-5	9.6*10^-4 (GenomePlex)	2.3–2.4*10^-4
Coverage non-uniformity	FREQUENT	LESS FREQUENT	LESS FREQUENT
ADO (allelic drop out)	(LESS) FREQUENT (12–65%)	FREQUENT (76% GenomePlex)	LEAST FREQUENT (12–28%)
Chimeric DNA molecules	2%	15%	13%
Preferred application	SNV analysis	CNV analysis	CNV analysis

Percentages and numbers are provided specifically for WGA products generated with the REPLI-g® kit for multiple displacement alignment (MDA), the Ampli1® or GenomePlex® kits for polymerase chain reaction (PCR)-based WGA, and the PicoPLEX® kit for displacement pre-amplification and PCR amplification (DA-PCR) WGA

Chimeric DNA Molecules

Throughout WGA, the structure of DNA can get altered from the original DNA molecule, forming chimeric DNA molecules [75], which interfere with the detection and interpretation of structural variants such as inversions and other intra- or interchromosomal rearrangements.

Types of WGA

Multiple Displacement Amplification (MDA)

When executing MDA, alkaline lysis disrupts the cell membrane and denatures the genomic material contained within a single cell into single-stranded DNA templates to which random hexamer primers can anneal followed by isothermal amplification at 30°C for several hours. During this synthesis phase, the polymerase at the 3′ end of a newly synthesized fragment will displace the 5′ end of an adjoining synthesized fragment on its template strand, causing this single-stranded DNA to become available for another round of primer annealing and DNA synthesis. Classically, Phi29 is the polymerase of choice for MDA since it is capable of performing strand displacement and proofreading and it has a high stability, allowing for the DNA synthesis to last for hours [76]. Because the polymerase employed in MDA has proofreading, less nucleotide errors are induced in comparison to other WGA methods. Nevertheless, there is a strong bias toward C:G > T:A transitions [73, 77]. However, single-cell multiple displacement amplification (SCMDA) by Dong et al. minimizes cytosine deamination artifacts by performing cell lysis and DNA denaturation on ice through alkaline lysis [78]. Furthermore, MDA provides the highest breadth of coverage (72–97%) following sequencing among the WGA methods and generates 1–2 μg amplicons of 12 kb or larger per cell [71–73, 79]. Those features make MDA the preferred method for investigating SNVs (Fig. 1). Because chimeric DNA molecules are frequently observed, using MDA for the identification of structural variants should be done with care [75]. This method is less suited for profiling CNVs due to nonuniformity in coverage [72, 73]. However, methods for paired-end sequence analysis developed by Voet et al. enable discrimination of single-cell copy number variants from artifacts introduced by MDA WGA [80]. A commonly used, commercially available kit for MDA-based WGA is Qiagen's REPLI-g® single-cell kit.

Whole-Genome Amplification through Polymerase Chain Reaction (PCR)

PCR amplification of complete genomes precedes the first attempts at single-cell sequencing. Various approaches are available, such as primer extension pre-amplification-PCR (PEP-PCR), degenerate oligonucleotide primed-PCR (DOP-PCR), and linker adaptor-PCR (LA-PCR). PEP-PCR amplifies the DNA from a single cell with *Taq* DNA polymerase by repeated primer extension with a mixture of random degenerate 15-base oligonucleotides. The PCR reaction is initiated at low annealing temperature followed by increasing annealing temperatures [81]. DOP-PCR is typically carried out in two stages. Semi-random priming with degenerate primers and low annealing and extension temperatures as a first stage generates many binding sites for nonrandom priming by primers with defined sequences at higher annealing temperature to specifically amplify the fragments tagged by the first round of priming in a second stage [82]. LA-PCR first fragments the input DNA through enzymatic treatment, followed by ligation of adapters with known sequence which serve as annealing sites for specific primers during the PCR amplification [83]. The performance of PCR-based WGA differs among the various approaches, though generally, the breadth of genome coverage is low, and the lower fidelity for base copying among PCR-based WGA methods makes them rather less suited for SNV analysis [71–73]. On the other hand, these methods provide a fairly deep and even coverage of the genome, which makes them the preferred option for copy number analysis (Fig. 1) [71–73]. Ampli1® is an example of a commercially available LA-PCR WGA method.

Displacement Pre-Amplification and PCR Amplification (DA-PCR)

DA-PCR methods such as Takara's PicoPLEX® WGA method (formerly Rubicon Genomics) runs through 12 cycles of displacement pre-amplification, generating full amplicons with recognition sites for the PCR primers at both sides for exponential amplification which eventually can yield 2–5 μg of 200–1000 bp DNA fragments per single cell [71–73]. Similar to PicoPLEX, during multiple annealing and looping-based amplification cycles (MALBAC), looping of the full amplicons takes place, forming hairpin structures, which are unavailable for further amplification [71, 84]. Because of the combination of rounds of displacement pre-amplification and PCR, DA-PCR provides a more even amplification and hence more uniform sequence coverage of the genome with a low ADO rate. Though a less broad coverage is obtained, as approximately only 52–72% of the genome is covered after PicoPLEX and MALBAC WGA, respectively. DA-PCR is a method of choice for CNV analysis (Fig. 1) [71–73].

Alternatives for WGA Amplification

To overcome amplification bias and false-positive CNV calls introduced by WGA, new methods were developed which avoid upfront WGA or conduct alternative WGA versions. Even though these methods provide a uniform coverage of the genome, which put them in favor for CNV analysis, the breadth of coverage is generally rather low, making them less suited for SNV analysis. As a first example, the ligation-based direct library preparation method of Van Den Bos et al. is based on StrandSeq from Falconer et al., where DNA is fragmented with micrococcal nuclease followed by end-repair, A-tailing, and ligation of Illumina PE forked adapters which allows for PCR amplification with barcoded primers [85, 86]. More recent direct library preparation methods use the Tn5 transposase to simultaneously fragment the DNA and ligate sequencing adaptors, after which PCR amplification with barcoded primers can take place (Fig. 1) [87–89]. Alternative whole-genome amplification methods include the linear amplification methods LIANTI, which combines Tn5 tagmentation and T7 in vitro transcription, and sci-L3, which conducts several rounds of barcoding followed by in vitro transcription [90, 91]. LIANTI provides a genome coverage of 97%, and the false-positive rate for single-nucleotide variants is $5.4*10^{-6}$, hereby outperforming MDA as a suited approach for single-cell SNV analysis. Ligation-based and tagmentation-based direct library preparation have been executed on brain tissues from postmortem control and AD brains and fresh mouse tissue respectively, allowing the detection of CNVs smaller than 1 Mb [86, 89].

Bioinformatic Analysis of Single-Cell Genome Sequencing Data

When enough start material is obtained from single neurons, tagmentation-based library preparation through, for example, Illumina's Nextera XT DNA library prep kit, is executed to generate sequencing-ready libraries for read-out of the genome on Illumina sequencing devices. Resulting DNA sequencing reads are trimmed for adaptor contamination and mapped to the reference genome prior to downstream analysis to define SNVs, CNVs, or structural variants through base variant, focal read-depth, and paired-end/split read analyses, respectively (Fig. 1). In more detail, at least three cells carrying the same base substitution, as detected through alignment with the reference genome, are often required to reliably call an SNV [84]. Additionally, bioinformatics pipelines such as single-cell-variant caller (SCcaller) and linked-read analysis (LiRA) are developed to accurately identify SNVs and differentiate true double-stranded SNVs from single-stranded variants and artifacts [22, 78]. For CNV detection, LogR values are computed based on the number of uniquely mapped reads per genomic bin, corrected for GC%-bias, normalized and segmented, after which an integer copy number is estimated, revealing de novo-acquired CNVs [80, 92]. Lastly, structural variants are detected by evaluating discordantly mapping read-pairs or split reads crossing a rearrangement, revealing inversions, translocations, and DNA-imbalanced structural variants in the neuronal genome [80, 93].

2 Single-Cell Epigenomics

The epigenome was first described by Conrad Waddington in 1942 as "the branch of biology which studies the causal interactions between genes and their products, which brings the phenotype into being" [94]. Epigenomics entails the study of heritable changes in gene expression without altering the DNA sequence, plays a crucial regulatory role in gene expression, and is involved in many biological processes such as genomic imprinting, cellular reprogramming, and X chromosome inactivation and development, but also in disease etiology [95]. Modifications to the epigenome comprise posttranslational histone modifications, DNA methylation variants, and alterations to chromosome conformation and chromatin accessibility. The epigenome can provide insights in regulatory mechanisms underlying cell type, state, and fate decisions in disease and health.

2.1 Burden of Epigenomic Errors Detected in Neuronal Tissue

Over the past years, DNA methylation has been associated with psychiatric disorders such as schizophrenia and depression [96] as well as neurodevelopmental disorders [97]. Wockner et al. identified 2929 differentially methylated genes in the frontal cortex of people with schizophrenia compared to unaffected controls. Furthermore, aberrant methylation in the prefrontal cortex during the fetal-to-postnatal transition can increase the risk of schizophrenia [96]. Recently, the dynamics of epigenomic alterations in the human and mouse brain have been studied both at bulk [98, 99] and single-cell resolution [100–103]. Firstly, single-nucleus methylcytosine sequencing (snmC-seq) of 6000 frontal cortex nuclei derived from both 8-week-old mouse brain and 25-year-old human brain has identified 16 mouse and 21 human neuronal subpopulations [102]. Secondly, aberrations in nuclear chromatin architecture have been linked to a variety of neuropsychiatric disorders such as Alzheimer's disease, schizophrenia, and other neurological diseases [104]. Open chromatin profiling has been performed at both bulk and single-cell resolution on the adult and developing human brain, respectively [98, 101]. More specifically, bulk ATAC-seq of 14 distinct regions of 5 adult human brains showed higher regional differences in chromatin state as compared to non-neuronal tissues [98], while analysis of 75,000 cells in the developing forebrain during midgestation disclosed thousands of loci undergoing extensive cell-specific chromatin accessibility changes, correlated with neuronal differentiation [101]. Additionally, scATAC-seq revealed novel distinctions between neural progenitor cells from different cortical areas, which is not observed in their transcriptomic profile, stressing the added value of epigenomic analysis during cell fate decisions. Lastly, studies of chromatin interactions through bulk Hi-C experiments of neurons and glia in the human brain have identified over 187,000 and 167,000 promoter interactions in neurons and glia, respectively [105]. Approximately, 37% of interactions occurred between enhancers and promoters, while ~23% occurred between two different promoters. Furthermore, more than 10,000 cell-type-specific H3K27ac peaks were assigned to approximately 7800 genes in neurons further illustrating gene expression regulation through chromatin architecture. scCUT&Tag, a method for histone modification profiling as described in more detail later, has been applied in glioblastoma to profile H3K27me3 before and after surgery and radiation therapy, identifying four distinct cell clusters within the relapse sample [106].

2.2 Methods for Single-Cell Epigenomic DNA Analysis in Brain

2.2.1 DNA Methylation

DNA methylation entails the covalent addition of a methyl group to the fifth carbon atom of a cytosine base (5mC) mainly in CpG context in mammals, though low levels of non-CpG methylation have been described in human embryonic stem cells and brain tissue [107]. In humans, 70–80% of CpG sites are methylated, in contrast to CpG islands (CGIs) which are typically hypomethylated [107]. CGIs are classically defined as regions in the genome with a GC content above 50%, an observed versus expected ratio of CpG sites larger than 0.6 and longer than 200 bp [108]. Promoter and enhancer methylation is usually linked with transcriptional repression, while gene body methylation positively correlates with transcription [107]. There are two mechanisms by which promoter methylation can induce transcriptional repression; firstly, the methyl group can inhibit binding of the transcription factor to the promoter, or secondly, methyl-binding proteins (MBPs) can bind the methylated promoter, thereby occupying the transcription factor binding site [109]. DNA methylation is a very versatile modification as the genome can be (de)methylated over time, e.g., during development or cell differentiation. Two main classes of DNA methyltransferases (DNMTs) are involved in methylation: de novo DNMTs (DNMT3A and DNMT3B) and maintenance DNMT (DNMT1) [110]. Demethylation can occur either passively, through, for example, a lack of maintenance during cell division, or actively, for example, enzymatic modification. Active demethylation can occur through oxidation of 5mC to 5-hydroxymethylcytosine (5hmC) to 5-formylcytosine (5fC) and finally 5-carboxylcytosine (5caC) by the ten-eleven translocation (TET) enzyme family [111]. During replication, 5fC and 5caC can be converted to cytosine following glycosylation through the thymidine DNA glycosylase (TDG) and base excision repair (BER) pathway [107].

The golden standard for single-cell DNA methylation analysis is based on sodium bisulfite conversion. DNA methylation can be profiled genome-wide or targeted to loci enriched for CpG dinucleotides. Examples of whole-genome bisulfite sequencing (WGBS) of individual cells are single-cell bisulfite sequencing (scBS-seq) and single-nucleus methylcytosine sequencing (snmC-seq2) (Fig. 2) [112, 113]. Single cells are treated with bisulfite, causing deamination of cytosines to uracils, while methylated cytosines are protected. After conversion, adapters are incorporated through random priming and extension, followed by PCR amplification, creating sequencing-ready libraries. Whole-genome scBS-seq is able to profile up to 48% of the CpG sites genome-wide in mouse, but has the disadvantage of lower data quality compared to conventional scDNA-seq methodologies, labor intensiveness, and

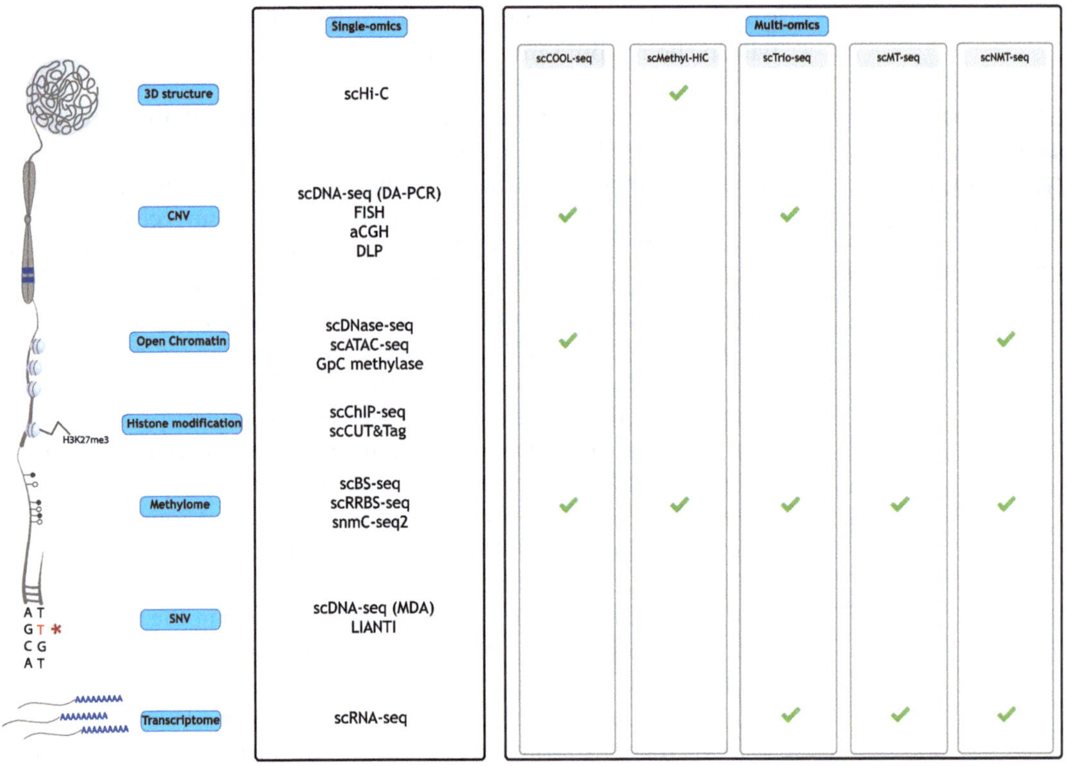

Fig. 2 Overview of single-cell (epi)genomic methods. This figure depicts the different hierarchies of (epi) genomic modifications together with the most commonly used single-cell techniques to analyze them. The method of choice for investigation of the 3D structure of the genome is scHi-C. CNV analysis can be achieved following scDNA-seq, with (DA-)PCR as the WGA method of choice, or alternatively via direct library preparation of the cell's DNA. Chromatin accessibility can be assessed using scATAC-seq or scDNase-seq, while posttranslational histone modifications can be profiled using scCUT&Tag or scChIP-seq. DNA methylation profiling can be achieved genome-wide (scBS-seq) or targeted at CpG sites (scRRBS). Finally, SNVs can be profiled using scDNA-seq after opting for MDA-based WGA or LIANTI, while scRNA-seq can be used for transcriptome analysis. Furthermore, three single-cell multi-omics methodologies are listed profiling the genome as well as epigenome; scCOOL-seq profiles a cell's copy number variations, open chromatin, and methylome; scMethyl-HiC profiles both the methylome and the 3D organization of the DNA; and finally scTrio-seq profiles a cell's copy number variations, methylome, and transcriptome. Lastly, two multi-omics techniques profiling epigenome and transcriptome, scNMT-seq, and scMT-seq are shown. scMT-seq allows methylome and transcriptome profiling, while scNMT-seq allows additional open chromatin profiling

high cost per cell. A cost-effective alternative is single-cell reduced representation bisulfite sequencing (scRRBS) (Fig. 2) [114] which employs a methylation-insensitive restriction enzyme (e.g., MspI) to digest the genome, resulting in fragments flanked by CpG dinucleotides [115]. Following end-repair and A-tailing, methylated adapters are added using a T4 ligase. After bisulfite conversion, the fragments are amplified. scRRBS is very time-consuming and has limited resolution. Both scBS-seq and scRRBS have the ability to profile DNA methylation at single-base resolution, however are unable to distinguish 5mC from 5hmC or to profile the other cytosine modifications, such as 5fC and 5caC. 5hmCs can be profiled using scAba-seq [116]. Following glucosylation of 5hmC, glucosylation-dependent digestion using AbaSI, and ligation of adapters carrying a T7 promoter, the DNA is amplified by in vitro transcription and subjected to directional RNA library preparation. Single-cell methylase-assisted bisulfite sequencing (scMAB-seq) allows 5fC and 5caC detection through methylation of all cytosines using a CpG methylase M.SssI, leaving 5fC and 5caC to be the only susceptible sites to C-to-T conversion after bisulfite treatment [117]. Furthermore, a high-throughput single-cell methylome technology, single-cell combinatorial indexing for methylation (sci-MET) has been developed in which cells are collected in pools of 2000 nuclei and tagmented with a unique barcode per pool, after which all nuclei are mixed and re-pooled to 22 nuclei per well using FANS. Next, the nuclei are bisulfite converted and subjected to library preparation according to standard WGBS procedure, ligating two additional indices [118], providing single-cell resolved sequencing data.

2.2.2 Chromatin Accessibility

DNA is compacted in the nucleus by wrapping it around nucleosomes, consisting of an octameric complex of four histones, connected by nucleosome-free linker DNA [119]. The chromatin state can be open or closed and is linked to transcriptional activity.

Techniques to analyze the chromatin accessibility can be divided into count-based and read-based methods. The count-based methods rely on comparing the number of reads at a specific locus to the average number of reads genome-wide. Examples of count-based methods are scATAC-seq and scDNase-seq (Fig. 2). Single-cell assay for transposase-accessible chromatin (scATAC-seq) uses a prokaryotic hyperactive transposase Tn5 to tagment accessible DNA [120]. After gap-fill, the adapter-ligated fragments are then PCR amplified and sequenced. DNase-seq uses a DNase I to digest the DNA, which is only possible at nucleosome-free DNA, i.e., DNase hypersensitivity sites (DHSs), and is followed by size selection and library preparation [121]. DNase-seq is applicable to single-cell inputs by means of using carrier DNA, i.e., circular plasmid DNA, preventing DNA loss during purification steps and library preparation.

Single-cell nucleosome occupancy and methylome sequencing (scNOMe-seq) is a read-based multi-omics technique allowing accessible chromatin detection using a GpC methyltransferase in vitro methylating cytosines exclusively in GpC context of nucleosome-depleted DNA [122, 123]. The DNA is then processed according to the scBS-seq procedure. The advantage of read-based methods is the ability to discriminate between inaccessible and missing data. As for count-based methods, true negatives (closed chromatin) are hard to distinguish from false negatives (loss of DNA). Additionally, scNOMe-seq is able to profile the chromatin state at single-base resolution and is not limited by the frequency of GpC sites as they occur very frequently in the human genome (1 in ~16 bp) [124]. However, GpCpG and CpCpG sites need to be excluded from the analysis due to the inability to discriminate between in vitro and in vivo methylation and the off-target effects of the enzyme, retaining 52% of all GpC sites.

2.2.3 Histone Modifications

The histones of the nucleosome are susceptible to posttranslational modifications, influencing chromatin dynamics and regulating gene expression [125, 126]. The most frequent modifications are histone acetylation, methylation, and phosphorylation. Histone acetylation entails the addition of an acetyl group to the ε-amino group of the lysine side chain and is facilitated or undone by histone acetyltransferases and histone deacetylases, respectively [125]. As such, the positive charge of lysine is neutralized, weakening the DNA-histone interactions. Histone phosphorylation entails the addition of a phosphor group to the hydroxyl group of a target amino acid side chain in the N-terminus of the histone tail and is regulated by kinases and phosphatases. This modification adds a negative charge to the histone affecting DNA-histone interactions. Lastly, histone methylation comprises the addition of a methyl group to the side chain of lysine and arginine mainly. Lysines can be mono-, di-, or tri-methylated, while arginines can be mono- or (a)symmetrically di-methylated. In contrast to acetylation and phosphorylation, methylation does not alter the electrical charge.

One of the most commonly used methods to study DNA modifications is chromatin immunoprecipitation followed by sequencing (ChIP-seq), which recently has been optimized for single cells (Fig. 2) [127]. scChIP-seq uses an antibody targeting the modified histone which is used to enrich the target loci. The nonspecific binding of the antibody can lead to noise, which is especially profound in single cells. However, Rotem et al. developed a microfluidics system, overcoming this issue by ligating a barcode to the chromatin from single cells after fragmentation and multiplexing many cells *prior* to immunoprecipitation. scChIP-seq shows high specificity, i.e., 50% of reads aligning to known positive sites, as well as obtaining nearly 100% accuracy in distinguishing cell types. Limitations of scChIP-seq can also be overcome by single-cell cleavage under targets and tagmentation (scCUT&Tag), using antibodies directed to chromatin proteins to tether a Tn5 transposase, tagmenting the DNA followed by subsequent library preparation (Fig. 2) [128]. Alternative methods to study histone modifications entail scDamID [129] and SCAN [130]. Single chromatin molecule analysis in nanochannels (SCAN) digests the native chromatin using a micrococcal nuclease (MNase) and adds a fluorophore-labeled antibody targeting a specific histone modification as well as a fluorescent intercalator after which the DNA is driven through a nanochannel and excitated [130, 131]. The chromatin fragments of interest can be recovered for further PCR analysis or sequencing. Single-cell DamID uses a fusion protein of *E. coli* deoxyadenosine methylase (Dam) and a protein targeting the modification of interest [129]. The Dam will methylate adenosines in proximity of the modification, followed by methylation-sensitive digestion using DpnI and subsequent library preparation.

2.2.4 Chromosome Conformation

Regulation of gene expression is not only accomplished by local (epi)genomic modifications but also through higher-order chromatin interactions [132]. The leading technique to study chromatin interactions is single-cell Hi-C [133]. After cross-linking chromatin interactions, enzymatic restriction digestion (e.g., BglII or DpnII) is performed, followed by biotin fill-in and religation. The religated molecules are pulled down using streptavidin-coated beads and digested with a second restriction enzyme AluI, fragmenting the DNA, followed by adapter ligation, PCR amplification, and sequencing.

3 Single-Cell Multi-omics

Integration of different single-omics technologies performed on different populations of single cells is susceptible to confounding factors such as interindividual and cellular heterogeneity. Multi-omics techniques that scrutinize multiple layers of information of the same single cell in parallel may circumvent this issue. However, single-cell multi-omics are labor-intensive and costly and integrative data analysis remains challenging. The field of single-cell multi-omics has dramatically evolved over the last decade to provide a wide range of combined multi-omics techniques.

3.1 Techniques Analyzing the Genome and Transcriptome

DNA and mRNA seq (DR-seq) [134] and genome-plus-transcriptome sequencing (G&T-seq) [135] allow parallel assessment of the genome and transcriptome. DR-seq starts with reverse transcription of the mRNA followed by quasilinear amplification of the gDNA and single-stranded cDNA in a single tube, preventing loss of material. After pre-amplification, the gDNA and cDNA are split where the gDNA is subjected to library preparation, while the cDNA is further amplified using in vitro transcription, followed by reverse transcription and PCR [134]. G&T-seq initiates with separation of the gDNA and mRNA contained within a single cell by means of oligodT-coated paramagnetic beads which capture poly-A-mRNA. The mRNA is reverse transcribed and pre-amplified following a modified smart-seq2 protocol [137, 138], followed by Nextera XT library preparation and sequencing. The gDNA is pre-amplified by means of WGA, supporting the downstream analysis (PicoPLEX for CNV and REPLI-g for SNV), which is followed by Nextera XT library preparation and sequencing. Recently, direct nuclear tagmentation and RNA sequencing (DNTR-seq) has been developed [139]. DNTR-seq is based on physical separation of the nucleus and cytosol by centrifugation, followed by a modified smart-seq2 protocol [137, 138], whereas the gDNA is subjected to direct library preparation, e.g., DLP(+), circumventing conventional WGA plus library preparation. As the abovementioned techniques, only profile mRNA, which is only a fraction of the total RNA, simultaneous isolation of genomic DNA, and total RNA (SIDR), has been developed [140]. Cells are coated with antibody-conjugated magnetic beads, after which hypotonic lysis is performed to release all cytoplasmic RNA, while the nucleus is pulled down using a magnet. Finally, Huang et al. have developed parallel RNA and DNA analysis after deep-sequencing (PRDD-seq) which employs single-cell MosaicHunter (scMH) to perform joint analysis of single-cell and bulk DNA sequencing to identify somatic SNVs and cell lineage markers to determine the neuronal cell type, cell lineage, and sequential neuronal formation of affected neurons [141].

In sum, methodologies for assessing the genome and transcriptome in parallel enable annotation of genetic cell lineage trees with cell type and state information, providing deeper insight in stem cell biology and differentiation, as well as the investigation of functional and phenotypic consequences of genomic aberrations acquired by individual cells.

3.2 Techniques Analyzing Multiple (Epi)genomic Layers

There are several techniques profiling multiple epigenomic layers or both (epi)genomic layers such as single-cell nucleosome occupancy and methylome sequencing (scNOMe-seq) [122] and single-cell chromatin overall omic-scale landscape sequencing (scCOOL-seq) (Fig. 2) [142]. During cell isolation, the DNA is treated with a GpC methyltransferase in vitro methylating cytosines exclusively in GpC context and nucleosome-depleted DNA. After isolation, the DNA is processed according to scBS-seq [112] inferring DNA methylation from C-to-T conversions in CpG context and chromatin accessibility from C-to-T conversions in GpC context. Additionally, scCOOL-seq allows inference of the cell's copy number variation and ploidy.

Recently, scMethyl-HiC, allowing parallel profiling of a single cell's chromosome conformation and methylome, has been developed, combining in situ Hi-C and bisulfite sequencing (Fig. 2) [143]. Following standard Hi-C procedure, bisulfite conversion is performed *prior* to library preparation. In general, over 16 million CpG sites could be profiled, with 20% less CpG sites covered compared to stand-alone bisulfite sequencing, while a higher CpG coverage in open chromatin was observed.

3.3 Techniques Analyzing the (Epi) Genome and Transcriptome

Examples of techniques that allow parallel investigation of the epigenome and transcriptome are single-cell methylome and transcriptome sequencing (scMT-seq and scM&T-seq) [136], single-cell chromatin accessibility and transcriptome sequencing (scCAT-seq) [144], single-cell Trio-seq2 [145], and single-cell nucleosome, methylome and transcriptome sequencing (scNMT-seq), which allows for profiling of a cell's open chromatin, methylome, and transcriptome through combining scNOMe-seq with the gDNA/mRNA separation of G&T-seq. However, scNMT-seq remains a labor-intensive and costly method with bisulfite-induced DNA damage, lowering data quality [124]. scMT-seq entails manual separation of the cell's cytosol and nucleus. The mRNA is processed according to smart-seq2 procedure [137, 138], while the DNA is processed according to scRRBS procedure [114]. In contrast, scM&T-seq automatically separates the gDNA and mRNA according to G&T-seq procedure [135, 136] after which the gDNA is processed according to scBS-seq procedure [112]. Single-cell Trio-seq2 – enabling DNA methylome, CNV, and RNA analyses – separates the nucleus from the cytoplasm using centrifugation, followed by reverse transcription, polyadenylation, PCR amplification, and library preparation [145], while the DNA is processed according to scBS-seq (Fig. 2) [112]. scCAT-seq commences by separation of the nucleus and cytoplasm which is processed according to smart-seq2 [137, 138], while the gDNA is processed according to scATAC-seq procedure [120]. Lastly, droplet-based single-nucleus chromatin accessibility and mRNA expression sequencing (SNARE) has been developed [146]. Nuclei are permeabilized and accessible chromatin is tagmented using a Tn5 transposase to be encapsulated in a droplet, together with their mRNA. Through heating, the tagmented chromatin and mRNA are released and mixed with a splint oligonucleotide, complementary to the transposase-inserted sequence allowing capture of the tagmented DNA, and carrying a poly(A)-tail allowing simultaneous capture of poly(A) mRNA and the tagmented DNA by an oligo(dT)-carrying barcoded bead. After capture, libraries for mRNA and gDNA are prepared. SNARE-seq was applied to mouse neonatal cerebral cortex, recovering 5000 nuclei, identifying 19 cell clusters capturing finer distinctions between closely related cell states.

Acknowledgments

We acknowledge the Research Foundation Flanders (FWO, G081318N, 1S24017N), KU Leuven C14/18/092, and the Wellcome Trust (Grant 206194).

References

1. Lupski JR (2007) Genomic rearrangements and sporadic disease. Nat Genet 39(7S): S43–S46. https://doi.org/10.1038/ng2084
2. Janssen A, Van Der Burg M, Szuhai K, Kops GJPL, Medema RH (2011) Chromosome segregation errors as a cause of DNA damage and structural chromosome aberrations. Science (80) 333(6051):1895–1898. https://doi.org/10.1126/science.1210214
3. Gelot C, Magdalou I, Lopez BS (2015) Replication stress in mammalian cells and its consequences for mitosis. Genes (Basel) 6(2): 267–298. https://doi.org/10.3390/genes6020267
4. Zhang F, Khajavi M, Connolly AM et al (2015) The DNA replication FoSTeS/MMBIR mechanism can generate genomic, genic and exonic complex rearrangements in humans Feng. Nat Genet 41(7):849–853. https://doi.org/10.1038/ng.399.The
5. McKinnon PJ (2009) DNA repair deficiency and neurological disease. Nat Rev Neurosci 10(2):100–112. https://doi.org/10.1038/nrn2559
6. Crasta K, Ganem NJ, Dagher R et al (2012) DNA breaks and chromosome pulverization from errors in mitosis. Nature 482(7383): 53–58. https://doi.org/10.1038/nature10802
7. Santaguida S, Amon A (2015) Short- and long-term effects of chromosome mis-segregation and aneuploidy. Nat Rev Mol Cell Biol 16(8):473–485. https://doi.org/10.1038/nrm4025
8. Stephens PJ, Greenman CD, Fu B et al (2011) Massive genomic rearrangement acquired in a single catastrophic event during cancer development. Cell 144(1):27–40. https://doi.org/10.1016/j.cell.2010.11.055
9. Zhang C, Leibowitz ML, Pellman D, Zhang C, Leibowitz ML, Pellman D (2013) Chromothripsis and beyond : rapid genome evolution from complex chromosomal rearrangements:2513–2530. https://doi.org/10.1101/gad.229559.113
10. Leija-Salazar M, Piette C, Proukakis C (2018) Review: somatic mutations in neurodegeneration. Neuropathol Appl Neurobiol 44(3): 267–285. https://doi.org/10.1111/nan.12465
11. Muotri AR, Chu VT, Marchetto MCN, Deng W, Moran JV, Gage FH (2005) Somatic mosaicism in neuronal precursor cells mediated by L1 retrotransposition. Nature 435(7044):903–910. https://doi.org/10.1038/nature03663
12. Madabhushi R, Pan L, Tsai LH (2014) DNA damage and its links to neurodegeneration. Neuron 83(2):266–282. https://doi.org/10.1016/j.neuron.2014.06.034
13. McConnell MJ, Moran JV, Abyzov A et al (2017) Intersection of diverse neuronal genomes and neuropsychiatric disease: the brain somatic mosaicism network. Science (80) 356 (6336):eaal1641. https://doi.org/10.1126/science.aal1641
14. Lodato MA, Woodworth MB, Lee S et al (2015) Somatic mutation in single human neurons tracks developmental and transcriptional history. Science (80). 350(6256): 94–98. https://doi.org/10.1126/science.aab1785
15. Strachan TRA (2018) Human molecular genetics, 5th edn. Taylor Fr Inc Pap, p ISBN: 9780815345893. https://doi.org/10.1038/nn1443
16. Rehen SK, Mcconnell MJ, Kaushal D, Kingsbury MA, Yang AH, Chun J (2001) Chromosomal variation in neurons of the developing and adult mammalian nervous system. PNAS 98(23):13361–13366
17. Iourov IY, Liehr T, Vorsanova SG, Kolotii AD, Yurov YB (2006) Visualization of interphase chromosomes in postmitotic cells of the human brain by multicolour banding (MCB):223–229
18. Yurov YB, Iourov IY, Vorsanova SG et al (2007) Aneuploidy and confined chromosomal mosaicism in the developing human brain. PLoS One 2(6):4–9. https://doi.org/10.1371/journal.pone.0000558
19. Mcconnell MJ, Lindberg MR, Brennand KJ et al (2014) Mosaic copy number variation in human neurons 342(6158):632–637. https://doi.org/10.1126/science.1243472. Mosaic
20. Chronister WD, Wierman MB, Burbulis IE et al (2018) CNV neurons are rare in aged human. Neocortex
21. Fischer H-G, Morawski M, Brü Ckner MK, Mittag A, Tarnok A Arendt T. Changes in neuronal DNA content variation in the human brain during aging. https://doi.org/10.1111/j.1474-9726.2012.00826.x
22. Lodato MA, Rodin RE, Bohrson CL et al (2017) Aging and neurodegeneration are associated with increased mutations in single human neurons. Science 559:1–8. https://doi.org/10.1126/science.aao4426

23. Knouse KA, Wu J, Whittaker CA, Amon A (2014) Single cell sequencing reveals low levels of aneuploidy across mammalian tissues. Proc Natl Acad Sci U S A 111(37): 13409–13414. https://doi.org/10.1073/pnas.1415287111
24. Cai X, Evrony GD, Lehmann HS et al (2014) Single-cell, genome-wide sequencing identifies clonal somatic copy-number variation in the human brain. Cell Rep 8(5):1280–1289. https://doi.org/10.1016/j.celrep.2014.07.043
25. Poduri A, Evrony GD, Cai X, Walsh CA Somatic mutation, genomic variation, and neurological disease. Science
26. Rodin RE, Walsh CA (2018) Somatic mutation in pediatric neurological diseases. Pediatr Neurol 87:20–22. https://doi.org/10.1016/j.pediatrneurol.2018.08.008
27. D'Gama AM, Pochareddy S, Li M et al (2015) Targeted DNA sequencing from autism Spectrum disorder brains implicates multiple genetic mechanisms. Neuron 88(5): 910–917. https://doi.org/10.1016/j.neuron.2015.11.009
28. Yurov YB, Vorsanova SG, Iourov IY et al (2007) Unexplained autism is frequently associated with low-level mosaic aneuploidy. J Med Genet 44(8):521–525. https://doi.org/10.1136/jmg.2007.049312
29. Oliveira G, Matoso E, Vicente A et al (2003) Partial tetrasomy of chromosome 3q and mosaicism in a child with autism. J Autism Dev Disord 33(2):177–185. https://doi.org/10.1023/A:1022943627660
30. Havlovicova M et al (2006) A girl with Neurofibromatosis type 1, atypical autism and mosaic ring chromosome 17. Am J Hum Genet 143(A):76–81. https://doi.org/10.1002/ajmg.a
31. D'Gama AM, Woodworth MB, Hossain AA et al (2017) Somatic mutations activating the mTOR pathway in dorsal Telencephalic progenitors cause a continuum of cortical dysplasias. Cell Rep 21(13):3754–3766. https://doi.org/10.1016/j.celrep.2017.11.106
32. Erickson RP (2010) Somatic gene mutation and human disease other than cancer: an update. Mutat Res - Rev Mutat Res 705(2): 96–106. https://doi.org/10.1016/j.mrrev.2010.04.002
33. Yurov YB, Vostrikov VM, Vorsanova SG, Monakhov VV, Iourov IY (2001) Multicolor fluorescent in situ hybridization on postmortem brain in schizophrenia as an approach for identification of low-level chromosomal aneuploidy in neuropsychiatric diseases. Brain and Development 23:186–190. https://doi.org/10.1016/S0387-7604(01)00363-1
34. Bundo M, Toyoshima M, Okada Y et al (2014) Increased L1 retrotransposition in the neuronal genome in schizophrenia. Neuron 81(2):306–313. https://doi.org/10.1016/j.neuron.2013.10.053
35. Perez-Rodriguez D, Kalyva M, Leija-Salazar M et al (2019) Investigation of somatic CNVs in brains of synucleinopathy cases using targeted SNCA analysis and single cell sequencing. Acta Neuropathol Commun 7(1):1–22. https://doi.org/10.1186/s40478-019-0873-5
36. Mokretar K, Pease D, Taanman J-W et al (2018) Somatic copy number gains of α-synuclein (SNCA) in Parkinson's disease and multiple system atrophy brains. Brain 141(8):2419–2431. https://doi.org/10.1093/brain/awy157
37. Granic A, Padmanabhan J, Norden M, Potter H (2010) Alzheimer Abeta peptide induces chromosome mis-segregation and aneuploidy, including trisomy 21: requirement for tau and APP. Mol Biol Cell 21(4): 511–520. https://doi.org/10.1091/mbc.E09-10-0850
38. Boeras DI, Granic A, Padmanabhan J, Crespo NC, Rojiani AM, Potter H (2008) Alzheimer's presenilin 1 causes chromosome missegregation and aneuploidy. Neurobiol Aging 29(3):319–328. https://doi.org/10.1016/j.neurobiolaging.2006.10.027
39. Beck JA, Poulter M, Campbell TA et al (2004) Somatic and germline mosaicism in sporadic early-onset Alzheimer's disease. Hum Mol Genet 13(12):1219–1224. https://doi.org/10.1093/hmg/ddh134
40. Bushman DM, Kaeser GE, Siddoway B et al (2015) Genomic mosaicism with increased amyloid precursor protein (APP) gene copy number in single neurons from sporadic Alzheimer's disease brains. elife. https://doi.org/10.7554/eLife.05116.001
41. Arendt T, Bruckner MK, Lasche A (2015) Regional mosaic genomic heterogeneity in the elderly and in Alzheimer disease as a correlate of neuronal vulnerability. Acta Neuropathol 130(4):501–510. https://doi.org/10.1007/s00401-015-1465-5
42. Arendt T, Brückner MK, Mosch B, Lösche A (2010) Selective cell death of hyperploid neurons in Alzheimer's disease. Am J Pathol 177(1):15–20. https://doi.org/10.2353/ajpath.2010.090955

43. Moorhead PS, Nowell PC, Mellman WJ, Battips DM, Hungerford DA (1960) Chromosome preparations of leukocytes cultured from human peripheral blood. Exp Cell Res 20(3):613–616. https://doi.org/10.1016/0014-4827(60)90138-5
44. Caspersson T, Zech L, Johansson C (1970) Differential binding of alkylating fluorochromes in human chromosomes. Exp Cell Res 60(3):315–319. https://doi.org/10.1016/0014-4827(70)90523-9
45. Rowley JD (1973) A new consistent chromosomal abnormality in chronic myelogenous leukaemia identified by quinacrine fluorescence and Giemsa staining. Nature 273: 290–293. https://doi.org/10.3389/fnins.2018.00081
46. Shaffer LG, Bejjani BA (2004) A cytogeneticist's perspective on genomic microarrays. Hum Reprod Update 10(3):221–226. https://doi.org/10.1093/humupd/dmh022
47. Pinkel D, Straume T, Gray JW (1986) Cytogenetic analysis using quantitative, high-sensitivity, fluorescence hybridization. Proc Natl Acad Sci U S A 83(9):2934–2938. https://doi.org/10.1073/pnas.83.9.2934
48. Pinkel D, Gray JW, Trask B et al (1986) Cytogenetic analysis by in situ hybridization with fluorescently labeled nucleic acid probes. Cold Spring Harb Symp Quant Biol 83: 2934–2938. https://doi.org/10.1002/dvg.20740
49. Liehr T, Heller A, Starke H et al (2002) Microdissection based high resolution multicolor banding for all 24 human chromosomes. Int J Mol Med 9:335–339. https://doi.org/10.1111/j.1471-4159.2009.06124.x
50. Parra I, Windle B (1993) High resolution visual mapping of stretched DNA by fluorescent hybridization. Nat Genet 5(1):17–21. https://doi.org/10.1038/ng0993-17
51. Raap AK, Florijn RJ, Blonden LAJ et al (1996) Fiber FISH as a DNA mapping tool. Methods A Companion Methods Enzymol 9(1):67–73. https://doi.org/10.1006/meth.1996.0009
52. Kallioniemi A, Kallioniemi OP, Sudar D et al (1992) Comparative genomic hybridization for molecular cytogenetic analysis of solid tumors. Science 258(5083):818–821. https://doi.org/10.1126/science.1359641
53. Speicher MR et al (1993) Molecular cytogenetic analysis of formalin-fixed, paraffin-embedded solid tumors by comparative genomic hybridization after universal DNA-amplification. Hum Mol Genet 2: 1907–1914. https://doi.org/10.1074/mcp.M700467-MCP200
54. Carvalho B, Ouwerkerk E, Meijer GA, Ylstra B (2004) High resolution microarray comparative genomic hybridisation analysis using spotted oligonucleotides. J Clin Pathol 57(6):644–646. https://doi.org/10.1136/jcp.2003.013029
55. Fiegler H, Geigl JB, Langer S et al (2007) High resolution array-CGH analysis of single cells. Nucleic Acids Res 35(3):1–10. https://doi.org/10.1093/nar/gkl1030
56. Vanneste E, Voet T, Le Caignec C et al (2009) Chromosome instability is common in human cleavage-stage embryos. Nat Med 15(5): 577–583. https://doi.org/10.1038/nm.1924
57. McConnell MJ, Lindberg MR, Brennand KJ et al (2013) Mosaic copy number variation in human neurons.-supplementary online material. Science 342(6158):632–637. https://doi.org/10.1126/science.1243472
58. Lander ES, Linton LM, Birren B et al (2001) Initial sequencing and analysis of the human genome. Nature 409(6822):860–921. https://doi.org/10.1038/s41593-019-0349-8.Single-cell
59. Nik-Zainal S, Alexandrov LB, Wedge DC et al (2012) Mutational processes molding the genomes of 21 breast cancers. Cell 149(5): 979–993. https://doi.org/10.1016/j.cell.2012.04.024
60. Zhong S, Zhang S, Fan X et al (2018) A single-cell RNA-seq survey of the developmental landscape of the human prefrontal cortex. Nature. https://doi.org/10.1038/nature25980
61. Davie K, Janssens J, Koldere D et al (2018) A single-cell transcriptome atlas of the aging drosophila brain. Cell:1–17. https://doi.org/10.1016/j.cell.2018.05.057
62. Bakken TE, Hodge RD, Miller JM et al (2018) Equivalent high-resolution identification of neuronal cell types with single-nucleus and single-cell RNA-sequencing. BioRxiv. 239749. https://doi.org/10.1101/239749
63. Lacar B, Linker SB, Jaeger BN et al (2016) Nuclear RNA-seq of single neurons reveals molecular signatures of activation. https://doi.org/10.1038/ncomms11022
64. Koenitzer JR, Wu H, Atkinson JJ, Brody SL, Humphreys BD. Single nucleus RNASeq profiling of mouse lung : reduced dissociation bias and improved detection of rare cell types compared with single cell RNASeq. 2020

65. Habib N, Li Y, Heidenreich M et al (2016) Div-Seq: single-nucleus RNA-Seq reveals dynamics of rare adult newborn neurons. Science 7:1–20. https://doi.org/10.1126/science.aad7038
66. Abdelmoez MN, Iida K, Oguchi Y et al (2018) SINC-seq : correlation of transient gene expressions between nucleus and cytoplasm reflects single-cell physiology. Genome Biol:1–11
67. Swiech L, Heidenreich M, Banerjee A et al (2015) In vivo interrogation of gene function in the mammalian brain using CRISPR-Cas9. Nat Biotechnol 33(1):102–106. https://doi.org/10.1038/nbt.3055
68. Slyper M, Porter CBM, Ashenberg O, et al. A single-cell and single-nucleus RNA-seq toolbox for fresh and frozen human tumors
69. Macaulay IC, Voet T (2014) Single cell genomics: advances and future perspectives. PLoS Genet 10(1). https://doi.org/10.1371/journal.pgen.1004126
70. Kalisky T, Blainey P, Quake SR (2011) Genomic analysis at the single-cell level. Annu Rev Genet 45(46):431–445. https://doi.org/10.1146/annurev-genet-102209-163607
71. Huang L, Ma F, Chapman A, Lu S, Xie XS (2015) Single-cell whole-genome amplification and sequencing: methodology and applications. Annu Rev Genomics Hum Genet 16(1):79–102. https://doi.org/10.1146/annurev-genom-090413-025352
72. Gawad C, Koh W, Quake SR (2016) Single-cell genome sequencing: current state of the science. Nat Rev Genet 17(3):175–188. https://doi.org/10.1038/nrg.2015.16
73. Hou Y, Wu K, Shi X et al (2015) Comparison of variations detection between whole-genome amplification methods used in single-cell resequencing. Gigascience 4(1): 37. https://doi.org/10.1186/s13742-015-0068-3
74. Findlay I, Ray P, Quirke P et al (1995) Diagnosis and preventing inherited disease: allelic drop-out and preferential amplification in single cells and human blastomeres: implications for preimplantation diagnosis of sex and cystic fibrosis. Hum Reprod 10(6):1609–1618. https://doi.org/10.1021/pr800207h
75. Lasken RS, Stockwell TB (2007) Mechanism of chimera formation during the multiple displacement amplification reaction. BMC Biotechnol 7(19). https://doi.org/10.1186/1472-6750-7-19
76. Dean FB, Nelson JR, Giesler TL, Lasken RS (2001) Rapid amplification of plasmid and phage DNA using Phi29 DNA polymerase and multiply-primed rolling circle amplification. Genome Res 11(6):1095–1099. https://doi.org/10.1101/gr.180501
77. Leung ML, Wang Y, Waters J, Navin NE (2015) SNES : single nucleus exome sequencing SNES : single nucleus exome sequencing. https://doi.org/10.1186/s13059-015-0616-2
78. Dong X, Zhang L, Milholland B et al (2017) Accurate identification of single-nucleotide variants in whole-genome-amplified single cells. Nat Methods 14(5):491–493. https://doi.org/10.1038/nmeth.4227
79. Dean FB, Hosono S, Fang L et al (2002) Comprehensive human genome amplification using multiple displacement amplification. Proc Natl Acad Sci U S A 99(8):5261–5266. https://doi.org/10.1073/pnas.082089499
80. Voet T, Kumar P, Van Loo P et al (2013) Single-cell paired-end genome sequencing reveals structural variation per cell cycle. Nucleic Acids Res 41(12):6119–6138. https://doi.org/10.1093/nar/gkt345
81. Zhang LIN, Cui X, Schmitt K, Hubert R, Navidit W, Arnheim N (1992) Whole genome amplification from a single cell: Implications for genetic analysis 89(July):5847–5851
82. Cheung VG, Nelson SF (1996) Whole genome amplification using a degenerate oligonucleotide primer allows hundreds of genotypes to be performed on less than one nanogram of genomic DNA. Proc Natl Acad Sci U S A 93(25):14676–14679. https://doi.org/10.1073/pnas.93.25.14676
83. Klein CA, Schmidt-Kittler O, Schardt JA, Pantel K, Speicher MR, Riethmüller G (1999) Comparative genomic hybridization, loss of heterozygosity, and DNA sequence analysis of single cells. Proc Natl Acad Sci U S A 96(8):4494–4499. https://doi.org/10.1073/pnas.96.8.4494
84. Zong C, Lu S, Chapman AR, Xie XS (2012) Genome-wide detection of single-nucleotide and copy-number variations of a single human cell. Science 338(6114):1622–1626. https://doi.org/10.1126/science.1229164
85. Falconer E, Hills M, Naumann U et al (2013) DNA template strand sequencing of single-cells maps genomic rearrangements at high resolution. Nat Methods 9(11):1107–1112. https://doi.org/10.1038/nmeth.2206.DNA
86. van den Bos H, Spierings DCJ, Taudt A et al (2016) Single-cell whole genome sequencing reveals no evidence for common aneuploidy in normal and Alzheimer's disease neurons.

Genome Biol 17(1):1–9. https://doi.org/10.1186/s13059-016-0976-2

87. Zahn H, Steif A, Laks E, Eirew P, Vaninsberghe M, Sohrab P (2017) Scalable whole-genome single-cell library preparation without pre-amplification. Nat Methods 2: 1–39. https://doi.org/10.1038/nmeth.4140
88. Laks E, McPherson A, Zahn H et al (2019) Clonal decomposition and DNA replication states defined by scaled single-cell genome sequencing. Cell 179(5):1207–1221.e22. https://doi.org/10.1016/j.cell.2019.10.026
89. Rohrback S, April C, Kaper F et al (2018) Submegabase copy number variations arise during cerebral cortical neurogenesis as revealed by single-cell whole-genome sequencing. Proc Natl Acad Sci U S A 5(16): 201812702. https://doi.org/10.1073/pnas.1812702115
90. Chen C, Xing D, Tan L et al (2017) Single-cell whole-genome analyses by linear amplification via transposon insertion (LIANTI). Science 356(6334):189–194. https://doi.org/10.1126/science.aak9787
91. Yin Y, Jiang Y, Lam KG et al (2019) High-throughput single-cell sequencing with linear amplification. Mol Cell:1–15. https://doi.org/10.1016/j.molcel.2019.08.002
92. Baslan T, Kendall J, Rodgers L et al (2012) Genome-wide copy number analysis of single cells. Nat Protoc 7(6):1024–1041. https://doi.org/10.1038/nprot.2012.039
93. Evrony GD, Cai X, Lee E et al (2012) Single-neuron sequencing analysis of l1 retrotransposition and somatic mutation in the human brain. Cell 151(3):483–496. https://doi.org/10.1016/j.cell.2012.09.035
94. Goldberg AD, Allis CD, Bernstein E (2007) Epigenetics: a landscape takes shape. Cell 128(4):635–638. https://doi.org/10.1016/j.cell.2007.02.006
95. Allis CD, Jenuwein T (2016) The molecular hallmarks of epigenetic control. Nat Rev Genet 17(8):487–500. https://doi.org/10.1038/nrg.2016.59
96. Wockner LF, Noble EP, Lawford BR et al Genome-wide DNA methylation analysis of human brain tissue from schizophrenia patients. Transl Psychiatry 2014(4):1–9. https://doi.org/10.1038/tp.2013.111
97. Tognini P, Napoli D, Pizzorusso T (2015) Dynamic DNA methylation in the brain: a new epigenetic mark for experience-dependent plasticity. Front Cell Neurosci 9: 1–11. https://doi.org/10.3389/fncel.2015.00331
98. Fullard JF, Hauberg ME, Bendl J et al (2018) An atlas of chromatin accessibility in the adult human brain. Genome Res 28(8): 1243–1252. https://doi.org/10.1101/gr.232488.117
99. Bryois J, Garrett ME, Song L et al (2018) Evaluation of chromatin accessibility in prefrontal cortex of individuals with schizophrenia. Nat Commun 9(1). https://doi.org/10.1038/s41467-018-05379-y
100. Sinnamon JR, Torkenczy KA, Linhoff MW et al (2019) The accessible chromatin landscape of the murine hippocampus at single-cell resolution. Genome Res 29(5):857–869. https://doi.org/10.1101/gr.243725.118
101. Ziffra R, Kim C, Wilfert A et al (2019) Single cell epigenomic atlas of the developing human brain and organoids. https://doi.org/10.1101/2019.12.30.891549
102. Luo C, Keown CL, Kurihara L et al (2017) Single-cell methylomes identify neuronal subtypes and regulatory elements in mammalian cortex. Science 4:357–600
103. Liu H, Zhou J, Tian W et al (2020) DNA methylation atlas of the mouse brain at single-cell resolution. Biorxiv
104. Medrano-Fernández A, Barco A (2016) Nuclear organization and 3D chromatin architecture in cognition and neuropsychiatric disorders. Mol Brain 9(1):1–12. https://doi.org/10.1186/s13041-016-0263-x
105. Hu B, Won H, Mah W et al (2020) Neuronal and glial 3D chromatin architecture illustrates cellular etiology of brain disorders. Bioarxiv. https://doi.org/10.1101/2020.05.14.096917
106. Wu SJ, Furlan SN, Mihalas AB, et al. Single-cell analysis of chromatin silencing programs in development and tumor progression. Biorxiv
107. Jang HS, Shin WJ, Lee JE, Do JT (2017) CpG and non-CpG methylation in epigenetic gene regulation and brain function. Genes (Basel) 8(6):2–20. https://doi.org/10.3390/genes8060148
108. Siegfried Z, Simon I (2010) DNA methylation and gene expression. Wiley Interdiscip Rev Syst Biol Med 2(3):362–371. https://doi.org/10.1002/wsbm.64
109. Tirado-Magallanes R, Rebbani K, Lim R, Pradhan S, Benoukraf T (2017) Whole genome DNA methylation: beyond genes silencing. Oncotarget 8(3):5629–5637. https://doi.org/10.18632/oncotarget.13562

110. Kohli RM, Zhang Y (2013) TET enzymes, TDG and the dynamics of DNA demethylation. Nature 502(7472):472–479. https://doi.org/10.1038/nature12750

111. Wu X, Zhang Y (2017) TET-mediated active DNA demethylation: mechanism, function and beyond. Nat Rev Genet 18(9):517–534. https://doi.org/10.1038/nrg.2017.33

112. Clark SJ, Smallwood SA, Lee HJ, Krueger F, Reik W, Kelsey G (2017) Genome-wide base-resolution mapping of DNA methylation in single cells using single-cell bisulfite sequencing (scBS-seq). Nat Protoc 12(3):534–547. https://doi.org/10.1038/nprot.2016.187

113. Luo C, Rivkin A, Zhou J et al (2018) Robust single-cell DNA methylome profiling with snmC-seq2. Nat Commun 9(1):1–7. https://doi.org/10.1038/s41467-018-06355-2

114. Guo H, Zhu P, Wu X, Li X, Wen L, Tang F (2013) Single-cell methylome landscapes of mouse embryonic stem cells and early embryos analyzed using reduced representation bisulfite sequencing. Genome Res 23(12):2126–2135. https://doi.org/10.1101/gr.161679.113

115. Gu H, Smith ZD, Bock C, Boyle P, Gnirke A, Meissner A (2011) Preparation of reduced representation bisulfite sequencing libraries for genome-scale DNA methylation profiling. Nat Protoc 6(4):468–481. https://doi.org/10.1038/nprot.2010.190

116. Mooijman D, Dey SS, Boisset JC, Crosetto N, Van Oudenaarden A (2016) Single-cell 5hmC sequencing reveals chromosome-wide cell-to-cell variability and enables lineage reconstruction. Nat Biotechnol 34(8):852–856. https://doi.org/10.1038/nbt.3598

117. Wu X, Inoue A, Suzuki T, Zhang Y (2017) Simultaneous mapping of active DNA demethylation and sister chromatid exchange in single cells. Genes Dev 31(5):511–523. https://doi.org/10.1101/gad.294843.116

118. Mulqueen RM, Pokholok D, Norberg SJ et al (2018) Highly scalable generation of DNA methylation profiles in single cells. Nat Biotechnol 36(5):428–431. https://doi.org/10.1038/nbt.4112

119. Estévez-Torresa ABD (2011) DNA compaction: fundamentals and applications. Soft Matter 11. https://doi.org/10.1016/j.bbadis.2016.01.002

120. Buenrostro JD, Wu B, Litzenburger UM et al (2015) Single-cell chromatin accessibility reveals principles of regulatory variation. Nature 523(7561):486–490. https://doi.org/10.1038/nature14590

121. Cooper J, Ding Y, Song J, Zhao K (2017) Genome-wide mapping of DNase I hypersensitive sites in rare cell populations using single-cell DNase sequencing. Nat Protoc 12(11):2342–2354. https://doi.org/10.1038/nprot.2017.099

122. Statham AL, Taberlay PC, Kelly TK, Jones PA, Clark SJ (2015) Genome-wide nucleosome occupancy and DNA methylation profiling of four human cell lines. Genomics Data 3:94–96. https://doi.org/10.1016/j.gdata.2014.11.012

123. Pott S (2017) Simultaneous measurement of chromatin accessibility, DNA methylation, and nucleosome phasing in single cells. elife 6:1–19. https://doi.org/10.7554/eLife.23203

124. Clark SJ, Argelaguet R, Kapourani CA et al (2018) ScNMT-seq enables joint profiling of chromatin accessibility DNA methylation and transcription in single cells e. Nat Commun 9(1):1–9. https://doi.org/10.1038/s41467-018-03149-4

125. Bannister AJ, Kouzarides T (2011) Regulation of chromatin by histone modifications. Cell Res 21(3):381–395. https://doi.org/10.1038/cr.2011.22

126. Lawrence M, Daujat S, Schneider R (2016) Lateral thinking: how histone modifications regulate gene expression. Trends Genet 32(1):42–56. https://doi.org/10.1016/j.tig.2015.10.007

127. Rotem A, Ram O, Shoresh N et al (2015) Single-cell ChIP-seq reveals cell subpopulations defined by chromatin state. Nat Biotechnol 33(11):1165–1172. https://doi.org/10.1038/nbt.3383

128. Kaya-Okur HS, Wu SJ, Codomo CA et al (2019) CUT&tag for efficient epigenomic profiling of small samples and single cells. Nat Commun 10(1). https://doi.org/10.1038/s41467-019-09982-5

129. Lo P-K, Zhou Q (2018) Emerging techniques in single-cell epigenomics and their applications to cancer research. J Clin genomics 1(1):1–16. https://doi.org/10.4172/JCG.1000103

130. Murphy PJ, Cipriany BR, Wallin CB et al (2013) Single-molecule analysis of combinatorial epigenomic states in normal and tumor cells. Proc Natl Acad Sci U S A 110(19):7772–7777. https://doi.org/10.1073/pnas.1218495110

131. Cipriany BR, Zhao R, Murphy PJ et al (2010) Single molecule epigenetic analysis in a

nanofluidic channel. Anal Chem 82(6): 2480–2487. https://doi.org/10.1021/ac9028642

132. Li G, Reinberg D (2011) Chromatin higher-order structures and gene regulation. Curr Opin Genet Dev 21(2):175–186. https://doi.org/10.1016/j.gde.2011.01.022
133. Nagano T, Lubling Y, Stevens TJ et al (2013) Single-cell hi-C reveals cell-to-cell variability in chromosome structure. Nature 502(7469): 59–64. https://doi.org/10.1038/nature12593
134. Dey SS, Kester L, Spanjaard B, Bienko M, van Oudenaarden A (2015) Integrated genome and transcriptome sequencing of the same cell. Nat Biotechnol 33(3):285–289. https://doi.org/10.1038/nbt.3129
135. Macaulay IC, Haerty W, Kumar P et al (2015) G&T-seq: parallel sequencing of single-cell genomes and transcriptomes. https://doi.org/10.1038/Nmeth.3370
136. Angermueller C, Clark SJ, Lee HJ et al (2016) Parallel single-cell sequencing links transcriptional and epigenetic heterogeneity. Nat Methods 13(3):229–232. https://doi.org/10.1038/nmeth.3728
137. Picelli S, Björklund ÅK, Faridani OR, Sagasser S, Winberg G, Sandberg R (2013) Smart-seq2 for sensitive full-length transcriptome profiling in single cells. Nat Methods 10:1096–1098. https://doi.org/10.1038/nmeth.2639
138. Picelli S, Bjrklund AK, Reinius B, Sagasser S, Winberg G, Sandberg R (2014) Tn5 transposase and tagmentation procedures for massively scaled sequencing projects. Genome Res 24(12):2033–2040. https://doi.org/10.1101/gr.177881.114
139. Zachariadis V, Cheng H, Andrews N, Enge M (2020) A highly scalable method for joint whole-genome sequencing and gene-expression profiling of single cells. Mol Cell 80(3):541–553.e5. https://doi.org/10.1016/j.molcel.2020.09.025
140. Han KY, Kim KT, Joung JG et al (2018) SIDR: simultaneous isolation and parallel sequencing of genomic DNA and total RNA from single cells. Genome Res 28(1): 75–87. https://doi.org/10.1101/gr.223263.117
141. Huang AY, Huang AY, Huang AY et al (2020) Parallel RNA and DNA analysis after deep sequencing (PRDD-seq) reveals cell type-specific lineage patterns in human brain. Proc Natl Acad Sci U S A 117(25): 13886–13895. https://doi.org/10.1073/pnas.2006163117
142. Li L, Guo F, Gao Y et al (2018) Single-cell multi-omics sequencing of human early embryos. Nat Cell Biol 20(7):847–858. https://doi.org/10.1038/s41556-018-0123-2
143. Li G, Liu Y, Zhang Y et al (2019) Joint profiling of DNA methylation and chromatin architecture in single cells. Nat Methods 16(10):991–993. https://doi.org/10.1038/s41592-019-0502-z
144. Liu L, Liu C, Quintero A et al (2019) Deconvolution of single-cell multi-omics layers reveals regulatory heterogeneity. Nat Commun 10(1). https://doi.org/10.1038/s41467-018-08205-7
145. Bian S, Hou Y, Zhou X et al (2018) Single-cell multiomics sequencing and analyses of human colorectal cancer. Science (80). 362(6418):1060–1063. https://doi.org/10.1126/science.aao3791
146. Chen S, Lake BB, Zhang K (2019) High-throughput sequencing of the transcriptome and chromatin accessibility in the same cell. Nat Biotechnol 37(12):1452–1457. https://doi.org/10.1038/s41587-019-0290-0
147. Van Loo P, Voet T (2014) Single cell analysis of cancer genomes. Curr Opin Genet Dev 24(1):82–91. https://doi.org/10.1016/j.gde.2013.12.004

Chapter 3

Single-Cell Proteome Profiling of Neuronal Cells

Santosh A. Misal and Ryan T. Kelly

Abstract

Characterizing neuronal heterogeneity at the single-cell level can improve our understanding of developmental and disease processes and lead to novel therapeutic strategies. While single-cell genomic and transcriptomic measurements have matured in recent years, mass spectrometry-based proteome profiling has lagged due to sensitivity limitations. However, recent progress in miniaturizing sample processing and separations have made it possible to profile >1000 proteins from single mammalian cells including neurons. Here, we present a detailed procedure that has provided the greatest depth of coverage for proteome profiling of single neurons to date. Single motor neurons and interneurons were isolated from human spinal tissue sections by laser capture microdissection, processed for proteomic analysis within nanoliter volumes and analyzed by ultralow-flow nanoLC-MS/MS. The protocol should be adaptable to a variety of applications and measurement platforms to further neuronal research.

Key words Single-cell proteomics, NanoPOTS, Neuroproteomics, Laser capture microdissection, High-resolution mass spectrometry, FAIMS

1 Introduction

The neuron is the fundamental structural and functional unit of the highly complex and orchestrated central nervous system (CNS) and is responsible for the transmission of signaling/sensory information across the human body. Various supporting cells, including microglia, astrocytes, ependymal cells, and oligodendrocytes, also contribute to neurophysiological activities. Neurons exhibit an enormous diversity of morphology and intercellular synaptic connections in different regions of the CNS, and their anatomical, physiological, and biochemical properties tend to be more complex than other cell types [1–3]. Decoding neuronal complexity and developmental stages at the single-cell level is important for understanding different biological contexts and functions in neurological research. For example, using single-cell technologies, underlying disease pathophysiologies and progressions can be precisely mapped, and changes in biochemical expression associated with

Jonathan V. Sweedler et al. (eds.), *Single Cell 'Omics of Neuronal Cells*,
Neuromethods, vol. 184, https://doi.org/10.1007/978-1-0716-2525-5_3,

neuron-neuron and neuron-environment interactions can potentially inform the development of novel precision therapeutic strategies.

Rapidly advancing single-cell omics technologies, including genomics, transcriptomics, metabolomics, and proteomics, have tremendous potential to uncover developmental trajectories and pathophysiological mechanisms of various neurodegenerative diseases [4–7]. Genomic and transcriptomic techniques are well developed and can unambiguously and robustly profile single cells [8]. However, cellular structure and function are largely determined by the proteome, i.e., the complement of expressed proteins [9–11], and single-cell proteomics measurements to date have been much less tractable. This is due largely to limitations in sensitivity and to the inability to amplify proteins above the detection limit of the measurement platform. Fortunately, a number of recent developments in experimental design, sample preparation, separations, and mass spectrometry have dramatically increased the measurement throughput and depth of coverage for such measurements [12, 13]. For example, miniaturized sample preparation approaches including iPAD [14], OAD [15], and our recently developed nanoPOTS (Nanodroplet Processing in One pot for Trace Samples) proteomic sample processing workflow have increased the efficiency and recovery of proteomic sample processing by downscaling volumes to the nanoliter scale [16, 17]. Such miniaturization minimizes nonspecific adsorption of proteins to reaction vessels and increases protein/peptide concentrations for more efficient chemical and enzymatic reactions [18, 19]. Besides miniaturizing sample processing, a number of other recent developments across the analytical workflow have increased the robustness, throughput, quantitative accuracy, and depth of coverage for single-cell proteomics workflows. Isobaric labeling using, e.g., tandem mass tag reagents enables multiple samples to be analyzed in a single experiment and can utilize a larger "carrier proteome" to improve detection of low-abundance species (e.g., in the SCoPE-MS workflow [20]). Isobaric labeling has also been adapted to the nanoPOTS platform for increased measurement sensitivity [21, 22]. Miniaturization of liquid chromatography is also beneficial for single-cell proteome profiling due to increased ionization efficiency at low--nanoliter-per-minute flow rates [19]. Finally, the additional selectivity provided by incorporating gas-phase ion fractionation technologies such as high-field asymmetric waveform ion mobility spectrometry (FAIMS) has pushed the achievable proteome coverage for single cells to >1000 proteins for label-free studies [23].

Using a combination of these state-of-the-art technologies, we have studied the neuronal proteomics using tissues derived from the rat brain [18] and human spinal tissue [23]. In this chapter, we provide a detailed protocol for the proteomic analysis of single neurons, following the workflow outlined in Fig. 1. The protocol,

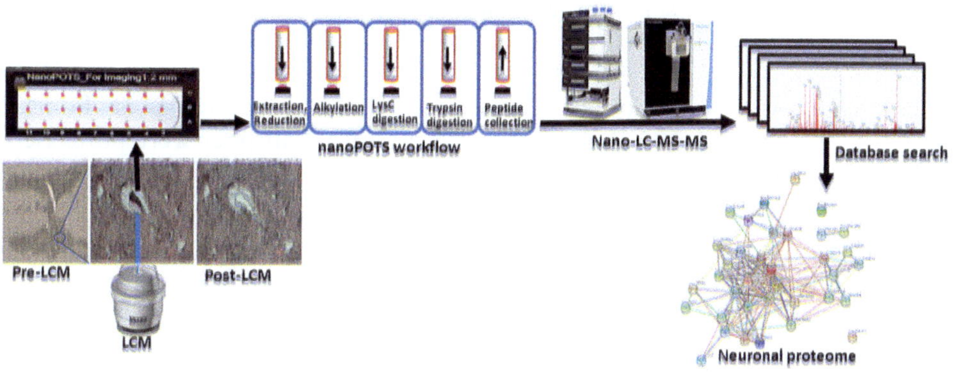

Fig. 1 Label-free proteome profiling of single neurons using LCM, nanoPOTS, and nanoLC-MS/MS

along with the notes at the end of this chapter, should enable the reader to reproduce the procedure or tailor it to a particular application to achieve an in-depth understanding of protein expression in CNS tissues with single-neuron resolution.

2 Materials

2.1 Tissue Preparation and Laser Microdissection

We sectioned frozen tissues to the desired thickness (typically 10–20 μm) using a TN50 cryostat from Tanner Scientific (Sarasota, FL) and using disposable microtome Platinum Blades purchased from Mercedes Scientific (Lakewood Ranch, FL). The specimen holder discs and other cryostat consumables were also from Mercedes Scientific.

Mayer's hematoxylin solution (MHS16–500 ml) and eosin Y solution (110,116–500 ml) purchased from Sigma-Aldrich (St. Louis, MO) were used for H&E staining. Scott's bluing reagent (6697–32) was from RICCA Chemical Company (Arlington, TX). Ethanol was from Decon Laboratories (King of Prussia, PA). We imaged the tissue sections and performed laser microdissection with a Zeiss PALM MicroBeam (Oberkochen, Germany) system equipped with a 40× objective. The Slide Collector 48 slide adapter and PEN membrane slides (415190–9041-000) were also purchased from Zeiss.

2.2 NanoPOTS Chip Fabrication

We patterned nanoPOTS chips in-house from 25 × 75 mm glass slides that were coated with chromium and photoresist as purchased from Telic Company (Valencia, CA). 5-inch glass photomasks (Telic) were prepared in-house on a pattern generator, although patterned photomasks may also be purchased from many suppliers. Custom 1-mm-thick glass spacer frames having outer dimensions of 74 × 23 mm and an inside space of 44 × 14 mm were purchased from Beijing Gin Koo MEMS Scientific & Technological Company (Beijing, China). Sylgard

184 silicone elastomer was a product of Dow Chemical (Midland, MI). Hydrochloric acid; (heptadecafluoro-1,1,2,2-tetrahydrodecyl) dimethylchlorosilane; 2,2,4-trimethylpentane; and 10:1 buffered oxide etchant were purchased from Sigma (St. Louis, MO). AZ 400 K developer and AZ 400 T stripper were purchased from Integrated Micro Materials (Argyle, TX), and CEP-200 chromium etchant was a product of Transene (Danvers, MA).

2.3 NanoPOTS Liquid Handler Construction

Linear stages with built-in controllers were purchased from Zaber Technologies (Vancouver, BC) and mounted on a T-slot aluminum scaffolding (80/20 Inc., Columbia City, IN) to create a three-axis stage with an additional stage serving as an integrated syringe pump. A Ximea MQ013MG-ON camera (Lakewood, CO) with a microscope adapter was mounted on the stage to visualize the position of a tapered fused silica capillary serving as pipet tip relative to the nanoPOTS chip. The three-axis stage was enclosed in a custom polycarbonate chamber, and humidified air was introduced into the chamber from a Crane Drop Ultrasonic Cool Mist Humidifier (Amazon.com). A fan (Delta DSB0912HH, Amazon.com) was installed to circulate the humidified air and a humidity controller served to power cycle the humidifier to maintain the chamber at 95% relative humidity (WILLHI, Model WH1436H, Amazon.com). A LabVIEW 2016 program was developed to operate and control the nanoPOTS system.

2.3.1 Sample Preparation and LC-MS Analysis

Dithiothreitol (DTT), iodoacetamide (IAA), LC-MS-grade formic acid (FA), and Pierce HeLa Protein Digest Standard were purchased from Thermo Fisher Scientific (Waltham, MA). CHROMASOLV LC-MS-grade water and acetonitrile were products of Honeywell (Charlotte, NC). MS-grade trypsin and Lys-C were purchased from Promega (Madison, WI). Solid-phase extraction (SPE) and analytical columns were packed in-house using C18-bonded 3-μm, 300-Å porous Jupiter particles from Phenomenex (Torrance, CA). Fused silica capillaries having an inner diameter of 20, 30, or 75 μm and an outer diameter of 360 μm were purchased from Polymicro Technologies (Phoenix, AZ). Column frits were prepared to a length of ~2 mm using Kasil 1, Kasil 1624, and formamide purchased from Next Advance (Troy, NY) according to manufacturer instructions. The other accessories used for column packing include 20,000 psi-rated needle valves, Swagelok fittings, tubing (HiP, Erie, PA), a Micro (Flea) magnetic spin bar (VWR), and an MS-188 Haskel pump (Burbank, CA). LC separations were performed using an UltiMate 3000 RSLCnano pump. Mass spectrometers employed for these studies include the Orbitrap Fusion Lumos, Orbitrap Eclipse, and Orbitrap Exploris 480 (Thermo Fisher Scientific).

3 Methods

3.1 Isolation of Single Neuronal Cells and NanoPOTS Workflow

1. Fresh frozen human spinal tissues obtained from ProteoGenex (Los Angeles, CA) were mounted on the specimen holder disc and securely embedded in purified water. Cryosectioning was performed at −16°C using a Platinum disposable microtome blade on the Tanner Scientific TN50 cryostat. Prior to use, the PEN membrane slides were exposed to UV light at 254 nm for 30 min to facilitate tissue adhesion and then precooled to −16°C in the cryostat. The tissues were then sectioned to a thickness of 10–20 μm depending on the application and deposited onto Zeiss PEN membrane microscope slides. Following the deposition of thin sections onto the slides, the tissue sections were briefly thawed to promote adhesion (*see* **Note 1**).
2. The tissue sections were immediately fixed with 70% ethanol for 15 min at 4°C, rapidly dried in a vacuum desiccator, and stored at −80°C until further use (*see* **Note 2**).
3. Before staining the tissue sections, the slides were slowly warmed from −80°C to −20°C and then to 4°C over the course of an hour. The tissue sections were H&E stained as follows. Tissue sections were rehydrated in deionized (DI) water for 30 s followed by immersion in Mayer's hematoxylin for 20 s. Excess hematoxylin was removed by dipping the slide in DI water for another 30 s. Bluing was done with Scott's tap water substitute reagent for 10 s followed by dehydration of tissues in 70% ethanol for 15 s. Tissue sections were then stained with a 20% eosin Y solution (v/v in ethanol) for 5 s rapidly dipped in 95% ethanol. The slides were further dehydrated by twice immersing in fresh 95% ethanol for 30 s and twice with 100% ethanol for another 30 s. The slides were then quickly dried in a vacuum desiccator. The sections were full scanned on the Zeiss PALM MicroBeam system at 40× resolution (*see* **Note 3**). The scanned images of the selected upper and lower ventral horn regions of the spinal tissue section were acquired in .jpg format as sequential tiles and stitched together using ImageJ to form a high-resolution image of the regions of interest.
4. Nanowells of the nanoPOTS chip were prefilled with 200 nL of DMSO using the nanopipetting robot, and the chip was mounted onto the LCM platform using the Slide Collector 48 adapter.
5. Single motor and interneurons were identified by a pathologist from the thin section images of the ventral horn regions of spinal tissue sections. These cells were excised from the tissue by LCM with two passes of the laser with the cutting energy set to 30–35 and a speed of 1% (*see* **Note 4**). The neurons were

then transferred to DMSO-filled nanowells of a nanoPOTS chip by laser-pressure catapulting with an energy of 20 and the "Center Robo LPC" mode in the PALMRobo V 4.6 control software. The process has been described in detail previously [18].

6. Following microdissection of each neuron, the target nanowell was imaged at 10× resolution to confirm the successful transfer of the cell. After cell collection, DMSO was evaporated to ~50 nL at 60°C (*see* **Note 5**).
7. Proteins from the neurons were extracted by adding 100 nL of 0.1% DDM and 5 mM DTT into the nanowells and incubating at 70°C for 1 h. To avoid drying out the samples in the oven, the nanoPOTS chip was sealed tightly with a PDMS-coated cover slide secured by binder clips and placed in a small humidity chamber. After incubation, the cover slide was removed from the chip, and the solvent was evaporated from the chip at the same temperature.
8. The samples were reconstituted with 150 nL of 10 mM IAA and incubated in the dark at room temperature for 30 min to alkylate the proteins.
9. For two-step enzyme digestion, 50 nL of 0.005 ng/nL Lys-C in 50 mM ammonium bicarbonate buffer was added to the samples and incubated for 4 h at 37°C. 50 nL of 0.005 ng/nL trypsin (0.25 ng) was then added and incubated overnight at 37°C (*see* **Note 6**).
10. Protease activity was quenched by adding 50 nL of 0.1% TFA and incubating for 1 h at room temperature. All reagent additions were performed with the nanoPOTS nanopipetting system.
11. For collection of prepared samples from nanowells, a 200 μm i.d. fused silica capillary was connected to the syringe of the nanopipettor and filled with buffer A (0.1% FA in water). The nanopipettor aspirated 50 nL of air followed by 200 nL of sample from the nanowell. Two additional washes of equivalent volume were also collected into the capillary from the same well to maximize sample recovery. This was done by aspirating mobile phase A (0.1% formic acid in water) into the capillary, dispensing it into the well, and repeating the above procedure. The sample and washes were separated from each other in the capillary by a ~ 50 nL air gap. The ~3 cm section of capillary containing sample and washes was cut from the remaining capillary and sealed with paraffin. The samples were stored at −20°C or 4°C until further use (*see* **Note** 7).

3.2 Liquid Chromatography-Mass Spectrometry Analysis

1. For manual sample loading, the sample-containing capillaries were brought to 4°C (if stored at −20°C) and connected to a 5-cm-long, 75-μm-i.d. solid-phase extraction (SPE) column on one end and the nanoLC pump on the other end. Mobile phase A was delivered by the pump at 500 nL/min for 10 min to transfer samples from the sample storage capillary to the SPE to ensure salts and debris were removed (*see* **Notes 7** and **8**).
2. The sample-containing SPE column was then manually connected to the 50-cm-long, 20-μm-i.d. analytical column with a zero-dead-volume union (Valco, Houston, TX) and the LC gradient was initiated. The LC flow rate was ~20 nL/min after being split from 250 nL/min programmed flow that was supplied by the ultimate 3000 nanoLC pump. The split column was an in-house-prepared 60-cm-long, 75-μm-i.d. column with the same packing material as the analytical column (*see* **Note 9**).
3. Peptides were separated with a gradient of mobile phase (MP) B (0.1% FA in acetonitrile) from 5 to 25% for 100 min. An additional gradient MP B from 25 to 45% was used to elute hydrophobic peptides for 20 min, and then MP B was ramped to 90% over 5 min and held for another 5 min to wash the column. Finally, MP B was reduced to 2% over 5 min and held for 20 min to re-equilibrate both the SPE and analytical columns (*see* **Note 10**).
4. During the LC run, the MS and MS/MS data were acquired on a Thermo Fisher Orbitrap Eclipse Tribrid mass spectrometer with a FAIMS Pro interface. The mass spectrometer was run in a peptide application mode with the Nanospray Flex ion source having the electrospray potential set at 2200 V. The FAIMS compensation voltages (CVs) were set at −55 and − 70 V to add selectivity and remove the singly charged contaminant ions. The ion transfer tube was set at 200°C for desolvation, and the ion funnel RF level was set to 40%. Full MS scans were acquired at 120,000 resolution (m/z 200) using the orbitrap detector in the m/z range of 375–1800. The AGC target and maximum injection time were set at 1E6/250 ms.
5. Data-dependent acquisition mode was selected to acquire the MS/MS spectra using the ion trap mass analyzer. MS/MS fragmentation was triggered after applying the MIPS filter, a minimum signal intensity of 1E4, and a dynamic exclusion window of 90 s. The maximum AGC target and injection time were set at 1E5/500 ms. Precursor ions having charge states of +2 to +7 were selected for fragmentation with normalized higher energy collision-induced dissociation (HCD) energy set to 30%. The full-cycle time for MS followed by MS/MS was set to 3 s.

3.3 Mass Spectrometry Data Analysis: MaxQuant, Proteome Discoverer, and Statistical Analysis

1. Raw data files were processed using MaxQuant (version 1.6.6.0) and Proteome Discoverer 2.5 (San Jose, CA) using the UniProtKB/Swiss-Prot human database (*see* **Note 11**).
2. For MaxQuant analysis, the raw files were analyzed using the following group-specific and global parameters. The enzyme digestion mode was set as a trypsin/P with three missed cleavages. The variable and fixed modifications were selected as oxidation (M) and acetylation (protein N-term) and carbamidomethylation (C), respectively. The mass tolerances for precursors and fragments were 5 and 20 ppm, respectively. The minimum and maximum peptide lengths were 6 and 60 amino acids, respectively. Proteins and PSMs were filtered with a maximum FDR of 0.01. The match between runs was enabled with the match time window of 0.7 min and alignment time window of 20 min. Label-free quantification was also enabled with the LFQ minimum ratio count of 2 and stricter requirements of MS/MS for LFQ comparisons. All other parameters were used as defaults.
3. Proteome Discoverer 2.5 analysis was carried out using three-stage processing workflow settings. In the first stage search, the SEQUEST HT search was enabled for full trypsin enzyme digestion mode with up to three missed cleavages. Dynamic modifications were selected as oxidation (M), acetylation (protein N-terminus), methionine loss from the protein N-terminus, and methionine loss plus acetylation (M). Carbamidomethylation (C) was selected as a static modification. The second stage SEQUEST search was enabled for semi-trypsin enzyme digestion mode with other settings used in the first search. The third stage SEQUEST search was done with the INFERYS rescoring node using full trypsin digestion with a peptide length of 7–30 amino acids. The dynamic and static modifications were set as oxidation (M) and carbamidomethylation (C), respectively. All three search stages were performed with a precursor mass tolerance of 5 ppm and fragment ion mass tolerance of 0.5 Da. The strict FDR target value was set to 0.01 and relaxed was set at 0.05. Search data were filtered using Comprehensive Enhanced Annotation LFQ and Precursor Quan consensus workflow. The low and medium confidence PSMs, peptide groups, and proteins were filtered out manually. Only high-confidence master proteins with more than one unique peptide were used for quantitation. The output files were transferred to Perseus (v1.6.5.0) for statistical analysis.
4. For analysis in Perseus, the data from MaxQuant or PD were imported as .CSV or tab-delimited text files. Protein raw abundances or normalized abundances were set as the main rows and categorized them as motor neuron (MN) and interneuron (IN). The data were filtered based on the presence of valid

values for at least two of three replicates in each group. The abundances were log2 transformed and missing values were imputed by a normal distribution (width 0.3 and downshift 1.8). A two-component PCA plot was generated using a p-value threshold of 0.05. To evaluate the differential detection of proteins in neuronal subtypes, the volcano plot was generated with the two-tailed t-test. The FDR of 0.05 with fold-change 2 was used.

5. From single IN and MN, an average of 1336 and 1225 high-confidence master proteins, and 4563 and 4042 peptide groups were identified, respectively, from three replicates using an FDR cutoff of 0.01 and without MS1-level feature matching (Fig. 2). In both cell types, 1524 quantifiable master proteins were found with additional data filters based on the presence of valid abundance values for at least two in three replicates in each cell type. Of these, 30 were significantly more abundant in MNs relative to INs, and 20 proteins are detected with greater abundance in IN relative to MN. The two-component principal component analysis readily differentiated the two cell subtypes using a p-value threshold of 0.05. However, the MN and IN cells also share 76% of identified proteins (Fig. 3). This nanoPOTS-based workflow provides the greatest depth of coverage for label-free analysis of single cells that have been developed to date.

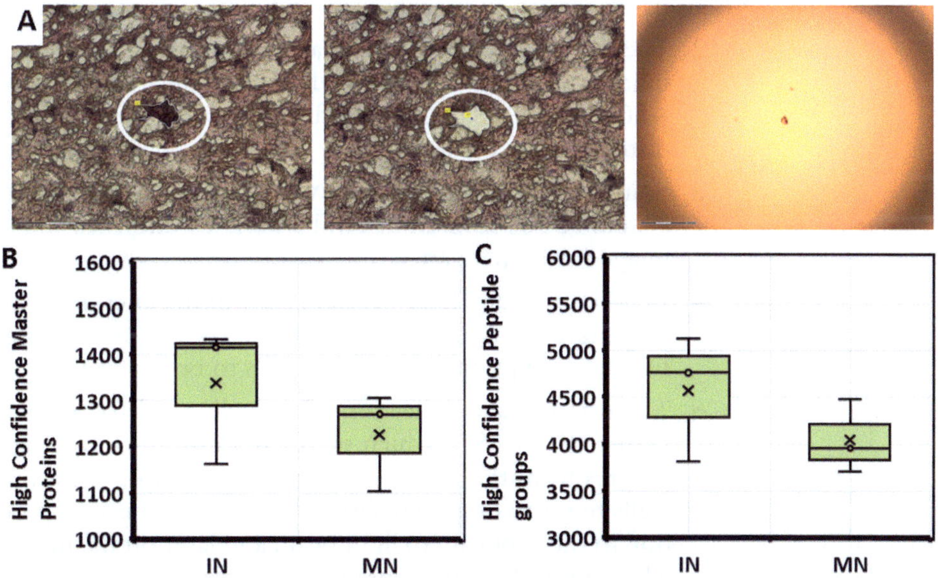

Fig. 2 (**a**) Isolation of single motor neurons by LCM and collection into nanowells for processing. (**b**) Approximately 1300 high-confidence master proteins and (**c**) >4000 peptide groups were identified using the Proteome Discoverer 2.5 SEQUEST/INFERYS two-stage search algorithm

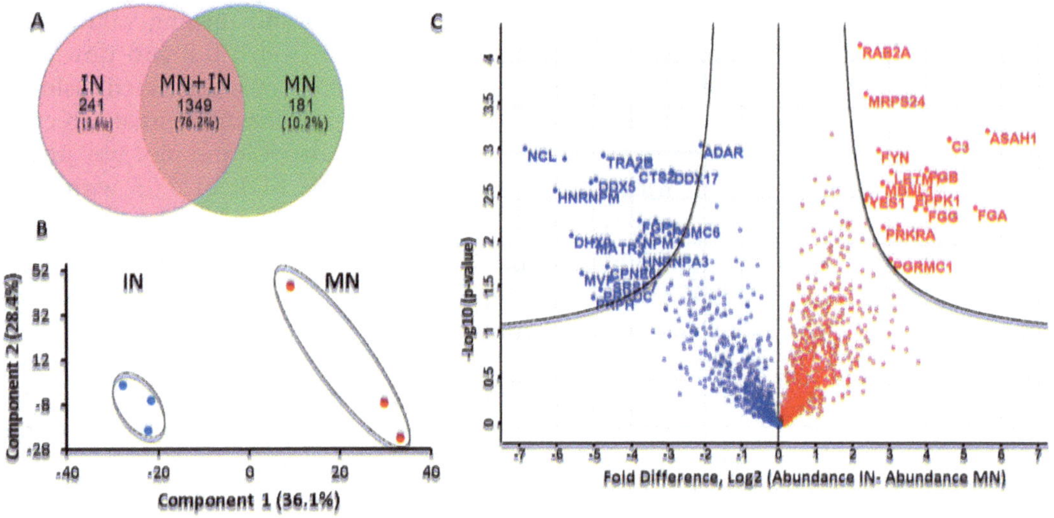

Fig. 3 (**a**) Venn diagram of high-confidence master proteins indicating overlap of identified proteins in interneurons (IN) and motor neurons (MN). Master proteins were identified with 1% FDR. (**b**) Differentiation of neuronal subtypes by principal component analysis based on 1524 quantifiable master proteins. (**c**) Volcano plot of differentially detected master proteins in IN and MNs ($p < 0.05$, fold difference ≥ 2)

4 Notes

1. To improve the tissue section adhesion to the slide, a warm finger can be placed on the back of the slide. This will allow the tissue section to quickly thaw and adhere to the slide.
2. The fresh-frozen human spinal tissue section slides need to be quickly dried in a vacuum desiccator and can be stored at −80°C with desiccant after fixing with 70% ethanol to avoid distorting the tissues.
3. Scanning the tissue sections with a Zeiss PALM MicroBeam system at 40x resolution can be done by autofocusing at each image for better clarity.
4. LCM cutting and laser-pressure catapulting energies need to be optimized for tissues of different thicknesses using test regions of the tissue prior to collecting cells to be analyzed.
5. The excision and collection of the cells need to be completed before the DMSO, which serves as a capture solvent and evaporates from the nanowells. The 200 nL DMSO in the nanowells tends to evaporate faster during cell collection procedure due to the exposure to light beam and laser on the microscope. The nanowell chip needs to be covered securely and may be stored at −20°C when not in use.
6. For proteomic sample analysis, always use protein low binding tubes for the storage of enzymes and other solutions.

7. The samples collected in the 200-μm-i.d. capillary and sealed on both ends can be stored at 4°C for up to 1 week or − 20°C for up to 1 month. For longer-term sample storage, storing in liquid nitrogen may improve sample recovery relative to storing at −80°C.
8. In manual sample loading, the SPE column needs to be equilibrated with mobile phase A prior to sample loading.
9. Check for leaks during the sample loading and separation by monitoring the observed pressure of the pump. The columns should not be disconnected while pressurized as this can reduce the efficiency of the column.
10. Electrospray stability can be checked on the tune software of the mass spectrometer. The emitter position relative to the ion transfer tube may be adjusted to maintain the stable spray during the LC gradient.
11. The raw data can be converted to mzXML or mgf format to analyze with the different search algorithms.

Acknowledgments

This work was supported by the National Institutes of Health under award numbers R33 CA225248 and R01 GM138931, and through a sponsored research agreement from Biogen, Inc.

References

1. Cadwell CR, Palasantza A, Jiang XL, Berens P, Deng QL, Yilmaz M, Reimer J, Shen S, Bethge M, Tolias KF, Sandberg R, Tolias AS (2016) Electrophysiological, transcriptomic and morphologic profiling of single neurons using patch-seq. Nat Biotechnol 34(2): 199–203. https://doi.org/10.1038/nbt.3445
2. Bean BP (2007) The action potential in mammalian central neurons. Nat Rev Neurosci 8(6): 451–465. https://doi.org/10.1038/nrn2148
3. Ko H, Hofer SB, Pichler B, Buchanan KA, Sjostrom PJ, Mrsic-Flogel TD (2011) Functional specificity of local synaptic connections in neocortical networks. Nature 473(7345): 87–91. https://doi.org/10.1038/nature09880
4. Kitchen RR, Rozowsky JS, Gerstein MB, Nairn AC (2014) Decoding neuroproteomics: integrating the genome, translatome and functional anatomy. Nat Neurosci 17(11): 1491–1499. https://doi.org/10.1038/nn.3829
5. Choi SB, Lombard-Banek C, Munoz LP, Manzini MC, Nemes P (2018) Enhanced peptide detection toward single-neuron proteomics by reversed-phase fractionation capillary electrophoresis mass spectrometry. J Am Soc Mass Spectrom 29(5):913–922. https://doi.org/10.1007/s13361-017-1838-1
6. Johnson MB, Wang PP, Atabay KD, Murphy EA, Doan RN, Hecht JL, Walsh CA (2015) Single-cell analysis reveals transcriptional heterogeneity of neural progenitors in human cortex. Nat Neurosci 18(5):637–646. https://doi.org/10.1038/nn.3980
7. Chappell L, Russell AJC, Voet T (2018) Single-cell (multi)omics technologies. Annu Rev Genomics Hum Genet 19:15–41. https://doi.org/10.1146/annurev-genom-091416-035324
8. Saikia M, Burnham P, Keshavjee SH, Wang MFZ, Heyang M, Moral-Lopez P, Hinchman MM, Danko CG, Parker JSL, De Vlaminck I (2019) Simultaneous multiplexed amplicon sequencing and transcriptome profiling in

single cells. Nat Methods 16(1):59–62. https://doi.org/10.1038/s41592-018-0259-9

9. Angel TE, Aryal UK, Hengel SM, Baker ES, Kelly RT, Robinson EW, Smith RD (2012) Mass spectrometry-based proteomics: existing capabilities and future directions. Chem Soc Rev 41(10):3912–3928. https://doi.org/10.1039/c2cs15331a
10. Blackstock WP, Weir MP (1999) Proteomics: quantitative and physical mapping of cellular proteins. Trends Biotechnol 17(3):121–127. https://doi.org/10.1016/s0167-7799(98)01245-1
11. Bludau I, Aebersold R (2020) Proteomic and interactomic insights into the molecular basis of cell functional diversity (vol 21, pg 327, 2020). Nat Rev Mol Cell Bio 21(6):353–353. https://doi.org/10.1038/s41580-020-0249-5
12. Kelly RT (2020) Single-cell proteomics: progress and prospects. Mol Cell Proteomics. https://doi.org/10.1074/mcp.R120.002234
13. Kelly RT, Zhu Y (2019) Ultrasmall sample biochemical analysis. Anal Bioanal Chem 411(21):5349–5350. https://doi.org/10.1007/s00216-019-01957-1
14. Shao X, Wang X, Guan S, Lin H, Yan G, Gao M, Deng C, Zhang X (2018) Integrated proteome analysis device for fast single-cell protein profiling. Anal Chem 90(23):14003–14010. https://doi.org/10.1021/acs.analchem.8b03692
15. Li ZY, Huang M, Wang XK, Zhu Y, Li JS, Wong CCL, Fang Q (2018) Nanoliter-scale oil-air-droplet Chip-based single cell proteomic analysis. Anal Chem 90(8):5430–5438. https://doi.org/10.1021/acs.analchem.8b00661
16. Zhu Y, Piehowski PD, Zhao R, Chen J, Shen Y, Moore RJ, Shukla AK, Petyuk VA, Campbell-Thompson M, Mathews CE, Smith RD, Qian WJ, Kelly RT (2018) Nanodroplet processing platform for deep and quantitative proteome profiling of 10-100 mammalian cells. Nat Commun 9(1):882. https://doi.org/10.1038/s41467-018-03367-w
17. Zhu Y, Clair G, Chrisler WB, Shen Y, Zhao R, Shukla AK, Moore RJ, Misra RS, Pryhuber GS, Smith RD, Ansong C, Kelly RT (2018) Proteomic analysis of single mammalian cells enabled by microfluidic Nanodroplet sample preparation and ultrasensitive NanoLC-MS. Angew Chem Int Ed Engl 57(38):12370–12374. https://doi.org/10.1002/anie.201802843
18. Zhu Y, Dou M, Piehowski PD, Liang Y, Wang F, Chu RK, Chrisler WB, Smith JN, Schwarz KC, Shen Y, Shukla AK, Moore RJ, Smith RD, Qian WJ, Kelly RT (2018) Spatially resolved proteome mapping of laser capture microdissected tissue with automated sample transfer to Nanodroplets. Mol Cell Proteomics 17(9):1864–1874. https://doi.org/10.1074/mcp.TIR118.000686
19. Cong YZ, Liang YR, Motamedchaboki K, Huguet R, Truong T, Zhao R, Shen YF, Lopez-Ferrer D, Zhu Y, Kelly RT (2020) Improved single-cell proteome coverage using narrow-bore packed NanoLC columns and ultrasensitive mass spectrometry. Anal Chem 92(3):2665–2671. https://doi.org/10.1021/acs.analchem.9b04631
20. Budnik B, Levy E, Harmange G, Slavov N (2018) SCoPE-MS: mass spectrometry of single mammalian cells quantifies proteome heterogeneity during cell differentiation. Genome Biol 19(1):161. https://doi.org/10.1186/s13059-018-1547-5
21. Dou MW, Clair G, Tsai CF, Xu KR, Chrisler WB, Sontag RL, Zhao R, Moore RJ, Liu T, Pasa-Tolic L, Smith RD, Shi TJ, Adkins JN, Qian WJ, Kelly RT, Ansong C, Zhu Y (2019) High-throughput single cell proteomics enabled by multiplex isobaric labeling in a Nanodroplet sample preparation platform. Anal Chem 91(20):13119–13127. https://doi.org/10.1021/acs.analchem.9b03349
22. Tsai CF, Zhao R, Williams SM, Moore RJ, Schultz K, Chrisler WB, Pasa-Tolic L, Rodland KD, Smith RD, Shi T, Zhu Y, Liu T (2020) An improved boosting to amplify signal with isobaric labeling (iBASIL) strategy for precise quantitative single-cell proteomics. Mol Cell Proteomics 19(5):828–838. https://doi.org/10.1074/mcp.RA119.001857
23. Cong Y, Motamedchaboki K, Misal SA, Liang Y, Guise AJ, Truong T, Huguet R, Plowey ED, Zhu Y, Lopez-Ferrer D, Kelly RT (2020) Ultrasensitive single-cell proteomics workflow identifies >1000 protein groups per mammalian cell. Biorxiv. https://doi.org/10.1101/2020.06.03.132449

Chapter 4

Joint Analysis of Transcriptome and Proteome Measurements in Single Cells with totalVI

Zoë Steier, Annie Maslan, and Aaron Streets

Abstract

The recent development of single-cell multi-omic measurement techniques necessitates analysis strategies to combine these datasets into a single view of a cell. Here we present totalVI, a computational framework for joint analysis of RNA and protein data. Unlike previous strategies that analyze each modality independently, totalVI jointly models the paired data to form a more comprehensive view of cell state and of the relationship between molecular layers within a cell. To do so, totalVI creates a joint, low-dimensional representation of the data and models technical factors such as batch effects, sampling noise, and protein background to uncover true biological signal. This chapter contains an overview of the experimental and analytical pipeline, from collecting paired RNA and protein data from single cells to preprocessing and filtering the sequencing data, running the totalVI model, and conducting downstream analysis with totalVI. This joint analysis of single-cell multi-omic data not only provides a richer definition of cell state but also has the potential to elucidate the dynamics and mechanisms of cellular processes by characterizing the relationship between molecular layers within a cell.

Key words Single-cell analysis, Transcriptomics, Proteomics, Multi-omics, CITE-seq, Deep generative model, scRNA-seq, Human cell atlas

1 Introduction

Single-cell measurements can reveal meaningful biological heterogeneity that is otherwise obscured by bulk analysis. Whole transcriptome analysis with single-cell RNA sequencing (scRNA-seq), for example, has been used to generate comprehensive maps of cell types and transcriptional states in both healthy and diseased primary tissues [1]. New technologies to measure multiple components of a single cell are now uncovering not only the variability within a given molecular layer, such as the transcriptome, but also the variability in the associations across layers, including the genome, the proteome, and the epigenome. Such multimodal measurements are distinguished in their ability to make simultaneous orthogonal measurements on the same single cell. There are

Jonathan V. Sweedler et al. (eds.), *Single Cell 'Omics of Neuronal Cells*,
Neuromethods, vol. 184, https://doi.org/10.1007/978-1-0716-2525-5_4,

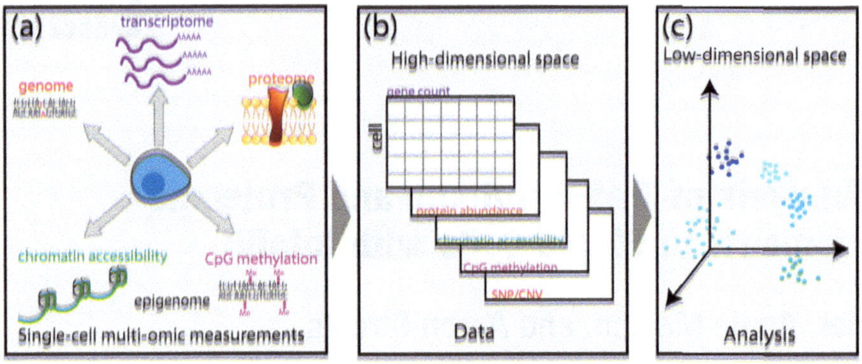

Fig. 1 Single-cell multi-omics overview. (**a**) Single-cell multi-omic measurements simultaneously profile multiple dimensions of the cell. Here, the major -omes currently used in multi-omic technologies are described: genome, epigenome, transcriptome, and proteome. (**b**) For each -ome, a cell by measurement matrix is produced. (**c**) These separate views can be combined to represent a cell in low-dimensional space

many emerging multimodal measurements that make use of tools such as imaging, sequencing, and other molecular assays [2–5]. The "multi-omic" methods that we discuss here probe large-scale and genome-wide features, typically by converting the molecular components into a sequenceable readout (Fig. 1a). As these single-cell multi-omic measurements mature, there is a growing need for analysis strategies to combine these datasets into a single view of a cell. Here we describe a computational pipeline for analyzing paired protein and RNA data. The approach to data analysis described here could, with modifications, be extended to other multi-omic measurements.

While high-throughput single-cell analysis technologies such as flow cytometry and scRNA-seq are independently powerful for identifying cell types and characterizing cellular heterogeneity, multi-omic analysis of single cells provides a more comprehensive view of cell state and of the relationship between molecular layers within a cell. For example, biologists have commonly relied upon an extensive literature of cell surface markers to define cell types and to sort cells for experiments, but recent cell atlas projects have amassed vast quantities of scRNA-seq data from which cells can be defined by their unique marker genes or gene signatures. The transcriptome offers a broader view of the cell than a limited protein panel; however, scRNA-seq has limited detection efficiency of genes expressed at low levels. Combining the breadth of the transcriptome with the depth and stability of the proteome creates a more complete view of a cell's state and allows the dynamics between the two modalities to be mapped. Beyond paired transcriptome and proteome measurements, other approaches for assessing multiple modalities in parallel have been developed. Many methods jointly measure the transcriptome and epigenome, such as with an epigenomic measurement of DNA methylation, accessibility, or genome organization [6–9].

Multi-omic approaches not only provide a more comprehensive view of cell state but also have the potential to illuminate the dynamics and mechanisms of cellular processes. For example, using joint measurements of chromatin accessibility and gene expression in single cells, Cao et al. predict gene expression from changes in chromatin accessibility at promoters and linked distal sites [7]. Similarly, using paired mRNA and protein data, Gorin et al. model protein translation kinetics to measure "protein velocity and acceleration" [10]. In particular, instances where mRNA and protein measurements are not correlated suggest an active layer of posttranslational regulation. Thus, multi-omic measurements can elucidate regulatory mechanisms and outcomes, whether that outcome is RNA expression, protein expression, or some other phenotype detected with imaging.

A major hurdle in working with these new types of multi-omic datasets is combining distinct data types into a single representation of a cell (Fig. 1b, c). Each measurement presents unique technical biases, such as technical noise, limited sensitivity, background, and sparsity. Distinguishing technical noise from meaningful biological variability across omic measurements with unique noise profiles therefore complicates analysis. Processing and normalizing different datasets and combining them into a single view of a cell presents a challenge, and much of the analysis to date has looked at modalities independently. However, methods such as MOFA, MOFA+, LIGER, and Seurat V3 have been developed to integrate multiple modalities [11–14]. Multi-omics factor analysis (MOFA) is a multi-view matrix factorization method that identifies factors that explain variation across multi-omic datasets and has been applied to paired datasets profiling mRNA, DNA methylation, and somatic mutations or chromatin accessibility [11, 15]. Unlike MOFA, MOFA+ accounts for additional information about the structure between cells, such as batch and experimental conditions [12]. LIGER uses an integrative nonnegative matrix factorization (iNMF) to create a low-dimensional space in which a cell is defined by dataset-specific factors and a set of shared factors [13]. Seurat V3 identifies pairwise correspondences, or "anchors," between single cells across datasets to transform the datasets into a shared space [14].

In this chapter, we focus on joint analysis of surface protein abundance and gene expression in single cells, a rapidly growing type of multi-omic data. Recently, CITE-seq and REAP-seq have demonstrated the ability to probe these two modalities simultaneously in single cells at high throughput with whole transcriptome profiling and the measurement of an expanded panel of proteins made possible by oligonucleotide-barcoded antibodies [16, 17].

We designed totalVI as a tool for conducting a joint analysis of RNA and protein data that addresses these assorted challenges of multi-omic analysis [18]. totalVI is a deep generative model that combines RNA and protein information into a single representation and leverages all available information (the gene and protein expression from each cell) to inform downstream analysis. Unlike previous strategies that analyze each modality independently and subsequently attempt to draw connections between the two, the joint analysis of totalVI combines this information to form a single, consistent view of cell state. Because of the direct connection between RNA and protein molecules through biology's central dogma, a single model that pools RNA and protein information is consistent with biological knowledge that these paired measurements were derived from molecules produced in the same single cell.

In its joint analysis, totalVI addresses the distinct sources of noise and technical bias in the RNA and protein measurements. While both RNA and protein data suffer from noise in the sampling of a limited number of molecules from each cell, other technical factors are specific to each modality. The RNA data tends to have limited sensitivity and variation in sequencing depth (where the total number of molecules detected per cell is known as the library size). The protein data is less sparse than the RNA data, but tends to be obscured by background that arises experimentally from a combination of unbound ambient antibodies and nonspecific antibody binding. These technical factors are modeled by totalVI so that analysis is based on the underlying biological signal rather than technical biases of the methods. totalVI can also correct technical differences between batches (known as batch effect correction) or between datasets (known as dataset integration). Unlike previous methods, totalVI is capable of integrating datasets with different protein panels (i.e., some or all proteins in a dataset are missing), and imputing the expression of the missing proteins.

To model single-cell sequencing data including noise and technical artifacts, totalVI draws conceptually on previous work modeling scRNA-seq data. To model RNA, totalVI uses a similar probabilistic modeling strategy as scVI to account for the sampling noise in the measurement [19]. For proteins, totalVI introduced a new model that addresses the issue of protein background by distinguishing the foreground and background components of the protein signal. totalVI combines RNA and protein data into a joint, low-dimensional representation, which is rooted in the notion that genes are often expressed in co-regulated networks or modules that can contain the information describing the high-dimensional transcriptome or proteome. This low-dimensional joint representation is derived using neural networks, which provide a powerful, nonlinear approach to accurately model the data. totalVI connects this model with downstream analysis tasks, providing an end-to-end framework for the analysis of RNA and protein data that both addresses the technical factors in each measurement and takes all information into account.

Combining RNA and protein information in a joint analysis could enable the identification of new cell types or cell states that were not easily identified from a single measurement. The additional information provided by combining these two modalities might enhance the ability to detect markers in either modality, which could be particularly useful in finding surface proteins that could serve as potential drug targets, surface markers to isolate a cell population of interest, or transcription factors and enzymes that might define or regulate cells with a known cell surface phenotype. totalVI can bridge the gap between these two related views to form a more comprehensive understanding of cell state. The paired measurement of RNA and protein in the same cell might also illuminate the connection between the expression levels of these molecules in different biological settings or regulatory dynamics of transcription and translation.

Conducting a joint analysis of RNA and protein data with totalVI consists of four main steps (Fig. 2):

1. Collecting a single-cell dataset with paired RNA and protein measurements.
2. Preprocessing and filtering the sequencing data.
3. Running the totalVI model.
4. Conducting downstream analysis with totalVI.

Practical guidance in completing each of these steps is discussed in detail below.

2 Experimental Considerations for Collecting Paired RNA and Protein Measurements in Single Cells

2.1 Methods for Generating a Single-Cell Dataset of Paired RNA and Protein Measurements

Droplet-based microfluidic platforms can be used to facilitate high-throughput scRNA-seq through rapid single-cell isolation and multiplexing using oligonucleotide barcodes, which capture mRNA transcripts with a region complementary to the poly-A tail, and identify the cell-of-origin of each captured molecule. Each barcode also contains a unique molecular identifier (UMI) that enables the quantification of the original number of captured molecules after amplification [20, 21]. Once barcoded, transcripts from thousands of single cells can be pooled together in the same tube, reverse transcribed, amplified, and prepared for sequencing in a process known as library generation. These libraries can be further pooled across samples, sequenced, and quantified, resulting in transcript counts for each gene in each single cell.

High-throughput droplet-based scRNA-seq can be extended to measure both RNA transcripts and surface proteins from the

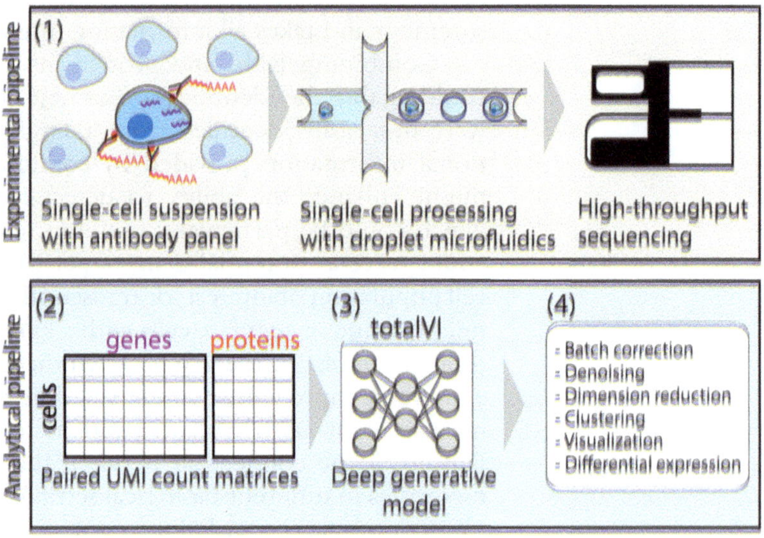

Fig. 2 totalVI experimental and analytical pipeline. This schematic shows the primary steps to take a single-cell sample through joint analysis of gene expression and protein abundance with totalVI. The experimental workflow for generating a dataset containing paired RNA and surface protein abundance in single cells is illustrated in the top panel. First, a single-cell suspension is stained with a panel of barcoded antibodies. Each antibody in the panel has a unique oligonucleotide barcode that identifies the target protein and contains a poly-A tail for capture along with the polyadenylated mRNA transcripts (1). Next, this sample is processed, typically using a droplet microfluidic platform, for isolation of single cells, and the reverse transcription and barcoding of captured molecules for multiplexed library preparation. After droplet processing, a single sequencing library is prepared and sequenced on a high-throughput sequencing platform. The bottom panel illustrates the totalVI analytical pipeline which requires first preprocessing and filtering the sequencing data to generate paired RNA and protein count matrices (2). Those count matrices are then used as input into the totalVI model (3). After running totalVI, downstream analysis allows for data interpretation (4)

same cell [16, 17]. Because proteins cannot be sequenced directly, these methods use antibodies specifically targeted to cell surface proteins, which are chemically conjugated to oligonucleotide barcodes that include polyadenylated tails (resembling the structure of mRNA transcripts). After staining cells with barcoded antibodies, cells are processed through droplet-generating devices, and both RNA transcripts and antibody barcodes can be captured, sequenced, and quantified in the same experiment.

Currently, this droplet-based approach is the state of the art for paired RNA and protein measurements. A number of companies have commercialized oligonucleotide-barcoded antibodies that target proteins on the cell surface, including BioLegend's TotalSeq and BD's AbSeq, which are compatible with the 10x Genomics platform [22] and the BD Rhapsody platform, respectively. The availability of these products has made the paired measurement of RNA and proteins in single cells increasingly accessible, and has enabled the application of this multi-omic technology to various cell types, organisms, and disease states [23–27].

2.2 Experimental Design

Step-by-step protocols for conducting paired RNA and protein measurements are extensively documented by the manufacturers of commercial platforms and in the original publications [16, 17]. However, there are a few aspects of experimental design that should be considered regardless of the experimental platform. These experimental decisions can affect the quality of the sequencing data collected and can have an impact on downstream analysis (*see* **Note 1**).

Cell Numbers

Droplet-based microfluidic platforms can process multiple thousands of cells per reaction. For robust downstream analysis with tools like totalVI, it is recommended that an experiment contains as many cells as possible (considering sequencing costs and sample availability), but at least around 2000 cells. Although cells can be loaded in numbers higher than those recommended by the manufacturer, caution should be taken since increasing cell numbers will increase the likelihood that two or more cells will be encapsulated in a single droplet, producing a "doublet" that should be filtered out from downstream analysis.

Cell Viability

The preparation of single-cell suspensions should be completed as quickly as possible to maintain cell viability. As cells die, RNA degrades more quickly than protein. Therefore, when analyzing sequencing data, low-quality RNA libraries could be an indication of poor cell viability. To preserve cell viability after tissue extraction, cells can be stained, washed, and loaded in media (including serum; see manufacturer's instructions for types of media that are incompatible with the platform in use). Commercial antibodies are formulated in buffers containing ingredients that might exacerbate cell death. To mitigate this, a buffer exchange can be performed on antibodies prior to cell staining to transfer antibodies into media or a more compatible buffer.

Cell Hashing

Cell hashing is the practice of staining a sample of cells with a uniquely barcoded antibody ("hashtag antibody") that targets an ubiquitous surface marker such that multiple samples receiving different hashtag antibodies can be pooled together in the same experiment and demultiplexed following sequencing [28]. Cell hashing is employed to save costs and mitigate batch effects by processing multiple samples in the same reaction. Cell hashing also provides a way to detect doublets between samples, facilitating the overloading of microfluidic devices with higher cell numbers. It is recommended to confirm that hashtag antibodies do indeed bind to the cell type of interest, since not all cells express so-called "ubiquitous" surface proteins.

Antibody Numbers

Because the available sequence space of oligonucleotide barcodes is virtually unlimited [16], the number of antibodies included per assay has thus far only been limited by the availability of barcoded antibodies. To date, experiments including hundreds [18] of barcoded antibodies have reported no negative effects due to high antibody numbers, raising the potential for a future proteome-wide assay. To reduce pipetting errors and improve consistency across experiments, some companies have begun producing pre-mixed "panels" or "cocktails" of barcoded antibodies. Caution should be taken to ensure that such pre-mixed panels have similar staining conditions (see *Antibody staining* below).

Antibody Titration

In some protocols, barcoded antibodies are added to the experiment in different concentrations (known as titrations). Traditionally, antibody titrations aimed to optimize the signal-to-noise ratio. Antibodies at too high a concentration might result in high levels of background due to nonspecific binding or excessive unbound antibodies in the ambient solution. Alternatively, antibodies at too low a concentration might be undetectable or indistinguishable from background noise. Finding the optimal concentration for each antibody can be challenging, since this concentration is expected to differ across tissues and cell types depending on the abundance of the target protein (i.e., high concentrations are often used to detect rare/lowly expressed proteins). A recent study of barcoded antibodies explored reducing antibody titrations relative to previously published protocols with the goal of achieving sufficient signal while reducing the high cost of sequencing [29]. For instance, reducing an antibody's titration might reduce background from excess antibodies, resulting in a lower required sequencing depth to recover the same level of signal over background. While reducing antibody concentrations might reduce background and lower sequencing costs, concentrations that are too low might inhibit the distinction of protein foreground from background and could reduce the power to detect differentially expressed proteins. Moreover, adding antibodies at different concentrations makes it challenging to draw comparisons between the absolute levels of two proteins.

Antibody Staining

Available protocols currently recommend uniform staining conditions for all antibodies in an experiment (e.g., add antibody panel to cell suspension and incubate for 30 minutes at 4 °C). It should be noted that antibodies for some proteins (e.g., chemokine receptors) typically require staining at higher temperatures and for longer times due to protein cycling onto and off of the cell surface. Therefore, pre-mixing antibody panels and conducting a single staining step might result in poor detection of some antibodies. It is recommended to consider the ideal staining conditions of each antibody included in the assay and to potentially conduct multiple rounds of staining to allow for the binding of all antibodies of interest.

Antibody Aggregation

Due to the potential of antibodies to aggregate in solution, most protocols recommend a high-speed centrifugation step prior to staining to remove antibody aggregates. It should be noted when processing sequencing results that cells with extremely high (outlier) numbers of protein counts could be caused by these antibody aggregates and should be filtered out prior to downstream analysis.

Antibody Washing

Following staining, washing is an important experimental step to remove unbound antibodies and only retain antibodies on the cell surface that have specifically bound their target protein. In principle, the wash volume and number of washes could be modulated to improve the signal-to-noise ratio of barcoded antibodies. While it is recommended to use no less than the recommended wash volume or number of washes in the appropriate protocol (which could result in higher levels of protein background), attempting to reduce protein background by increasing the number of washes would come with the trade-off of reducing the overall cell number (since cells are lost in each round of centrifugation).

Sequencing Depth

Sequencing depth is a major experimental consideration that can impact the quality of downstream analysis. Sequencing single-cell libraries to saturation (i.e., counting each captured molecule at least once) provides the maximum information in a single experiment, but can be cost prohibitive. Previous studies and current protocols make recommendations far below the point of saturation, but this can come at the cost of failing to detect some molecules at all (known as dropout events) or inaccurately quantifying relative molecule counts. Because RNA and protein libraries are generated separately following upstream processing steps, these libraries can be sequenced at different depths depending on the relative amounts of information provided by each molecule in a given experiment. The choice of sequencing depth typically depends on the complexity of each library (i.e., the number of unique transcripts in a cell and the total number of antibodies included in the antibody panel; see Fig. 3); cost considerations, including the

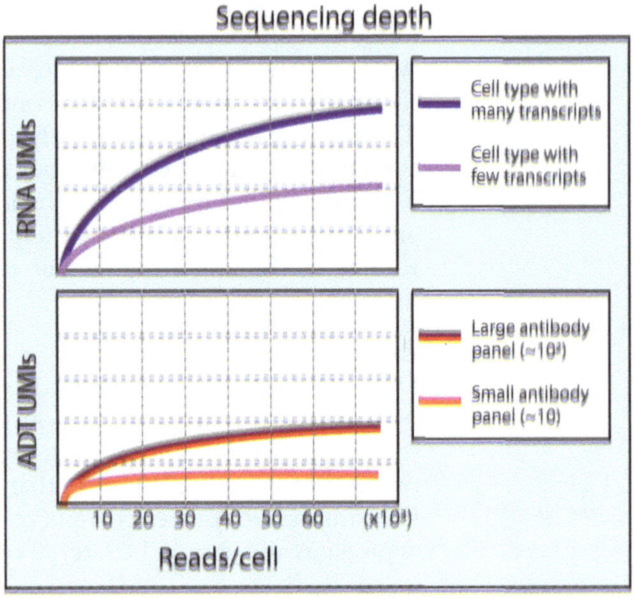

Fig. 3 Sequencing depth. This schematic illustrates the relationship between sequencing depth and recovered information as measured by average uniquely captured molecules (UMIs) corresponding to RNA transcripts (top) or surface proteins (bottom). In both RNA and protein libraries, the complexity of the system will determine the sequencing depth needed before all of the available information is collected. Cells with higher RNA content, such as HEK cells, saturate at higher sequencing depths than cells with lower RNA content, such as PBMCs

number of cells analyzed; and the downstream analysis goals. For example, accurate clustering and cell type identification can be achieved at relatively low sequencing depths [30, 31], but detailed downstream analysis, differential expression testing, and the detection of rare cell types or molecules might suffer at low depths. Particularly for protein analysis, sequencing at lower depth can make it more challenging to distinguish protein foreground from background. While a determination of the desired sequencing depth will depend on the cell type, proteins, and analysis of interest, it is recommended that initial experiments sequence at relatively high depth, and to reduce the sequencing depth in subsequent experiments if appropriate. For RNA libraries, a starting point is typically around 50,00–100,000 reads/cell depending on prior knowledge of low/high RNA content per cell. For protein libraries, a starting point is around 5000–25,000 reads/cell depending on the number of antibodies in the panel (e.g., 10–100+ antibodies). RNA and protein libraries can be pooled at the desired ratio in the same sequencing run to modulate the relative sequencing depth of each library.

3 Joint Analysis of Paired RNA and Protein Measurements in Single Cells with totalVI

3.1 Preprocessing and Filtering the Data

Sequencing of RNA and protein libraries will produce reads in FASTQ format. For all analysis, FASTQ files must first be processed into a UMI count matrix, which should be filtered for quality prior to analysis with totalVI (*see* **Note 1**).

Generating Count Matrices

UMI count matrices can be generated from FASTQ files using tools such as Cell Ranger [22] or kallisto [32], which align sequencing reads to a reference genome (for RNA) or to a reference of predefined oligonucleotide barcode sequences (for protein). The resulting matrix will contain a row for each feature (i.e., a gene or a protein) and a column for each cell. Each entry indicates the number of captured molecules (e.g., transcripts of a given gene or antibodies binding a target protein) in a given cell.

Doublet Filtering

Doublets, which are droplets that contain two or more cells rather than a single cell, can confound analysis (potentially appearing as nonexistent, hybrid cell types) and should be removed. Doublets composed of different cell types can be identified and removed by a number of existing packages [33, 34]. These doublet detection packages use similar algorithms to detect doublets composed of different cell types. Caution should be taken when applying these methods to datasets containing homogeneous populations or continuous developmental trajectories since these algorithms are not expected to perform well in this setting and could produce misleading results. If cell hashing was used, cross-sample doublets as determined by hashtag demultiplexing [28] should also be removed at this stage.

Cell Filtering by Quality

It is recommended to remove dead cells, low-quality cells, and doublets before analysis. For convenience, this stage of preprocessing can be performed using common single-cell analysis pipelines like Scanpy [35] or Seurat [14]. Typically, dead/low-quality cells can be identified by a high percentage of RNA UMI counts derived from mitochondrial genes (e.g., >15%) and low numbers of genes/proteins detected (the number will depend on sequencing depth per experiment). It is also recommended to remove outlier cells with extremely high protein UMI counts, since this is likely caused by antibody aggregation. However, it is not recommended to remove outlier cells with high RNA UMI content in an attempt to remove potential doublets, since these high UMI counts could be driven by biological differences that are of interest for analysis (e.g. large cell size, cell cycle stage).

Feature Filtering

Before running totalVI, it is recommended to conduct gene filtering. Users can decide how many genes to include in the model, but 5000 highly variable genes is a reasonable number in most settings. Using fewer than 500 genes might result in lower accuracy or less informative analysis, and using greater than 10,000 genes might lead to longer run-times without any substantial improvement in performance. Hashtag antibodies and isotype control antibodies (which have no specific protein target) should also be removed prior to running totalVI.

3.2 Running the totalVI Model

The totalVI Software

totalVI is available as open-source, Python-based software as a part of the scvi-tools software package (https://scvi-tools.org/) [36]. The package includes tutorials and documentation describing how to perform the analysis pipeline described below.

Model Inputs

totalVI takes as input the filtered UMI count matrix including both RNA and protein features. totalVI can also include a categorical covariate (called "batch"). Batch labels can be applied to remove batch effects, for example, differences across lanes of a 10x Genomics experiment, patient samples, or sequencing runs. Batch labels can also be applied to integrate datasets derived from different experiments or labs. totalVI is capable of integrating datasets with different protein panels (Fig. 4). The user can choose whether to

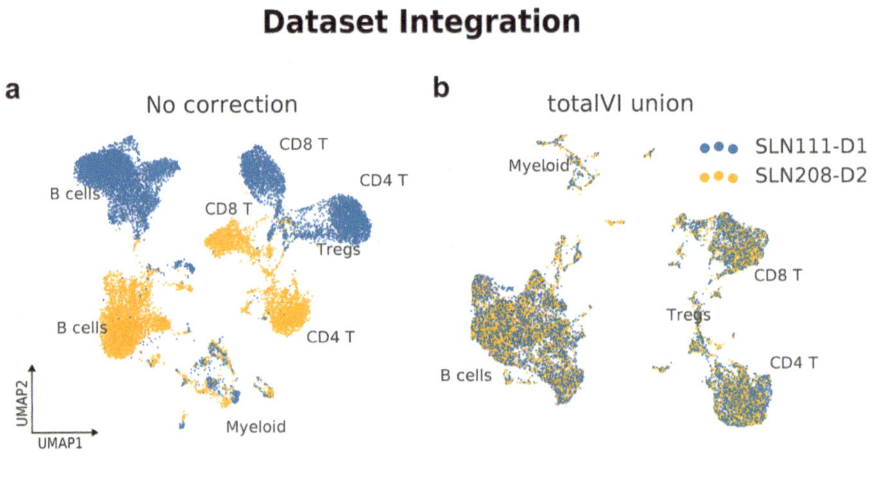

Fig. 4 Dataset integration with totalVI. (**a**) UMAP plot of two CITE-seq datasets of spleen and lymph node cells with panels containing either 111 or 208 antibodies (SLN111-D1, SLN208-D2). Linear dimensionality reduction with principal component analysis (PCA, 30 components) followed by UMAP reveals batch as a major source of variation. (**b**) UMAP plot of the totalVI latent space (totalVI-union) in which the batch effect between the datasets has been corrected. (Material from Gayoso et al. [18] Springer Nature)

run the model on only those proteins included in both datasets (totalVI-intersect) or to use all available protein information (totalVI-union), which will impute the missing protein values. totalVI-union can be used to integrate a scRNA-seq dataset (containing no protein information) with a CITE-seq dataset (containing both RNA and protein). In this case, totalVI can predict protein expression in the RNA-only data using the given CITE-seq data.

Running the Model

In general, it is recommended that users run the totalVI model with default hyperparameters (e.g., number of latent dimensions, learning rate, etc.). These parameters have demonstrated accurate model fits on multiple cell types and datasets and are a good starting point for any analysis (*see* **Note 2**). The larger the size of the dataset (number of cells and number of features), the longer it will take to run the totalVI model. The model will run considerably faster on a GPU (highly recommended), which can be accessed on cloud computing platforms. After training, the model can be saved for downstream analysis at a later time.

3.3 Downstream Analysis

Once trained, the totalVI model can be used for a wide variety of common analysis tasks. Here we present a common pipeline for downstream analysis with totalVI and show how totalVI can be used to learn about cells from the combination of their RNA and protein expression.

Model Outputs

The totalVI model contains two different components that can be used for downstream analysis. First, totalVI contains a low-dimensional representation of cells, known as the latent space, which contains information from both the RNA and protein data while controlling for noise and technical artifacts in each data type. The low-dimensional latent space can be used to visualize and cluster cell populations. Second, totalVI contains the parameters that describe the probabilistic distributions of each observed RNA and protein measurement. These distributions account for the noise and technical artifacts in the observed data and can be used to denoise the high-dimensional RNA and protein expression data. These distributions can also be used to test differential expression and perform other analysis tasks while accounting for the noise and technical artifacts in the RNA and protein measurements. Below, we describe how these components of totalVI can be used to conduct downstream analysis.

Dimension Reduction for Visualization

The totalVI model encodes the RNA and protein information from each cell in a single low-dimensional representation known as the latent space. Because the latent space includes a combination of

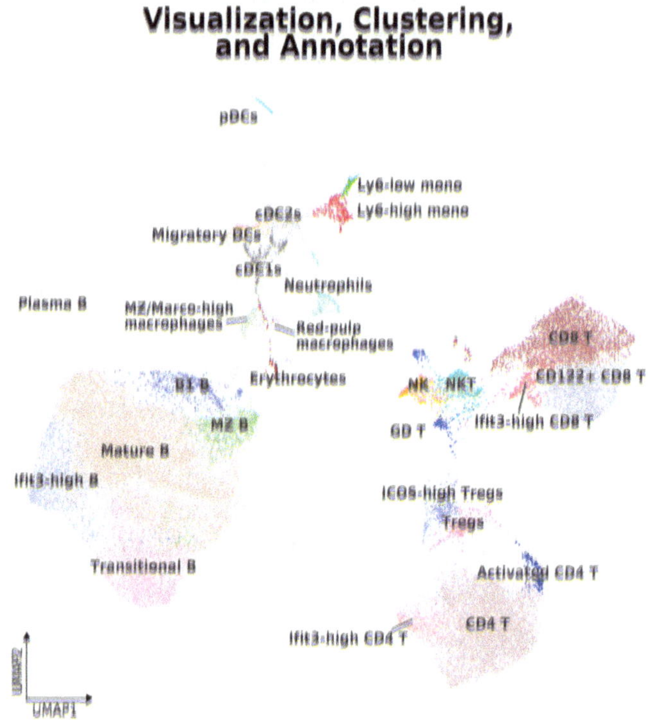

Fig. 5 Visualization, clustering, and annotation of cells in the totalVI latent space. The low-dimensional latent space learned by totalVI represents the RNA and protein information from each cell. The latent space of four integrated spleen and lymph node datasets (totalVI-intersect) is visualized in two dimensions with UMAP. The latent space is clustered by the Leiden algorithm, and clusters are annotated based on expression of RNA and protein markers of cell type or cell state. (Material from Gayoso et al. [18] Springer Nature)

RNA and protein information and has removed technical artifacts like batch and RNA library size, a cell's location in the latent space can be interpreted as representing the cell's underlying biological state. By applying a dimension reduction method like UMAP [37] to the latent space, users can visualize cells in two dimensions and interpret cell-cell similarities based on the combination of cells' RNA and protein expression profiles (Fig. 5).

Clustering Cells

Cells can be clustered using any clustering algorithm, such as the Leiden algorithm [38], based on cell-cell similarity in the latent space (Fig. 5). Therefore, clusters based on the totalVI latent space take into account both the RNA and protein information from each cell. These clusters can be used to identify and annotate cell types.

Denoising Features for Visualization

totalVI can produce a matrix of denoised expression values for RNA and protein that remove technical effects like RNA library size,

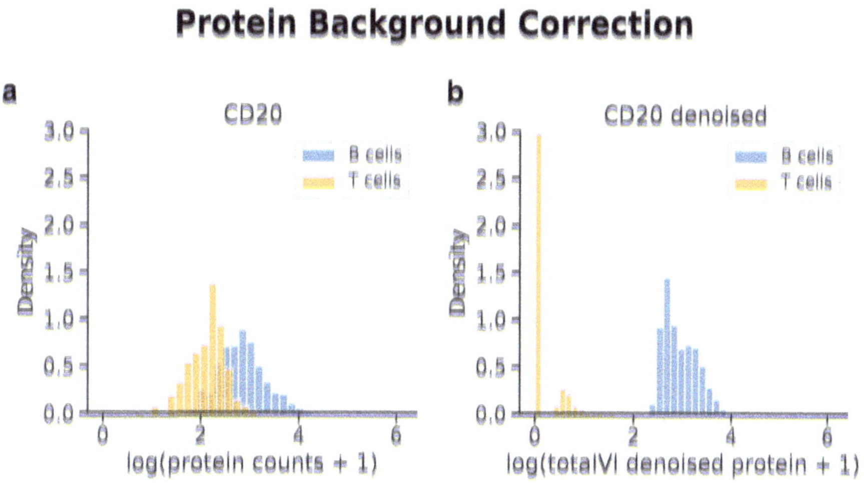

Fig. 6 totalVI corrects protein background. (**a**) For CD20 protein, which is a marker of B cells, the distributions of raw counts are difficult to distinguish between true expression in B cells and background in T cells in the SLN111-D1 dataset. (**b**) By denoising protein data with totalVI, the expected amount of background is effectively subtracted from the overall expected expression, resulting in a clear separation in CD20 expression levels between B cells and T cells. (Material from Gayoso et al. [18] Springer Nature)

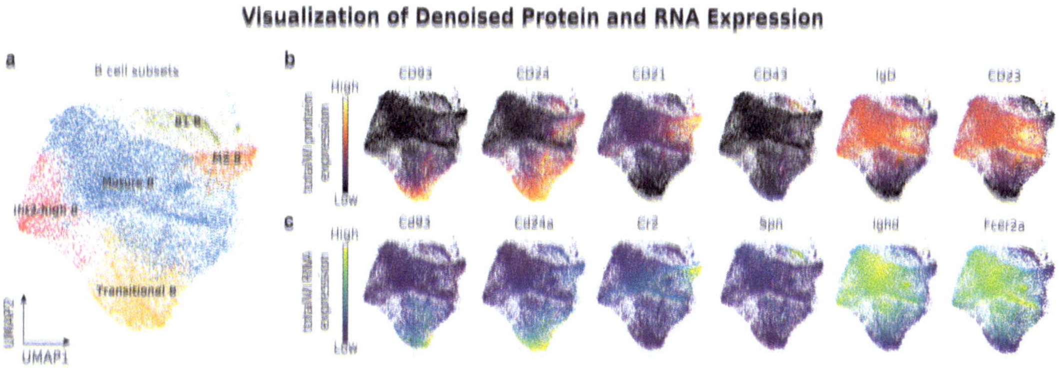

Fig. 7 totalVI denoises protein and RNA expression. (**a**) B cell clusters subset from the UMAP plot of the SLN-all dataset depicted in Fig. 5. (**b**) totalVI denoised expression of six proteins that are markers of different types of B cells visualized on the UMAP plot of the totalVI latent space. (**c**) totalVI denoised RNA expression of the six genes that encode the corresponding proteins in (**b**). (Material from Gayoso et al. [18] Springer Nature)

protein background, and sampling noise (Fig. 6). These denoised values can be visualized to observe expression patterns across cells (Fig. 7). Note that denoised values can be helpful for visualization, but downstream statistical analysis should not be applied to these values directly. Instead, for downstream analyses like differential expression testing and the calculation of correlations between features, totalVI contains statistical methods that account for technical artifacts (*see* **Note 3**). In the case of totalVI-union when a dataset might be missing data from some or all proteins, imputed protein values can be visualized to observe the predicted expression patterns of these proteins (*see* **Note 4**).

Interpreting Protein Foreground Probabilities
To accurately analyze protein data, the totalVI model learns to distinguish between protein foreground (a measurement likely derived from true biological signal) and background (a measurement likely due to ambient antibodies or nonspecific antibody binding). For each protein measurement, totalVI estimates the probability that the measurement was due to foreground signal. totalVI uses these foreground probabilities to remove protein background when denoising protein data and conducting other downstream analysis tasks so that they are not biased by this technical artifact in the protein measurement.

Users might be interested in visualizing these foreground probabilities to gain intuition about which cells are "on" or "off" for a particular protein. However, it is still recommended to use denoised protein expression (rather than foreground probabilities) for interpretation of cell type expression, since denoised values account for foreground probability while preserving the dynamic range of the protein measurement. Thus, users should not be concerned if a protein that is expected to be off has a higher-than-expected foreground probability, since the magnitude of denoised expression might still be very low. If the model is uncertain for a particular protein (i.e., if many cells have intermediate foreground probabilities near 0.5), it might be an indication that the targeting antibody was poorly titrated, making it challenging for the model to distinguish foreground from background.

Differential Expression Testing
totalVI can be used to test differential expression for both RNA and proteins (Fig. 8). totalVI includes a number of commonly performed differential expression tests. In a one-vs-one test, differential expression of all features is tested between two groups of cells (e.g., testing between two clusters). In a one-vs-all test, each cluster is compared with all remaining cells. This test can be particularly useful for cell type annotation or identifying unique markers for a cell type. When comparing between two conditions (e.g., case vs control), the within-cluster DE test can identify differences between two labeled populations within each cluster (e.g., to identify differences between diseased and healthy T cells that are mixed within the T cell cluster).

The output of the totalVI differential expression test includes an estimate of the log-fold-change of each feature between the comparison groups (indicating the magnitude of the difference) and a Bayes factor, which can be interpreted as the significance of the difference, where a higher Bayes factor means more significant.

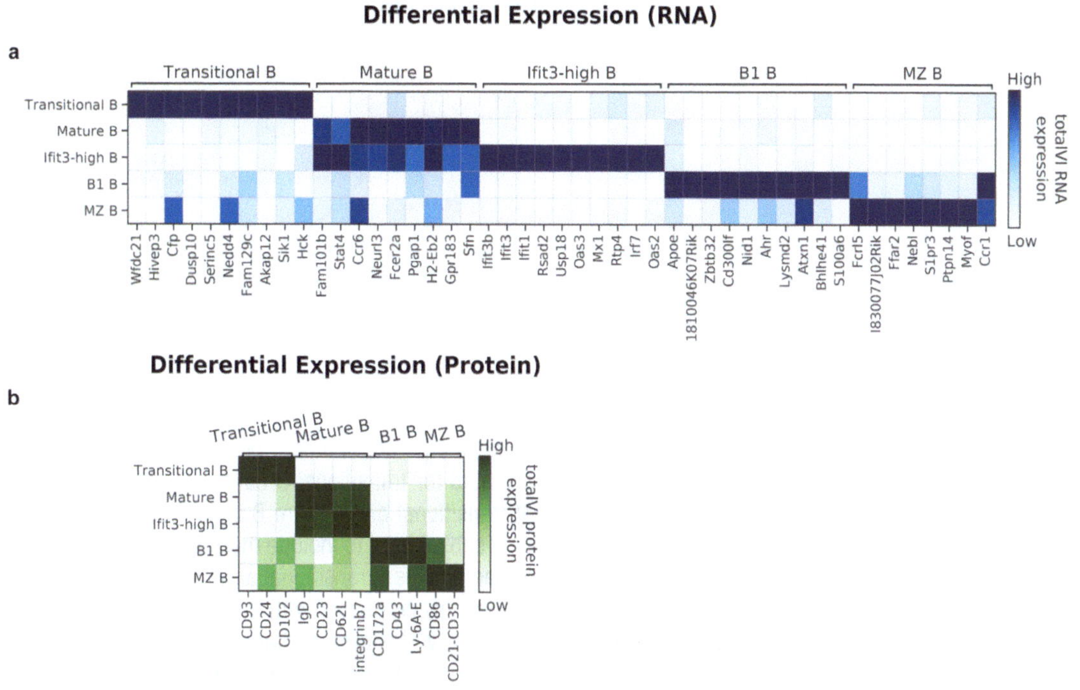

Fig. 8 totalVI tests differential expression for both RNA and proteins. Results of a one-vs-all differential expression test between the B cell subsets shown in Fig. 6. Results were filtered for significance (log Bayes factor > 0.7), ranked by median log fold change, and filtered to retain genes with nonzero UMI counts in at least 10% of the tested subset. (**a**) The top ten differentially expressed genes per subset. (**b**) The top three differentially expressed proteins per subset. (Material from Gayoso et al. [18] Springer Nature)

Correlations Between Features

Beyond the standard set of analyses described above, users might be interested in calculating correlations between features. totalVI has no explicit information about which RNA transcript encodes which protein, so any correlation learned by the model is not predetermined by known gene-protein relationships. totalVI can calculate a denoised correlation matrix (either Pearson or Spearman) between all features that removes batch effects and technical artifacts (e.g., protein background). From the denoised matrix, users can extract a correlation of interest (e.g., between a particular gene and the protein it encodes). Correlations can also be used to cluster features into modules, which might provide insight into regulatory networks (Fig. 9). It should be noted that correlations computed across an entire dataset might differ substantially from correlations computed within a cell type, since dataset-wide correlations might largely be driven by differences across cell types that either do or do not express a given feature.

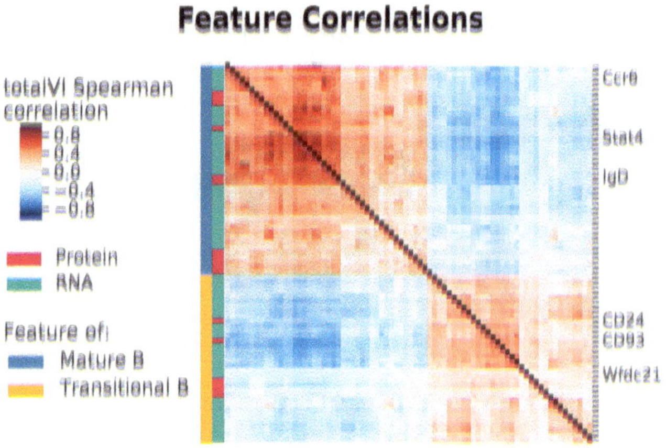

Fig. 9 totalVI calculates denoised correlations between features. totalVI Spearman correlations calculated for transitional B cells in the SLN-all dataset identify feature modules of transitional and mature B cells. The full correlation matrix between all features is subset here to include marker genes and proteins that are differentially expressed between transitional and mature B cells as well as additional features that are highly correlated with the differentially expressed features. Features are hierarchically clustered and are labeled as either RNA or protein, and by the cell type with which the feature is associated. (Material from Gayoso et al. [18] Springer Nature)

Interpretation of the Latent Space

Unlike many "black-box" deep-learning models that are difficult to interpret, the latent space of totalVI has interpretable latent dimensions. Users who are interested in exploring the meaning of the latent dimensions can visualize each dimension to observe which cells have high values (Fig. 10). A latent dimension could be interpreted as describing a particular cell type (e.g., B cells) or as describing a trend that spans cell types (e.g., cell cycle). Further exploration of the latent dimensions could include archetype analysis [39], which identifies which genes and which proteins contribute the most to each latent dimension.

Beyond Standard Analysis

In addition to the characterization of cell types and standard downstream tasks, totalVI can be used for a number of other types of analysis. For instance, totalVI could be used to inform aspects of experimental design such as identifying optimal antibody titrations to improve the detection of foreground and background in protein measurements. totalVI could also be used to explore the information gained per RNA or protein measurement in various analysis tasks, which could inform the desired sequencing depth of each library balanced with the cost of sequencing. Taking advantage of both RNA and protein information, totalVI could also be used to investigate basic biological questions on the relationship between

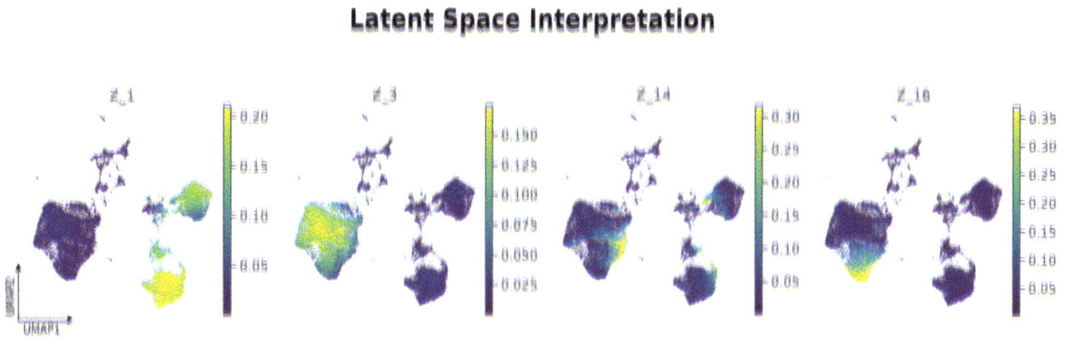

Fig. 10 Interpretation of totalVI latent space. UMAP plots of the totalVI latent space for the SLN-all dataset colored by four of the 20 latent dimensions. High values in these latent dimensions highlight (from left to right) T cells, B cells, cells undergoing the cell cycle, and transitional B cells. (Material from Gayoso et al. [18] Springer Nature)

RNA and protein levels, including the dynamics and regulatory processes governing transcription and translation in various cell types and experimental conditions.

4 Conclusions

totalVI provides a framework for end-to-end analysis of paired RNA and protein measurements in single cells. By combining RNA and protein information into a single model, totalVI creates a representation of cell state that forms a more complete description of a cell's phenotype than either modality alone. Moreover, the combination of both pieces of information about a single cell strengthens the interpretation of either modality alone. For example, the use of RNA information in the totalVI model helps to distinguish protein foreground from background, improving the accuracy of protein analysis. In addition, having protein measurements associated with single-cell transcriptomes can help relate RNA sequencing data to prior knowledge of cell types defined by surface marker expression.

In a standard analysis with totalVI, dimensionality reduction, clustering, and visualization of denoised genes and proteins can be applied to annotate cells, potentially discovering new cell types or cell states. Through differential expression, new RNA or protein markers for these cell types could be revealed, and cells can be defined by the combination of their RNA and protein expression profiles. totalVI also provides methods to investigate correlations between features. totalVI can conduct these analyses on multiple batches or datasets through integration even if datasets did not measure identical protein panels. The totalVI model accounts for technical artifacts in each measurement, like protein background, batch effects, and sampling noise, improving the accuracy of downstream analysis.

totalVI might be extended to perform additional tasks using RNA and protein data. One potential extension could perform an automated annotation of cell types based on both RNA and protein information. In the challenging problem of learning gene regulatory networks, totalVI could provide a connection between cell surface receptors receiving signals that are transduced to produce transcriptional changes, or connecting transcriptional events with changes in surface protein expression. totalVI could eventually serve as a useful tool to query and integrate data with various published cell atlases, which contain a combination of RNA and protein information with different protein panels across tissues.

The totalVI model was designed specifically to handle the nuances of analysis of paired transcriptome and surface protein data. In the future, totalVI could be applied to additional measurements and data types. For example, as more barcoded antibodies become available, totalVI can easily scale to include the analysis of larger protein panels. It is also expected that totalVI will readily apply to the analysis of intracellular proteins measured with barcoded antibodies when these methods become available. Beyond RNA and protein measurements, the flexible framework of the totalVI model could be extended to include additional multi-omic measurements such as chromatin accessibility. For each additional modality included, the specific technical artifacts and nuances of the modality must be considered in order to properly model and address them.

5 Notes

1. Experimental and preprocessing choices affect downstream analysis. In particular, for data that appears low quality, the most common explanation is experimental. For RNA data that appears low quality, a suggested place to begin troubleshooting is cell viability. In the cell processing stage, even cells that are negative for markers of cell death (e.g., propidium iodide or annexin V) might have begun degrading RNA due to stress or death, so methods can be applied to preserve cell viability (as described above). Poor cell viability could also explain the capture of fewer cells than expected, since dying cells with degraded RNA often fail to meet the minimum UMI-count or gene-count thresholds for high-quality cells. For proteins that are not detected or are detected at lower-than-expected levels, it should be considered whether the antibody titration and sequencing depth are high enough to detect a signal, or if this antibody requires different staining

conditions. After the data has been collected, skipping recommended preprocessing steps can result in a lower-quality analysis. Even though it might be challenging to determine whether a certain filtering threshold is correct for a given dataset, users should not be alarmed if a few low-quality cells, dead cells, or doublets are missed by the filters. Often, after running totalVI, these cells will form separate clusters that can be filtered out at a later stage, or filtering thresholds can be updated, and the model can be run again on more stringently filtered data.

2. It is highly recommended to start by running totalVI with default parameters. Advanced users might be interested in conducting parameter tuning for a particular dataset to improve performance. However, in our experience, running totalVI with default parameters provides sufficient performance across a variety of datasets that include different numbers of genes, proteins, and cells, different sequencing depths, heterogeneous cell types or more homogeneous continuous development, and different types and severity of batch effects.
3. Caution should be taken with downstream analysis on denoised values. Denoised values can be extremely helpful for visualizing RNA and protein expression after the removal of technical artifacts and sampling noise. However, due to the denoising process, denoised values might contain bias caused by spurious relationships between features. Therefore, it is not recommended for users to attempt to calculate correlations between these denoised values or to use these denoised values in a separate differential expression test. totalVI has statistical methods that control for nuisance variation while avoiding potential denoising-induced bias, which should be used for analysis testing the relationship between features and for differential expression.
4. Caution should be taken when drawing conclusions from imputed protein values (e.g., when observing predicted protein expression from a scRNA-seq dataset that has been integrated with a CITE-seq dataset). These predictions can be useful for making hypotheses to be further validated experimentally and have been shown to be highly accurate in some settings [18], but are expected to be less informative for cells that don't have high overlap with data where the observed proteins are present.

Acknowledgments

Research reported in this chapter was supported by the National Institute of General Medical Sciences of the National Institutes of Health under award number R35GM124916. Zoë Steier was supported by the National Science Foundation Graduate Research

Fellowship and the Siebel Scholarship. Annie Maslan was supported by the National Science Foundation Graduate Research Fellowship. Aaron Streets is a Chan Zuckerberg Biohub investigator. The authors would like to acknowledge Nir Yosef and Adam Gayoso for co-development of totalVI and support with the writing of this chapter.

References

1. Regev A, Teichmann SA, Lander ES et al (2017) The human cell atlas. eLife 6. https://doi.org/10.7554/ELIFE.27041
2. Chen KH, Boettiger AN, Moffitt JR et al (2015) Spatially resolved, highly multiplexed RNA profiling in single cells. Science (80) 348:aaa6090. https://doi.org/10.1126/science.aaa6090
3. Paul F, Arkin Y, Giladi A et al (2015) Transcriptional heterogeneity and lineage commitment in myeloid progenitors. Cell 163: 1663–1677. https://doi.org/10.1016/j.cell.2015.11.013
4. Wilson NK, Kent DG, Buettner F et al (2015) Combined single-cell functional and gene expression analysis resolves heterogeneity within stem cell populations. Cell Stem Cell 16:712–724. https://doi.org/10.1016/j.stem.2015.04.004
5. Chen TN, Gupta A, Zalavadia MD, Streets A (2020) μCB-seq: microfluidic cell barcoding and sequencing for high-resolution imaging and sequencing of single cells. Lab Chip 20: 3899–3913. https://doi.org/10.1039/D0LC00169D
6. Clark SJ, Argelaguet R, Kapourani CA et al (2018) ScNMT-seq enables joint profiling of chromatin accessibility DNA methylation and transcription in single cells e. Nat Commun 9: 1–9. https://doi.org/10.1038/s41467-018-03149-4
7. Cao J, Cusanovich DA, Ramani V et al (2018) Joint profiling of chromatin accessibility and gene expression in thousands of single cells. Science (80) 361:1380–1385. https://doi.org/10.1126/science.aau0730
8. Chen S, Lake BB, Zhang K (2019) High-throughput sequencing of the transcriptome and chromatin accessibility in the same cell. Nat Biotechnol 37:1452–1457. https://doi.org/10.1038/s41587-019-0290-0
9. Rooijers K, Markodimitraki CM, Rang FJ et al (2019) Simultaneous quantification of protein–DNA contacts and transcriptomes in single cells. Nat Biotechnol 37:766–772. https://doi.org/10.1038/s41587-019-0150-y
10. Gorin G, Svensson V, Pachter L (2020) Protein velocity and acceleration from single-cell multiomics experiments. Genome Biol 21:39. https://doi.org/10.1186/s13059-020-1945-3
11. Argelaguet R, Velten B, Arnol D et al (2018) Multi-Omics Factor Analysis—a framework for unsupervised integration of multi-omics data sets. Mol Syst Biol:14. https://doi.org/10.15252/msb.20178124
12. Argelaguet R, Arnol D, Bredikhin D et al (2020) MOFA+: a statistical framework for comprehensive integration of multi-modal single-cell data. Genome Biol 21:111. https://doi.org/10.1186/s13059-020-02015-1
13. Welch JD, Kozareva V, Ferreira A et al (2019) Single-cell multi-omic integration compares and contrasts features of brain cell identity. Cell 177:1873–1887.e17. https://doi.org/10.1016/j.cell.2019.05.006
14. Stuart T, Butler A, Hoffman P et al (2019) Comprehensive integration of single-cell data. Cell 177:1888–1902.e21. https://doi.org/10.1016/j.cell.2019.05.031
15. Argelaguet R, Clark SJ, Mohammed H et al (2019) Multi-omics profiling of mouse gastrulation at single-cell resolution. Nature 576: 487–491. https://doi.org/10.1038/s41586-019-1825-8
16. Stoeckius M, Hafemeister C, Stephenson W et al (2017) Simultaneous epitope and transcriptome measurement in single cells. Nat Methods 14:865–868. https://doi.org/10.1038/nmeth.4380
17. Peterson VM, Zhang KX, Kumar N et al (2017) Multiplexed quantification of proteins and transcripts in single cells. Nat Biotechnol 35:936–939. https://doi.org/10.1038/nbt.3973
18. Gayoso A, Steier Z, Lopez R et al (2021) Joint probabilistic modeling of single-cell multiomic data with totalVI. Nat Methods:1–11. https://doi.org/10.1038/s41592-020-01050-x
19. Lopez R, Regier J, Cole MB et al (2018) Deep generative modeling for single-cell

transcriptomics. Nat Methods 1512(15):1053–1058. https://doi.org/10.1038/s41592-018-0229-2
20. Macosko EZ, Basu A, Satija R et al (2015) Highly parallel genome-wide expression profiling of individual cells using Nanoliter droplets. Cell 161:1202–1214. https://doi.org/10.1016/J.CELL.2015.05.002
21. Klein AM, Mazutis L, Akartuna I et al (2015) Droplet barcoding for single-cell transcriptomics applied to embryonic stem cells. Cell 161:1187–1201. https://doi.org/10.1016/j.cell.2015.04.044
22. Zheng GXY, Terry JM, Belgrader P et al (2017) Massively parallel digital transcriptional profiling of single cells. Nat Commun 81(8):1–12. https://doi.org/10.1038/ncomms14049
23. Granja JM, Klemm S, McGinnis LM et al (2019) Single-cell multiomic analysis identifies regulatory programs in mixed-phenotype acute leukemia. Nat Biotechnol 37:1458–1465. https://doi.org/10.1038/s41587-019-0332-7
24. Praktiknjo SD, Obermayer B, Zhu Q et al (2020) Tracing tumorigenesis in a solid tumor model at single-cell resolution. Nat Commun 111(11):1–12. https://doi.org/10.1038/s41467-020-14777-0
25. Kotliarov Y, Sparks R, Martins AJ, et al (2020) Broad immune activation underlies shared set point signatures for vaccine responsiveness in healthy individuals and disease activity in patients with lupus. Nat Med 1–12. https://doi.org/10.1038/s41591-020-0769-8
26. Lavaert M, Liang KL, Vandamme N et al (2020) Integrated scRNA-Seq identifies human postnatal thymus seeding progenitors and regulatory dynamics of differentiating immature thymocytes. Immunity 52:1088–1104.e6. https://doi.org/10.1016/J.IMMUNI.2020.03.019
27. Muench DE, Olsson A, Ferchen K et al (2020) Mouse models of neutropenia reveal progenitor-stage-specific defects. Nat 5827810(582):109–114. https://doi.org/10.1038/s41586-020-2227-7
28. Stoeckius M, Zheng S, Houck-Loomis B et al (2018) Cell hashing with barcoded antibodies enables multiplexing and doublet detection for single cell genomics. Genome Biol 19:1–12. https://doi.org/10.1186/S13059-018-1603-1
29. Buus TB, Herrera A, Ivanova E et al (2021) Improving oligo-conjugated antibody signal in multimodal single-cell analysis. eLife:10. https://doi.org/10.7554/ELIFE.61973
30. Heimberg G, Bhatnagar R, El-Samad H, Thomson M (2016) Low dimensionality in gene expression data enables the accurate extraction of transcriptional programs from shallow sequencing. Cell Syst 2:239–250. https://doi.org/10.1016/J.CELS.2016.04.001
31. Svensson V, da Beltrame EV, Pachter L (2019) Quantifying the tradeoff between sequencing depth and cell number in single-cell RNA-seq. Biorxiv:762773. https://doi.org/10.1101/762773
32. Melsted P, Booeshaghi AS, Liu L et al (2021) Modular, efficient and constant-memory single-cell RNA-seq preprocessing. Nat Biotechnol 397(39):813–818. https://doi.org/10.1038/s41587-021-00870-2
33. Wolock SL, Lopez R, Klein AM (2019) Scrublet: computational identification of cell doublets in single-cell transcriptomic data. Cell Syst 8:281–291.e9. https://doi.org/10.1016/J.CELS.2018.11.005
34. Gayoso A, Shor J, Carr AJ et al (2019) GitHub. DoubletDetection. https://doi.org/10.5281/ZENODO.2678042
35. Wolf FA, Angerer P, Theis FJ (2018) SCANPY: large-scale single-cell gene expression data analysis. Genome Biol 191(19):1–5. https://doi.org/10.1186/S13059-017-1382-0
36. Gayoso A, Lopez R, Xing G, et al (2022) A Python library for probabilistic analysis of single-cell omics data. Nat Biotechnol 40(2): 163–166. https://doi.org/10.1038/s41587-021-01206-w
37. McInnes L, Healy J, Melville J (2018) UMAP: uniform manifold approximation and projection for dimension reduction. arXiv:1802.03426. https://doi.org/10.48550/arXiv.1802.03426
38. Traag VA, Waltman L, van Eck NJ (2019) From Louvain to Leiden: guaranteeing well-connected communities. Sci Reports 91(9):1–12. https://doi.org/10.1038/s41598-019-41695-z
39. Cutler A, Breiman L (1994) Archetypal analysis. Technometrics. https://doi.org/10.1080/00401706.1994.10485840

Chapter 5

Single-Cell Mass Spectrometry of Metabolites and Proteins for Systems and Functional Biology

Erika P. Portero, Leena R. Pade, Jie Li, Sam B. Choi, and Peter Nemes

Abstract

Molecular composition is intricately intertwined with cellular function, and elucidation of this relationship is essential for understanding life processes and developing next-generational therapeutics. Technological innovations in capillary electrophoresis (CE) and liquid chromatography (LC)-mass spectrometry (MS) provide previously unavailable insights into cellular biochemistry by allowing for the unbiased detection and quantification of molecules with high specificity. This chapter presents our validated protocols integrating ultrasensitive MS with classical tools of cell, developmental, and neurobiology to assess the biological function of important biomolecules. We use CE- and LC-MS to measure hundreds of metabolites and thousands of proteins in single cells or limited populations of tissues in chordate embryos and mammalian neurons, revealing molecular heterogeneity between identified cells. By pairing microinjection and optical microscopy, we enable cell lineage tracing and testing of the roles that dysregulated molecules play in the formation and maintenance of cell heterogeneity and tissue specification in frog embryos (*Xenopus laevis*). Electrophysiology extends our workflows to characterizing neuronal activity in sections of mammalian brain tissues. The information obtained from these studies mutually strengthen chemistry and biology and highlight the importance of interdisciplinary research to advance basic knowledge and translational applications forward.

Key words Single cell, Mass spectrometry, Functional biology, Proteomics, Metabolomics, Cell and developmental biology, Neurobiology, *Xenopus laevis*, Zebrafish, Mouse

1 Introduction

Modern 'omics provide an unprecedented set of technologies to study connections between molecular composition and biological function. By enabling the detection and quantification of genes, transcripts, proteins, peptides, and metabolites, valuable information is gained on the molecular underpinnings of biological processes. High-throughput sequencing allows for routine profiling of gene expression in limited to large populations of single cells [1–3].

Erika P. Portero and Leena R. Pade contributed equally with all other contributors.

Jonathan V. Sweedler et al. (eds.), *Single Cell 'Omics of Neuronal Cells*,
Neuromethods, vol. 184, https://doi.org/10.1007/978-1-0716-2525-5_5,

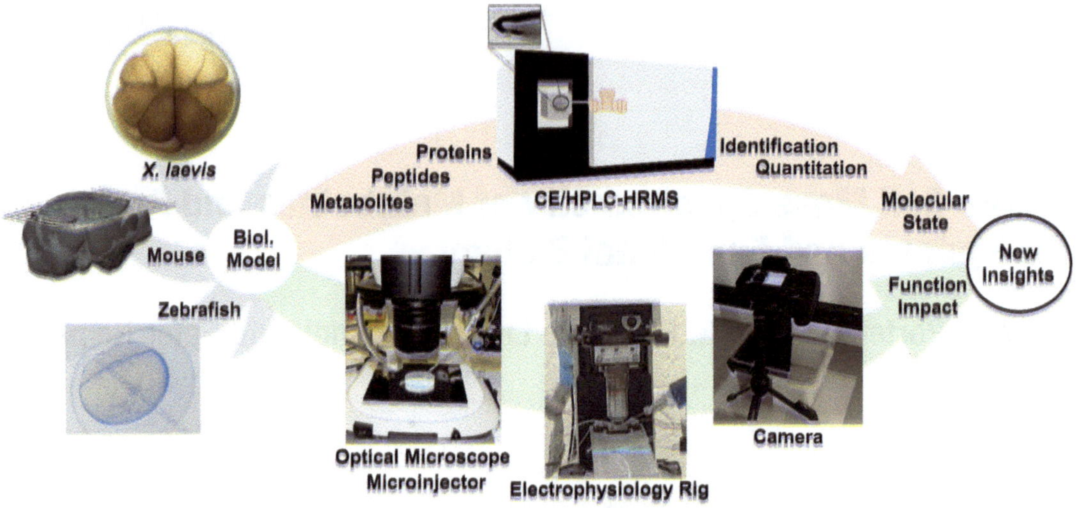

Fig. 1 Our approach for assessing chemistry and function in biological models. The protocols discussed in this chapter were developed and validated using embryos from the frog *Xenopus laevis* and zebrafish and sections from mouse brain. (Adapted with permission from [23] Copyright 2017 American Chemical Society. Adapted with permission from [25] Copyright 2019 American Chemical Society)

Identification and quantification of molecular markers from discovery 'omics help develop hypotheses and design experiments to evaluate their biological function.

Functional biological experiments leverage diverse types of approaches and technologies. Functional tests borrow tools of molecular biology to knock down or out genes, transcripts, or proteins, e.g., by using transcription or translation blocking morpholinos and CRISPR-Cas9. Electrophysiology allows for eavesdropping on neurons. Different types of microscopies, typically optical to electron, enable characterization of cell morphology, phenotype, and anatomy (Fig. 1). Single-cell RNA sequencing with single-molecule fluorescence in situ hybridization (sm-FISH) recently captured molecular cell heterogeneity in the brain in high spatial and molecular resolution [4], supplementing classical knowledge of brain anatomy with molecular information [5]. By comparing single thalamic neurons projecting to motor, somatosensory, and visual cortices in the mouse brain, this approach uncovered several cell types within each projection. Technologies from modern 'omics empower classical biology and neuroscience with new investigative capabilities.

Single-cell mass spectrometry (MS) supports systems biological studies with information on the molecular state of cells. It complements single-cell transcriptomics by directly measuring proteomic, peptidomic, and metabolomic composition (Fig. 1). Detection with excellent molecular specificity facilitates identification without the requirement for functional probes. For example, MS does not necessitate antibodies for detection. High sensitivity and a broad linear dynamic range permit quantification of molecules at endogenous concentrations. Stringent reporting guidelines [6, 7] and public repositories hosting data for reuse, reanalysis, and exchange (e.g., PRIDE [8] and Metabolomics Workbench [9]) promote scientific rigor and accountability in MS-based research. The current state of the field of single-cell MS was the focus of several reviews covering technology and application [10–19]. Single-cell MS is adaptable to broad types of molecules, can be made sufficiently sensitive to quantify physiological concentrations, and is compatible with cells of broad dimensions and types, as well as different model systems used in biology and health studies.

Optical microscopy with single-cell MS integrates morphological and molecular information. Recent reviews provide a comprehensive discussion of single-cell MS [12, 18–20]. Figure 1 illustrates our protocol for optically guided single-cell MS in embryos of the South African clawed frog (*X. laevis*) and zebrafish and sections of mouse brain. Cells were identified and dissected [21, 22] or their contents directly microaspirated [23–25] for metabolomic and proteomic analysis using capillary electrophoresis (CE) or liquid chromatography (LC)-electrospray ionization (ESI)-high-resolution MS (HRMS) (reviewed in [26–29]). Microscopy with single-cell MS enabled detection of ~1500 proteins in 2- to 50-cell *X. laevis* embryos and orthogonal validation using immunohistochemistry [30, 31]. Single-cell proteomics by MS (SCoPE MS) quantified 3000+ proteins from 1490 cells [32, 33], and a single-cell printer with liquid vortex capture enabled rapid metabolomics (~25 cells/min) [34]. With low attomole sensitivity, HPLC columns of narrow bore dimension or with a stationary phase supported on a monolith or a porous layer open tubular (PLOT) format allowed for the identification of ~1300–4000 proteins from 50 to 200 cells via magnetophoretic isolation from whole blood [35–37]. Nanodroplet processing in one pot for trace samples (NanoPOTS) identified over ~1500 proteins from 10 HeLa cells and ~2400 proteins from 100 pancreatic islet cells, supporting profiling across clinical samples [38]. These and other leaps in technology expanded the classical toolbox of cell biology, as was discussed in our recent review of the field [26–29].

We and others built CE-MS platforms to study biomolecules and their role in cell and neurobiological processes. CE renders several fundamental advantages for single-cell analyses. The physical dimensions of fused silica capillaries are amenable to the limited amounts of sample that are contained in single cells. CE provides several methods for concentration enrichment in the capillary to boost sensitivity to low-abundance molecules (e.g., reviewed in [39]). An exquisite separation power and various data alignment strategies permit reproducible identifications [40, 41]. CE-ESI interfaces offer various designs to help hyphenate CE with HRMS for sensitivity, robustness, and reproducibility [42, 43].

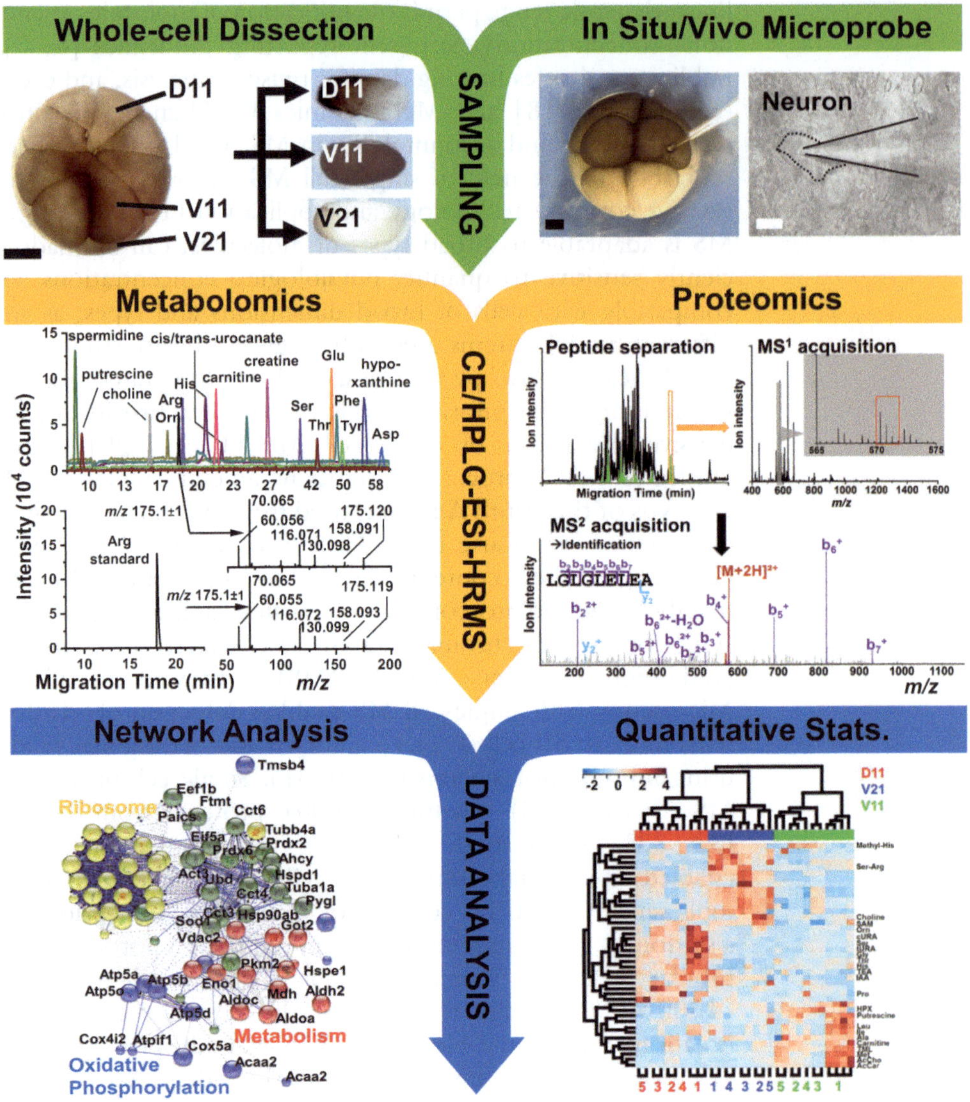

Fig. 2 Our general single-cell MS workflow enabling the analysis of metabolites and proteins in single cells of *Xenopus laevis* or zebrafish and single neurons in a section of the mouse brain. Scale bars: 250 μm (black), 20 μm (white). (Adapted with permission from Ref. [57]; Adapted with permission from Ref. [21]. Copyright 2015 National Academy of Sciences; Adapted with permission from [23] Copyright 2017 American Chemical Society; Adapted with permission from Ref. [26])

These custom-built CE-ESI-HRMS platforms revealed previously unknown details on cellular biochemistry. Proteins or metabolites were measured in single cells of *X. laevis* and zebrafish embryos (Figs. 1 and 2) [21–23, 25], single neurons dissected from *Aplysia californica* [44, 45] and mouse [40, 46], and single HeLa [47] cells. Our CE-ESI interface enabled the identification of hundreds of cationic and anionic metabolites (Fig. 3) [48] and ~700 proteins from ~5 ng protein digest from single *X. laevis*

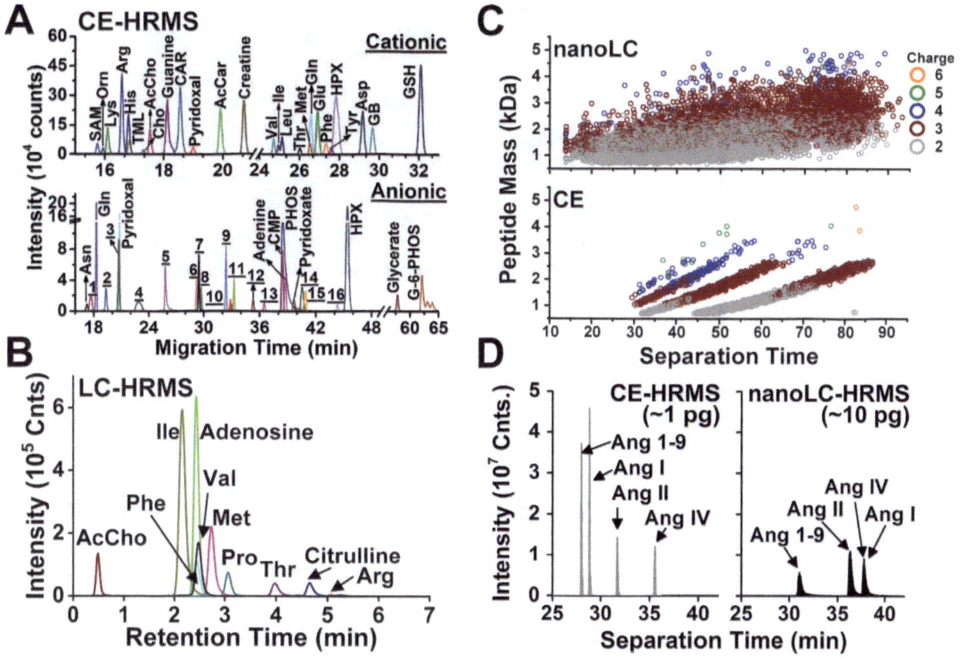

Fig. 3 Representative detection of metabolites and proteins by CE and HPLC ESI-HRMS. (**a**) Chemical profiling of anionic and cationic metabolites in a single *X. laevis* cell using CE-ESI-MS. (**b**) HILIC LC-MS of polar metabolites from limited populations of cells. (**c**) Comparison of peptide identifications by CE-ESI-HRMS and nanoLC-nanoESI-HRMS. (**d**) Targeted detection of angiotensin peptides in the PVN and SFO of the mouse hypothalamus using CE and nanoLC. (Reproduced from Ref. [48, 49] with permission from The Royal Society of Chemistry and Springer Nature, Copyright 2019)

cells [25]. Targeted neuropeptides were detected in record sensitivity in the subfornical organ (SFO) and the paraventricular nucleus (PVN) of the mouse hypothalamus (Fig. 3d) [49]. With a 200-zmol lower limit of detection, this technology also identified ~500 proteins from ~1 ng and ~225 proteins from ~500 pg protein digest, which estimates to a single neuron [46, 50, 51]. Our second-generation CE-ESI HRMS design employing a microprobe capillary enabled the in situ and in vivo analysis of single identified cells in live embryos [23, 25, 52] and mouse brain [50, 53, 54] (Fig. 2). As shown in Fig. 4, its integration with cell labeling and stereomicroscopy permitted the tracing of tissue developmental trajectory. These instrumental capabilities revealed differences in the proteomic [22, 25, 55] and metabolic [21, 23] state of cells in *X. laevis* and zebrafish embryos, including those occupying the dorsal-ventral [21, 22, 56], animal-vegetal [25, 57], and left-right [58, 59] developmental axes in the frog. They also led to discovering metabolite-induced cell fate changes [21, 60] and metabolic communication between neighboring cells in *X. laevis* [61]. Further, the approach can be extended to patch-clamp electrophysiology, permitting the metabolomic [62] and proteomic [53, 54, 63]

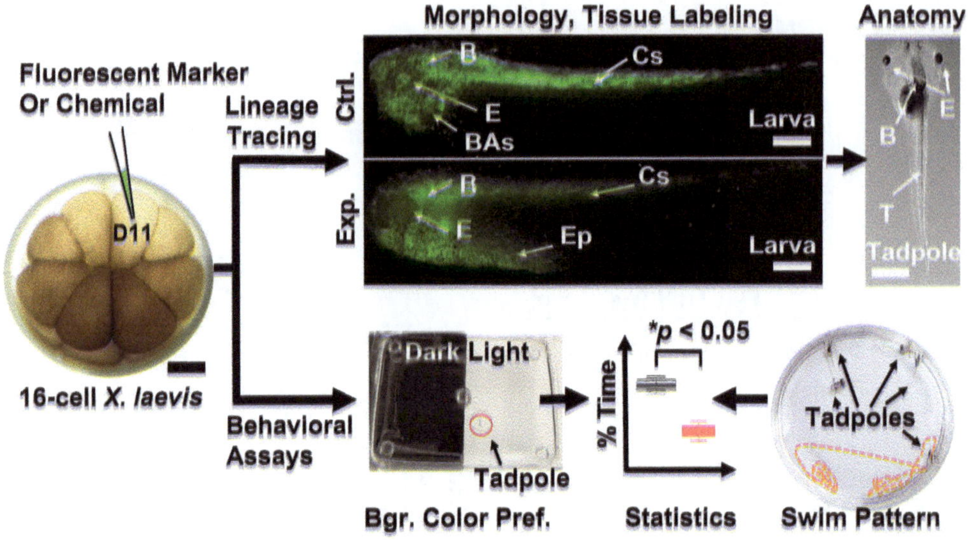

Fig. 4 Techniques to investigate chemistry and function during development. (Top panel) Analysis of cell fate, morphology, and anatomy following fluorescence lineage tracing of the left dorsal-animal midline cell (D11) in control (Ctrl.) and experimental (Exp.) *X. laevis* larvae. (Bottom panel) Background color preference and swim pattern assays evaluating behavior in *X. laevis* tadpoles. Key: B, brain; BAs, branchial arches; Cs, central somites; E, eye; Ep, epidermis; T, tail. Scale bars = ~250 μm (embryo, larvae), ~1.5 mm (tadpole). (Adapted with permission from [23] Copyright 2017 American Chemical Society)

characterization of identified neurons in mouse brain. Most recently, we also integrated single-cell CE MS with functional assays measuring background color preference and swim pattern to assess the impact of metabolic perturbation on organismal behavior [52, 61].

This chapter presents our protocol enabling functional biological studies with insights to the proteomic and metabolomic state of cells in chordate embryos and mammalian neurons (Fig. 1). We overview required consumables and instruments (see *Materials* and *Methods*) and discuss the experimental workflow (Fig. 2). After identifying cells based on optical microscopy or electrophysiology and cell sampling by dissection or microprobe aspiration, the collected material is processed, and the resulting metabolites, peptides, or proteins are measured using LC and CE. Statistical analysis of signal abundances detected by HRMS-MS/MS allowed us to identify compounds for biological investigations. As an example, we present approaches to test the effect of select compounds on tissue development (Fig. 4). Representative examples are discussed with references to data showing the integration of single-cell 'omics (Fig. 3) with functional biology (Fig. 4). The *Notes* section advises on troubleshooting from the vantage point of an experimentalist, thus hoping to promote the combined use of single-cell HRMS-based proteomics/metabolomics with functional biology in other cell types and biological models.

2 Materials

2.1 Culturing Embryos and Neurons

1. Animals: Adult male and female *Xenopus laevis* frogs (e.g., from Xenopus1, Dexter, MI); adult male and female zebrafish (e.g., from Zebrafish International Resource Center, Eugene, OR); adult male mouse (e.g., from Charles River Laboratories, Wilmington, MA) (*see* **Note 1**).
2. Equipment: Incubators set to 14°C and 18°C; stereomicroscope.
3. Solutions:
   (a) Dejellying solution (2% cysteine): Dissolve 4 g of cysteine in 200 mL deionized (DI) water. Add 10 M sodium hydroxide dropwise to adjust pH to 8.
   (b) 100% Steinberg's solution (SS): Mix 3.4 g sodium chloride (NaCl), 0.05 g potassium chloride (KCl), 0.08 g calcium nitrate ($Ca(NO_3)_2 \times 4\ H_2O$), 0.205 g magnesium sulfate ($MgSO_4 \times 7\ H_2O$), 0.66 g Trizma hydrochloride, 0.075 g Trizma base in 1 L of DI water. Adjust the pH to 7.4 by adding Trizma base and autoclave the solution. Store at 4–14°C.
   (c) Anesthetic solution (ketamine, 20 mg/mL, and dexmedetomidine, 0.1 mg/mL): Prepare by mixing 200 μL of ketamine stock solution (100 mg/mL) and 200 μL of dexmedetomidine stock solution (0.5 mg/mL) with 800 μL of 0.9% saline solution for injection (e.g., part no. NDC0409–4888-06, Hospira, Inc., Lake Forest, IL).
   (d) HEPES ringer solution: Prepare by mixing the following reagents to the following final concentrations: 86 mM NaCl, 2.5 mM KCl, 1.2 mM sodium phosphate (NaH_2PO_4), 35 mM sodium bicarbonate ($NaHCO_3$), 20 mM HEPES, 25 mM glucose, 5 mM sodium ascorbate, 2 mM thiourea, 3 mM sodium pyruvate, 1 mM magnesium sulfate ($MgSO_4$), 2 mM calcium chloride ($CaCl_2$).
   (e) Perfusion solution for mouse brain slices: Prepare by mixing the following reagents to final concentration: 126 mM NaCl, 21.4 mM $NaHCO_3$, 2.5 mM KCl, 1.2 mM NaH_2PO_4, 2.4 mM $CaCl_2$, 1.0 mM $MgSO_4$, and 11.1 mM glucose.

2.2 Embryology

1. Equipment:
   (a) Stereomicroscope for embryology to identify, inject, and microsample single cells (e.g., SMZ1270, Nikon, Melville, NY) and phenotype embryos, larvae, and tadpoles (e.g., SMZ18, Nikon).

   (b) Microinjector to inject into or aspirate contents from identified cells (e.g., PLI-100A, Warner Instruments, Hamden, CT).
   (c) Micromanipulator (e.g., MM-33, Warner Instruments).
   (d) Capillary pullers (e.g., P-1000 for fused silica capillaries and P-2000 for borosilicate capillaries, Sutter Instrument, Novato, CA).
   (e) Ancillary equipment: Heat block; centrifuge (e.g., Refrigerated 5430R, Eppendorf).

2. Materials and Solutions:
   (a) Fine sharp forceps (e.g., Dumont #5, Fine Science Tools).
   (b) Centrifuge vials (e.g., 0.2–0.5 mL LoBind tubes, Eppendorf).
   (c) Borosilicate glass capillary (e.g., 0.5/1.0 mm inner/outer diameter, Sutter Instrument).
   (d) Hair loop: Place both ends of a fine hair (~10 cm length) into a 6-inch Pasteur pipette to form a 2 mm loop. Secure it in place with melted paraffin. Sterilize before use by dipping in 70% ethanol and air-drying.
   (e) Injection/Dissection dish: Cover the bottom of a Petri dish (35 or 60 mm) with nontoxic modeling clay (e.g., Claytoon was tested in *Xenopus* labs [64]). Make 3–5 wells of ~1.5 mm diameter using a cool glass bead to hold the embryos in place.
   (f) Injection dish: Cover the bottom of a Petri dish (35 or 60 mm) with nontoxic modeling clay. Make several ~1.5 mm wells using a cool glass bead across the dish to hold ~20–30 embryos.
   (g) 50% Steinberg's solution: Dilute 500 mL of 100% SS with 500 mL of DI water.
   (h) 20% Steinberg's solution: Dilute 200 mL of 100% SS with 800 mL of DI water.

2.3 Sample Processing

1. Equipment: Vacuum concentrator (e.g., CentriVap, Lab-Conco, Kansas City, MO).
2. Reagents: All reagents are LC-MS grade to reduce chemical interference during MS detection. Methanol, anhydrous acetonitrile (ACN), water, acetic acid, 100 mM TEAB, 5% hydroxylamine, isobaric labeling kit (e.g., TMTsixplex, Thermo Fisher Scientific or iTRAQ, AB Sciex).
3. Solutions:
   (a) Metabolite extraction solvent: 50% (v/v) methanol in water containing 0.5% acetic acid.

(b) Proteomic digestion buffer: 50 mM ammonium bicarbonate containing protease inhibitor (1 protease inhibitor cocktail tablet per 10 mL).

(c) Trypsin (0.5 μg/μL) in 50 mM acetic acid.

(d) Patch-clamp solution: We use 50 mM ammonium bicarbonate in water as a compromise between sensitivity and function.

2.4 Mass Spectrometry

1. Instrument and Materials:

   (a) CE system (e.g., laboratory-built following [21, 40] or CESI, AB Sciex, Toronto, Canada).

   (b) High-resolution tandem mass spectrometer (e.g., quadrupole time-of-flight, Impact HD or timsTOF, Bruker Scientific, Billerica, MA, or quadrupole orbitrap Q-Exactive Plus or Orbitrap Fusion Lumos, Thermo Scientific, Fig. 1).

   (c) HPLC (e.g., Acquity I-class UPLC, Waters, Milford, MA, and Dionex Ultimate 3000, Thermo Scientific).

   (d) Separation CE capillary (e.g., 40/100 μm inner/outer diameter fused silica, Polymicro Technologies, Phoenix, AZ).

   (e) LC column (e.g., Acquity UPLC BEH Amide Column, 1.7 μm, 1 mm × 100 mm, and Acclaim PepMap C18 column, 3 μm, 0.075 mm × 250 mm, Waters).

2. Solutions: All solvents and reagents are LC-MS grade.

   (a) CE background electrolyte solution (BGE): 1% (v/v) formic acid in water for metabolomics; 25% (v/v) acetonitrile in water with 1 M formic acid for proteomics.

   (b) CE-ESI sheath solution: 0.1% (v/v) formic acid and 50% methanol for metabolomics; 10% (v/v) acetonitrile in water with 0.05% acetic acid for proteomics.

   (c) LC mobile phase for metabolomics using hydrophilic interaction LC (HILIC): For cationic separation, mobile phase A is aqueous 0.1% formic acid and B is acetonitrile containing 0.1% formic acid; for anionic separation, mobile phase A is aqueous 5% acetonitrile with 10 mM ammonium bicarbonate (pH 9) and B is aqueous 95% acetonitrile with 10 mM ammonium bicarbonate (pH 9).

   (d) LC mobile phase for proteomics using reversed-phase LC (RPLC): Mobile phase A is aqueous 0.1% formic acid and B is acetonitrile containing 0.1% formic acid.

2.5 Functional Studies

1. Instrument and Equipment:
   (a) Epifluorescence stereomicroscope (e.g., SMZ18, Nikon, Fig. 1).
   (b) Inverted microscope (e.g., Eclipse Ti-U, Nikon).
   (c) Microinjector (e.g., PLI-100A, Warner Instruments).
   (d) Micromanipulator (e.g., MM-33, Warner Instruments).
   (e) Camera with tripod (e.g., ESO70D, Canon, Fig. 1).
   (f) Software for processing movies (e.g., Windows media player software).
   (g) Ancillary: Incubator set to 14°C; Nutator rotator.
   (h) Patch amplifier for electrophysiology (e.g., Sutter Instrument).
2. Materials:
   (a) Tadpole food (e.g., part no. 7010, Xenopus1).
   (b) Transfer pipets.
   (c) Black electrical tape.
   (d) 26 G needle (e.g., part no. BD305115, Fisher Scientific).
   (e) 1/2-gallon tank (e.g., part no. SB19271M, Nasco).
   (f) Fine sharp forceps (e.g., Dumont #5).
   (g) Inoculating turntable (e.g., part no. 50809–022, VWR).
3. Reagents:
   (a) 200 proof ethanol.
   (b) Gentamicin antibiotic (e.g., part no. 17-528Z, Fisher Scientific).
   (c) Sylgard 184 silicone elastomer (e.g., part no. NC9285739, Fisher Scientific).
   (d) Ficoll 400 (Sigma-Aldrich, St. Louis, MO).
   (e) Benzocaine (part no. E1501-500G, Sigma-Aldrich).
   (f) 1X phosphate-buffered saline (PBS) (Fisher Scientific).
   (g) Fluorescent lineage tracer (e.g., fluorescent dextran 10,000 MW lysine fixable or mRNA lineage tracer, Invitrogen, Carlsbad, CA).
4. Solutions:
   (a) 3% Ficoll in 100% SS: Prepare by mixing 3 g of Ficoll in 100 mL of 100% SS.
   (b) 4% paraformaldehyde: Prepare by mixing 4 g of paraformaldehyde in 40 mL DI water at 60°C. Add a few drops of 1 N NaOH to adjust pH to 7.4. Add DI water to a total volume of 100 mL.
   (c) DEPC water: Add 1 mL diethyl pyrocarbonate to 1 L DI water. Autoclave the solution, seal, and store at room temperature.

(d) 10% benzocaine: Weight out 2 g of benzocaine and place in a glass beaker. Add 20 mL 200 proof ethanol and stir using a magnetic bar in a stirrer.

(e) 2% benzocaine in 20% SS: Add 2 mL of 10% benzocaine to 8 mL of 20% SS dropwise, ensuring complete dissolution.

3 Methods

3.1 Culturing

This step lays out our methodology to culture *X. laevis* embryos to larvae or tadpoles and primary neurons from the mouse on the basis of established protocols [65, 66]. *X. laevis* embryos require additional dejellying step for manipulation and functional experiments, as described below.

3.1.1 Frog Embryos

1. Obtain fertilized eggs by gonadotropin-induced natural mating of adult *X. laevis* frogs or in vitro fertilization as detailed elsewhere [65]. See **Note 1** on working with live vertebrate animals.
2. Remove the jelly coat from fertilized eggs as follows:
   (a) Remove excess media from dishes containing fertilized eggs.
   (b) Add dejellying solution and keep embryos unperturbed for 2 min.
   (c) Gently swirl dishes over a 2-min period and immediately decant excess dejellying solution after confirming removal of the jelly coating under a stereomicroscope.
   (d) Transfer the embryos to a 250 mL beaker and add 10% SS. Gently swirl for ~30 s and decant excess liquid.
3. Rinse embryos four times with 10% SS to remove remaining dejellying solution.
   (a) Transfer ~300–500 embryos into individual 100 mm Petri dishes containing 100% SS. Place dishes in the 14°C incubator (*see* **Note 1**).
4. Collect two-cell embryos:
   (a) Under a stereomicroscope, identify two-cell embryos that display stereotypical pigmentation to accurately mark the dorsal-ventral axis, in reference to established cell fate maps [67].
   (b) Place the selected two-cell embryos into a 100 mm Petri dish containing 100% SS and incubate at 14°C until the desired developmental stage.

3.1.2 Brain Section

1. Brain sections were collected following established protocols [68]:
   (a) Anesthetize male mice aged postnatal day (PND) 21–35 with an intraperitoneal injection of anesthetic solution and perfuse with ice-cold HEPES ringer solution.
   (b) After perfusion, dissect the brain rapidly in horizontal slices (220 μm) prepared in HEPES ringer solution using a vibratome.
   (c) Recover slices for 1 h at 34°C in oxygenated HEPES holding solution. Then, place slide in the same solution at room temperature until use.

4 Sample Collection from Single Cells

The goal of this step is to collect material from targeted single cells. In what follows, we give an example for *Xenopus*. The workflow starts with the identification of single cells in the *X. laevis* embryo in reference to established cell fate maps [69, 70]. As shown in Fig. 2, our laboratory established orthogonal strategies to collect single-cell samples. We dissected identified whole cells from *X. laevis* embryos [21, 26] or used fabricated microcapillaries as microprobes to aspirate portions of single identified cells from the embryo [23, 24, 28] or electrophysiologically characterized mouse neurons [53, 63].

4.1 Whole-Cell Dissection from X. Laevis *Embryos*

1. Under a stereomicroscope, identify the cell of interest based on stereotypical cleavage and pigmentation, and then dissect it manually as follows:
   (a) Transfer the embryo into a dissection dish containing 50% SS.
   (b) Using a hair loop, place the embryo of interest in a groove.
   (c) Gently remove the vitelline membrane surrounding the embryo using sharp forceps.
   (d) Use forceps to hold the embryo, preferably at the opposite side of the cell of interest. Lightly pull away the selected cell from the rest of the embryo.
2. Transfer the isolated cell using a pipette into a LoBind Eppendorf vial containing chilled 10 μL methanol (~4°C) or digestion buffer (*see* **Note 3**).
3. Cool the Eppendorf vial (on ice) to preserve sample at low temperature and store samples at −80°C until analysis and up to 3 months without detectable degradation.

4.2 In Situ/Vivo Microsampling

1. Fabricate the microprobe by pulling a borosilicate glass capillary to create a fine tip. We use a capillary puller (P-1000 Sutter Instrument) with custom settings: heat = 355, pull = 65, velocity = 80, time = 150.
2. Using sharp forceps, cut the needle tip to an aperture of ~10–20 μm (*see* **Note 4**).
3. Mount the microprobe into a capillary holder on a three-axis micromanipulator and connect its distal end to a microinjector.
4. Transfer *X. laevis* embryos into an injection dish containing 50% SS. This protocol does not require removal of vitelline membrane.
5. Use a stereomicroscope to aid viewing and manipulation of the embryo. Use a hair loop to immobilize the embryo of interest into a well in preparation for microsampling. Identify the cell of interest following protocols established elsewhere [69, 70].
6. Using a micromanipulator, guide the tip of the microprobe into the targeted cell to pierce through the membrane of the cell. Withdraw ~10–15 nL (or as needed) volume from the targeted cell by applying negative pressure to the microprobe using the connected microinjector.
7. To end the microsampling, reduce pressure and retract the microprobe from the cell.
8. Transfer the collected content into a LoBind Eppendorf vial containing 4 μL of metabolite extraction solvent or digestion buffer (*see* **Note 5**). We usually inject the collected material from the capillary by inserting the tip into the solvent and applying a positive pressure pulse.
9. Cool the Eppendorf vial (on ice) to preserve sample at low temperature and immediately process the samples via metabolomics or proteomics workflows to prevent molecular degradation.

5 Sample Processing for MS-Based 'Omics

This section discusses protocols to process the collected materials for HRMS analysis. We use LC and CE to separate biomolecules in complementary performance prior to ESI-HRMS.

5.1 Metabolomics Workflow

1. Extract metabolites from dissected single cells [21, 23] or aspirates collected by microprobe sampling [23, 48] as follows:
   (a) For dissected single cells: Retrieve the single-cell samples stored in 100% methanol from −80°C freezer and vacuum-dry them at 4°C. Add 10 μL metabolite extraction solvent. Vortex-mix the vials for 30 s at room temperature to facilitate cell lysis and extraction of

metabolites. Sonicate the sample vials for 3 min in an ice bath, followed by vortex-mixing for 1 min at room temperature.

(b) For microprobe aspirated samples: Retrieve single-cell samples collected in metabolite extraction solvent preserved on ice. Vortex-mix the vials for 1 min at room temperature to facilitate extraction of metabolites.

(c) For CE-ESI-HRMS, centrifuge the samples (dissected or microsamples) for 5 min at 8000 × *g* at 4°C to pellet cellular debris. Proper pelleting is important to avoid CE capillaries from getting clogged. We usually store the aliquot together with the pelleted debris to avoid sample losses. The samples are kept at −80°C until analysis.

(d) For HILIC-ESI-HRMS, centrifuge the samples at 13,000 × *g* for 10 min at 4°C and transfer the supernatant into a microvial and vacuum-dry the samples at 4°C. Reconstitute the samples in 10 μL 95% (v/v) acetonitrile in water and centrifuge the samples at 13,000 × *g* for 10 min at 4°C to pellet potential debris that could clog the column (*see* **Note 6**). Transfer the supernatant into an LC vial and store the sample at −80°C until analysis.

5.2 Proteomics Workflow

1. Lyse the collected cell or aspirate by sonication for 5 min. Heat the sample to 60°C for ~15 min to denature proteins. We usually skip reduction and alkylation steps for our single-cell samples for higher sensitivity [25].
2. For one-step digestion, add ~50 ng of trypsin protease to the protein extract and incubate the mixture at 37°C for 5–6 h. For neurons yielding less starting protein amounts, add ~2 ng trypsin and digest at 60°C for 1 h.
3. Vacuum-dry the resulting protein digest and store it at −80°C until analysis.
4. (Optional) To enable multiplexing relative quantification, barcode the dried protein digests. We use TMT isobaric labeling following the vendor's protocol (*see* **Note 7**), downscaled to the total amount of protein/peptide contained in the sample:

   (a) Reconstitute the dried protein digest in 10 μL of 100 mM TEAB and tag it with 1 μL of 85 mM TMT reagent.

   (b) Incubate each sample for 1 h at room temperature.

   (c) Quench the reaction with 2 μL of 5% hydroxylamine and incubate the mixture for 15 min at room temperature.

   (d) Mix the multiple tagged samples, vacuum-dry the mixture, and store it at −80°C for up to 1 month until analysis.

6 High-Resolution Mass Spectrometry

In this step, biomolecules in the resulting samples are separated and detected using ESI-HRMS. We present protocols for separation based on partition chromatography (LC) and electrophoresis (CE). These separation techniques provide complementary benefits in sensitivity, throughput, and molecular coverage (Fig. 3). The resulting data are processed using established approaches in bioinformatics, including but not limited to statistics, multivariate data analysis (e.g., principal component analysis and hierarchical cluster analysis), or machine learning (e.g., Trace [71, 72]). These technologies and related protocols (Fig. 2) allowed us to document metabolic and proteomic differences between cells in embryos of *X. laevis* and zebrafish [21, 22, 25, 58] and single neurons in the mouse brain [53, 54, 63].

6.1 CE-ESI-HRMS

1. Construct the CE-ESI interface following protocols established by us and others [40, 44, 50, 73]. A simplified procedure to build a blunt-tip CE-ESI interface follows:
   (a) Cleave a 1 m long piece of fused silica CE capillary.
   (b) On the outlet end of the CE capillary, burn off ~1.5 mm of polyimide coating and clean using isopropanol. Before proceeding to the next step, ensure that the capillary end is clean of burned residues to avoid the leaching of interfering ion signals from the burned residue.
   (c) Feed the CE capillary outlet-end into a T-junction connected to a sheath-flow capillary.
   (d) Mount the CE capillary into the T-junction to feed the CE capillary through the emitter, allowing the capillary to protrude ~40–50 μm past the emitter.
2. Hydrate the CE capillary by flushing with LC-MS grade water overnight.
3. Position the tip of the CE-ESI interface ~5 mm from the inlet orifice of the mass spectrometer.
4. Fill CE capillary with BGE and flush sheath-flow capillary with sheath solution.
5. Initiate the electrospray as follows:
   (a) Using a translation stage, fine-position the electrospray emitter tip ~2–3 mm from the mass spectrometer orifice to operate the electrospray in the cone-jet regime (ESI voltage 1.8–2.0 kV).
   (b) Monitor the electrospray using a stereomicroscope.

(c) Observe the stability of total ion current (TIC) for ~10–15 min to ensure stable operation before analyzing a sample (*see* **Note 8**).

6. Inject ~10–15 nL from the metabolite or protein extract hydrodynamically into the CE capillary following previously described protocols [24, 40].
7. Gradually increase CE separation voltage from ground (0 V) to ~20–22 kV. Sudden application of high voltage may break the capillary.
8. Load the MS method as described in Table 1 (see CE-HRMS) and start data acquisition (*see* **Note 9**).

6.2 LC-ESI-HRMS

For Metabolomics:

1. We use the following LC parameters: column temperature, 35°C; autosampler temperature, 4°C; injection volume, 1 μL; flow rate, 130 μL/min. Positive ion mode gradient: 0–0.5 min 95% B, 0.5–10 min 95–40% B, 10–13 min 40% B, 13–15 min 40–95% B, 15–22 min 95% B; negative ion mode gradient: 0–0.5 min 99% B, 0.5–2.5 min 99–82.5% B, 2.5–6.5 min 82.5–68% B, 6.5–10 min 68–30% B, 10–13 min 30% B, 13–15 min 30–99% B, 15–22 min 99% B.
2. Select MS method parameters described in Table 1 (see LC-HRMS).

For Proteomics:

1. LC parameters: 0–5 min 2% B, 5–85 min 2–35% B, 86–90 min 70% B, 91–120 min 2% B; autosampler temperature, 4°C; injection volume, 1 μL; flow rate, 300 nL/min.
2. Load the MS method parameters and start data acquisition. Our typical parameters are listed in Table 1. Adjust ion source settings to get a stable nanospray (*see* **Notes 8** and **9**).

7 Data Processing

Metabolomics:

1. Survey the MS-MS/MS data for molecular features (signals with unique *m/z* and separation time) using available software packages. For example, we employ MetaboScape Version 4.0.4 (Bruker Daltonics) using the following settings: intensity threshold, 1000 counts; minimum peak length, 5 spectra.
2. Annotate metabolites based on the accurate mass, isotopic distribution pattern, and tandem MS spectra against reference spectra available in MS-MS/MS databases, including but not limited to METLIN [74], EMBL (http://curatr.mcf.embl.de/), mzCloud

Table 1
Our typical instrumental settings for detecting metabolites and proteins in CE and LC ESI-HRMS using the positive (+) and negative (−) ion mode

Parameters	CE-ESI-HRMS		LC-ESI-HRMS	
Instrument Configuration	Q-TOF	Q-OT	Q-TOF	Q-OT
Compounds	Metabolites	Proteins	Metabolites	Proteins
MS survey scan frequency	MS^1–MS^2, 2 Hz	MS^1, 7 Hz; MS^2, 13 Hz	MS^1–MS^2, 2 Hz	MS^1, 7 Hz; MS^2, 13 Hz
Mass range (*m/z*) and spectral resolution	50–550 at 40,000 FWHM	400–1700 at 35,000 FWHM for MS^1 and 17,500 FWHM for MS^2	50–1300 at 40,000 FWHM	400–1700 at 35,000 FWHM for MS^1 and 17,500 FWHM for MS^2
Dry gas	2 L/min (N_2)	–	4 L/min (N_2)	–
Nebulizer gas	–	–	0.4 bar (+); 1 bar (−)	–
Dry temperature	100°C (+); 150°C (−)	275°C	220°C	275°C
Ion-transfer capillary voltage	−1700 V (+); +2100 V (−)	2300 V	−4500 V (+); +4000 V (−)	1800 V
Collision energy	18 eV (CID)	36% (normalized, HCD)	15–35 eV (CID)	36% (normalized, HCD)
Isolation m/z window	1.5 Da	1.5 Da	1.5 Da	1.5 Da
Dynamic exclusion mass tolerance	5.0 ppm	5.0 ppm	5.0 ppm	5.0 ppm
Dynamic exclusion	9 s	9 s	13 s	13 s
High-intensity ion signal threshold (counts)	1.5×10^6	1.5×10^6	1.5×10^6	1.5×10^6
AGC target (counts)	–	1×10^6	–	1×10^6
Minimum AGC target (ion counts)	–	9.2×10^2	–	9.2×10^2
Maximum injection time (ms)	–	50	–	50
DDA top N	5	20	5	20

(https://www.mzcloud.org/), MassBank of North America (https://mona.fiehnlab.ucdavis.edu/), and the Human Metabolome Database [75]. For example, we use METLIN with an annotation tolerance ≤10 ppm mass accuracy and MS/MS score ≥ 700–900.

3. Perform relative/absolute quantification using under-the-peak-areas (label-free quantification) or ion signal abundances (multiplexing quantification) serving as a proxy for metabolite abundance.
4. Perform statistics and multivariate data analysis to select molecules for follow-up functional studies (see Fig. 2).

Proteomics:

1. Identify proteins using established bioinformatics software packages broadly available for bottom-up proteomics. For example, we analyze the MS-MS/MS data in Proteome Discoverer (Thermo Fisher Scientific) or MaxQuant (Max Planck Institute of Biochemistry) against the mouse or *Xenopus* proteome (e.g., downloaded from UniProt [76] or Xenbase [77]) with the following search parameters: digestion enzyme, trypsin; missed cleavages, up to 2; variable modification, methionine oxidation; precursor mass tolerance, 10 ppm; fragment mass tolerance, 4.5 ppm; minimum peptide length, 5. Peptides are filtered to <1% false discovery rate (FDR), calculated against a reversed-sequence decoy database. The reported proteins are grouped based on the closest parsimony principle. We remove common contaminants from the final list of protein identifications by manually annotating for common contaminant proteins (downloaded from UniProt) .
2. Employ label-free or label-based strategies and software packages from the proteomics community to compare protein levels between single cells. For example, we used MaxQuant Version 1.5.5.1 [78] or Proteome Discoverer (Thermo Scientific) to quantify the proteomic state of single embryonic cells and neurons by calculating label-free quantitative indexes (LFQ values) [25, 57] or relative reporter ion signal abundances from TMTs [22].

8 Functional Studies

The goal of this step is to link chemistry with biological function. As an example, we describe a protocol to prepare brain tissues to record neuronal activity followed by in vitro single-cell proteomics

[53, 63]. In the context of cell differentiation, we trace cell fates to understand how cells divide to form specific tissues and organs. Because cell fates are reproducible in *X. laevis* [79, 80], it is possible to inject molecules into identified cells to determine their developmental impact on tissue specification and organogenesis [21, 60, 61]. Figure 4 shows an example, in which the vehicle or select metabolites were injected into specific cells while fluorescently monitoring their tissue clone via the co-injection of the green fluorescent dextran. Alternatively, fluorescent proteins can be expressed in the cell, for example, by injecting the corresponding mRNA [21, 64]. In *X. laevis* tadpoles, it is also possible to perform behavioral assays to assess sensory (e.g., visual), muscular, cognitive, and other functions. As an example, we adopted the background color preference assay [81] to test the behavior of pre-metamorphic *X. laevis* tadpoles after performing in vivo single-cell MS on the precursor embryo [52].

8.1 Single-Neuron Electrophysiology and Capillary Microsampling

1. Perfuse midbrain slices continuously at 1.5–2 mL/min with perfusion solution at 28–32°C, following established protocols [82].
2. Backfill the patch pipettes for recording (2–4 MΩ) with ~20 μL of 50 mM ammonium bicarbonate in water (*see* **Note 10**).
3. Putatively identify neuron type of interest. For example, dopaminergic neurons can be identified based on their location in the lateral portion of the substantia nigra and their size. Consult with brain anatomy atlases to improve the accuracy of tissue identifications.
4. Neuron identification may be aided by electrophysiology. Detection of a slow pacemaker firing pattern (>2 ms action potential) indicates dopaminergic neurons [83].
5. Obtain a giga-ohm seal and record action potential for 60 s in a cell-attached configuration. A patch-clamped neuron is demonstrated in Fig. 2. We used the SutterPatch software to control the devices [53, 54, 63].
6. Following electrophysical analysis, apply a steady negative pressure at the outlet end of pipette with a syringe to aspirate a portion of neuronal soma (*see* **Note 11**).
7. Under an inverted microscope (40× magnification), visually inspect the neuron during microaspiration. A slight reduction in neural soma size is anticipated from successful patching. Figure 2 shows a neuron during sampling after electrophysiological recording.
8. Gently withdraw the pipette from the cell and then expel the collected contents into an LoBind microvial containing 5 μL protein digestion solution chilled on ice. After a 1-h digestion at 60 °C, store samples at −80 °C for up to 1 month without detectable degradation, until analysis.

8.2 Cell Lineage Tracing

Cell Labeling by Microinjection

1. Prepare fluorescent tracer solution:
   (a) Prepare a 100 μL solution containing 0.5% fluorescent dextran in DEPC water.
   (b) (Optional) Synthesize capped mRNA using an in vitro transcription kit. Prepare a working solution containing ~50–100 pg/nL mRNA pellet in DEPC-treated water.
2. Prepare injection needles following steps described earlier. Calibrate the volume of injection by injecting water droplet into mineral oil and measuring the diameter of the water droplet.
3. Fill the injection needle with ~0.5–1 μL fluorescent tracer solution using a microinjector in "fill" mode following established protocols [64].
4. Transfer the embryo into an injection dish containing 3% Ficoll in 100% SS.
5. Under a stereomicroscope, use a hair loop to gently angle the embryo to orient the cell of interest for facile access for injection (see Fig. 4).
6. Using a calibrated micromanipulator, guide the tip of the injection capillary into the targeted cell. Inject ~1–5 nL of the sample by applying +40 psi on the capillary for ~300–500 ms using a microinjector (*see* **Note 12**).
7. Culture the injected embryos in 3% Ficoll in 100% SS for 3–4 h to allow the cell membrane to heal. Transfer the embryos to 50% SS and culture at 14–22°C until the larval stages 32–34 (*see* **Note 13**).

Fixing and Imaging of the Larvae/Tadpoles

1. Anesthetize the larvae or tadpoles on ice or using 0.5% benzocaine in 100% SS. Ensure the success of anesthesia by gently touching the tadpole with a capillary, anticipating no response if successful. Increase the concentration of benzocaine if necessary. Ensure tadpoles are handled and treated humanely so that the organisms do not suffer or feel pain (*see* **Note 1**).
2. Fix the specimens in 4% paraformaldehyde for 1 h on a rotator.
3. Rinse the tadpoles with 1X PBS twice. Store the tadpoles in 1X PBS at 4°C.
4. For imaging, mount the specimens in a 30 mm dish containing ~1–2 mL of 1X PBS. Image the specimens using epifluorescence microscopy (e.g., SMZ18, Nikon) (*see* **Note 14**).
5. Acquire images using a microscope following the manufacturer's instructions.

6. Conduct lineage analysis by determining the relative contribution of fluorescent cells to tissues and organs (see protocols in References [69, 70, 79]).

8.3 Behavioral Assay

Tadpole Preparation

1. Obtain and culture embryos as described earlier.
2. Inject identified cells with the test compounds as described earlier. Prepare negative control by injecting identified cells with DEPC-treated water. Use non-injected embryos for later use as the wild-type control group.
3. Place the embryos in the 14°C incubator until they reach gastrula stage. Transfer the embryos to room temperature in 90 mm Petri dishes containing 20% SS and change media every 2 days and culture the tadpoles until the feeding stage (Stage 45) (*see* **Note 1**).
4. Feed tadpoles every other day as follows:
   (a) Mix food with 20% SS solution to form a paste.
   (b) Place the paste in a corner of the Petri dish containing the tadpoles.
   (c) Provide more food as needed.
5. Maintain the tadpoles under a 12 h light/dark light cycle.

Background Color Preference Assay

1. The color preference assay is performed in a setup consisting of nested tanks following an established protocol [81]. The inner tank holds the tadpoles. The outer tank provides the background colors with half covered with a black tape and the other half covered with a white paper. Ensure both tanks are water-leveled to aid visual inspection of tadpole behavior.
2. Fill the inner test tank with 20% SS to the 5 cm water mark from the top and insert the inner tank inside the outer tank.
3. Mount the camera on a tripod to record the entire tank from above (*see* **Note 15**).
4. Transfer a single tadpole on the white background of the inner tank. Limit this experiment to one tadpole at a time to avoid interactions between tadpoles which may confound behavioral phenotypes.
5. Record the swim pattern of tadpole for 2 min.
6. After 2 min, carefully lift the outer tank, rotate the outer tank, and return the inner tank into the outer tank. This step helps minimize the impact environmental factors may have on behavior. Start recording immediately and set the timer for 2 min.

7. Place the tadpole back to a holding tank containing 20% SS. Record two trials for each of ~10–15 tadpoles.
8. Repeat the assay on the next day on the same tadpoles to test for reproducibility and enhance statistical evaluation of the results by obtaining more data.

Euthanasia

1. Place tadpoles in a 90 mm Petri dish containing 2% benzocaine in 20% SS for ~15–20 min.
2. Monitor the tadpole's reflex by gently touching with a hair loop. Anesthetized tadpole cannot swim or respond to mechanical stimuli (gentle touching). Only proceed to the next step if the tadpole is anesthetized.
3. Freeze the larvae/tadpoles at −20 °C overnight.
4. Dispose of the tadpoles following protocols approved by the relevant institutional and federal authorities (*see* **Note 1**).

9 Notes

1. *X. laevis*, mouse, and zebrafish are sentient and vertebrate animals; therefore, protocols pertaining to the care and handling of the animals must be approved by institutional and federal agencies. The work presented in this chapter was approved by the Institutional Animal Care and Use Committee of the University of Maryland (approval numbers R-DEC-17-57, R-FEB-21-07, and R-JUN-20-31) and/or The George Washington University (approval numbers #A311 and #A283).
2. Development is temperature-dependent in *Xenopus* [77, 84], thus providing a helpful tool to time biological and chemical experiments. Low temperatures slow down the speed of cell cleavage, extending the time to select embryos for experiments.
3. For metabolomics, place cells in ice-cold 100% methanol immediately upon collection. Methanol denatures enzymes and low temperatures slow down chemical reactions, thus minimizing metabolic changes.
4. Capillaries with too large or too small diameters challenge microsampling or microinjection. In *Xenopus*, we find apertures larger than ~20 μm tend to cause substantial damage to the cell membrane. Without the membrane being able to heal, the cytoplasm may leak into the media and the cell may not be able to continue division for functional experiments. Conversely, apertures below ~10 μm may clog with yolk and cytoplasmic content, requiring refabrication of the microprobe.
5. Tailor the composition of the metabolite extraction solvent to the type of metabolites of interest in a particular study. To study

polar metabolites, we use aqueous 40% (v/v) acetonitrile with 40% (v/v) methanol as the extraction solution. Theoretical predictions based on partition and distribution coefficients can help experimental design [58].

6. Debris may clog CE capillaries or LC columns. We find that centrifugation of samples before analysis prevents clogging. The efficiency of centrifugation depends on the field force and time of centrifugation. We typically use 13,000 × *g* for 10 min.
7. To select the spectral resolution appropriate for the multiplexing relative quantification, refer to instructions from the manufacturer of the reagents.
8. We consider the electrospray to be stable when the total ion current exhibits less than ~15% relative standard deviation over ~40 min of separation.
9. Tailor MS-MS/MS experimental settings to the chosen separation technology. For example, we adjust the number of targeted molecular features and the duration of dynamic exclusion depending on typical peak widths and the complexity of the sample in LC and CE experiments (see Table 1).
10. It is imperative to optimize the composition of the intracellular solution used for patch-clamp electrophysiology to the osmolarity of the neuron. Although potassium gluconate is commonly used in electrophysiological recordings, we use ammonium bicarbonate to minimize spectral and ionization interferences during MS caused by involatile salts.
11. Aid sample collection by applying consistent negative pressure and continuously monitoring the size of the neuron under the inverted microscope. Shrinking neuronal soma is an indication of successful sampling. Be careful not to aspirate the media surrounding the neuron to avoid sample dilution and interferences due to salt during MS analysis.
12. Limit damage to the cell membrane by carefully withdrawing the needle tip from the cell. With negligible damage, the cell membrane heals, and the embryo continues its development. Damage to the cell membrane can cause leakage of cytoplasmic content, which can result in low survival rates or lethality. Take extra care to also avoid damaging the neighboring cells to facilitate development.
13. Based on our experience and other protocols [80], culturing at lower temperature (14–16°C) improves survival rates.
14. Clearing agents (e.g., benzyl alcohol/benzyl benzoate, BA/BB) may be used in *X. laevis* embryos and tadpoles to improve fluorescent imaging [85]. Tissue clearing is recommended for imaging deep in intact embryos/tadpoles and sections.

15. The locations of the eyes in tadpoles help accurately determine crossings between the white and black backgrounds (see Fig. 4). Therefore, the camera used to record video trials should have sufficient optical resolution and frame rate to clearly identify the eyes and monitor fast tadpole swimming. We typically use ~30 fps to monitor tadpoles during both the background color preference and swim assays.

Acknowledgments

The work presented here was in part sponsored by an Arnold and Mabel Beckman Foundation Beckman Young Investigator award (to P.N.), COSMOS Club Foundation Awards (to E.P.P., L.P., and J.L.), the National Institute of General Medical Sciences award no. 1R35GM124755 (to P.N.), and the National Science Foundation award no. IOS-1832968 (to P.N.). The authors thank Abigail Polter (The George Washington University) for permitting to photograph the electrophysiology station in her laboratory.

References

1. Briggs JA, Weinreb C, Wagner DE, Megason S, Peshkin L, Kirschner MW, Klein AM (2018) The dynamics of gene expression in vertebrate embryogenesis at single-cell resolution. Science 360(6392):1–9
2. Tan TZ, Heong V, Ye J, Lim D, Low J, Choolani M, Scott C, Tan DSP, Huang RY-J (2019) Decoding transcriptomic intra-tumour heterogeneity to guide personalised medicine in ovarian cancer. J Pathol 247(3):305–319
3. Rohrback S, Siddoway B, Liu CS, Chun J (2018) Genomic mosaicism in the developing and adult brain. Dev Neurobiol 78(11): 1026–1048
4. Cembrowski MS (2019) Single-cell transcriptomics as a framework and roadmap for understanding the brain. J Neurosci Methods 326: 1–7
5. Phillips JW, Schulmann A, Hara E, Winnubst J, Liu C, Valakh V, Wang L, Shields BC, Korff W, Chandrashekar J, Lemire AL, Mensh B, Dudman JT, Nelson SB, Hantman AW (2019) A repeated molecular architecture across thalamic pathways. Nat Neurosci 22(11):1925–1935
6. Taylor CF, Paton NW, Lilley KS, Binz P-A, Julian RK, Jones AR, Zhu W, Apweiler R, Aebersold R, Deutsch EW, Dunn MJ, Heck AJR, Leitner A, Macht M, Mann M, Martens L, Neubert TA, Patterson SD, Ping P, Seymour SL, Souda P, Tsugita A, Vandekerckhove J, Vondriska TM, Whitelegge JP, Wilkins MR, Xenarios I, Yates JR, Hermjakob H (2007) The minimum information about a proteomics experiment (MIAPE). Nat Biotechnol 25(8):887–893
7. Sumner LW, Amberg A, Barrett D, Beale MH, Beger R, Daykin CA, Fan TWM, Fiehn O, Goodacre R, Griffin JL, Hankemeier T, Hardy N, Harnly J, Higashi R, Kopka J, Lane AN, Lindon JC, Marriott P, Nicholls AW, Reily MD, Thaden JJ, Viant MR (2007) Proposed minimum reporting standards for chemical analysis. Metabolomics 3(3):211–221
8. Perez-Riverol Y, Csordas A, Bai J, Bernal-Llinares M, Hewapathirana S, Kundu DJ, Inuganti A, Griss J, Mayer G, Eisenacher M, Pérez E, Uszkoreit J, Pfeuffer J, Sachsenberg T, Yilmaz S, Tiwary S, Cox J, Audain E, Walzer M, Jarnuczak AF, Ternent T, Brazma A, Vizcaíno JA (2019) The PRIDE database and related tools and resources in 2019: improving support for quantification data. Nucleic Acids Res 47 (D1):D442–D450
9. Sud M, Fahy E, Cotter D, Azam K, Vadivelu I, Burant C, Edison A, Fiehn O, Higashi R, Nair SK, Sumner S, Subramaniam S (2016) Metabolomics workbench: an international repository for metabolomics data and metadata, metabolite standards, protocols, tutorials and training, and analysis tools. Nucleic Acids Res 44(D1):D463–D470

10. Duncan KD, Fyrestam J, Lanekoff I (2019) Advances in mass spectrometry based single-cell metabolomics. Analyst 144(3):782–793
11. Yin L, Zhang Z, Liu Y, Gao Y, Gu J (2019) Recent advances in single-cell analysis by mass spectrometry. Analyst 144(3):824–845
12. Comi TJ, Do TD, Rubakhin SS, Sweedler JV (2017) Categorizing cells on the basis of their chemical profiles: Progress in single-cell mass spectrometry. J Am Chem Soc 139(11): 3920–3929
13. Couvillion SP, Zhu Y, Nagy G, Adkins JN, Ansong C, Renslow RS, Piehowski PD, Ibrahim YM, Kelly RT, Metz TO (2019) New mass spectrometry technologies contributing towards comprehensive and high throughput omics analyses of single cells. Analyst 144(3): 794–807
14. Zhang L, Vertes A (2018) Single-cell mass spectrometry approaches to explore cellular heterogeneity. Angew Chem Int Ed 57(17): 4466–4477
15. Yang Y, Huang Y, Wu J, Liu N, Deng J, Luan T (2017) Single-cell analysis by ambient mass spectrometry. Trends Anal Chem 90:14–26
16. Qi M, Philip MC, Yang N, Sweedler JV (2018) Single cell neurometabolomics. ACS Chem Neurosci 9(1):40–50
17. Rubakhin SS, Romanova EV, Nemes P, Sweedler JV (2011) Profiling metabolites and peptides in single cells. Nat Methods 8(4): S20–S29
18. Evers TMJ, Hochane M, Tans SJ, Heeren RMA, Semrau S, Nemes P, Mashaghi A (2019) Deciphering metabolic heterogeneity by single-cell analysis. Anal Chem 91(21): 13314–13323
19. Armbrecht L, Dittrich PS (2017) Recent advances in the analysis of single cells. Anal Chem 89(1):2–21
20. Zhang L, Sevinsky CJ, Davis BM, Vertes A (2018) Single-cell mass spectrometry of subpopulations selected by fluorescence microscopy. Anal Chem 90(7):4626–4634
21. Onjiko RM, Moody SA, Nemes P (2015) Single-cell mass spectrometry reveals small molecules that affect cell fates in the 16-cell embryo. Proc Natl Acad Sci U S A 112(21): 6545–6550
22. Lombard-Banek C, Moody SA, Nemes P (2016) Single-cell mass spectrometry for discovery proteomics: quantifying translational cell heterogeneity in the 16-cell frog (*Xenopus*) embryo. Angew Chem Int Ed 55(7): 2454–2458
23. Onjiko RM, Portero EP, Moody SA, Nemes P (2017) *In situ* microprobe single-cell capillary electrophoresis mass spectrometry: metabolic reorganization in single differentiating cells in the live vertebrate (*Xenopus laevis*) embryo. Anal Chem 89(13):7069–7076
24. Onjiko RM, Portero EP, Moody SA, Nemes P (2017) Microprobe capillary electrophoresis mass spectrometry for single-cell metabolomics in live frog (*Xenopus laevis*) embryos. J Vis Exp 130:e56956
25. Lombard-Banek C, Moody SA, Manzini MC, Nemes P (2019) Microsampling capillary electrophoresis mass spectrometry enables single-cell proteomics in complex tissues: developing cell clones in live *Xenopus laevis* and zebrafish embryos. Anal Chem 91(7):4797–4805
26. Lombard-Banek C, Moody SA, Nemes P (2016) High-sensitivity mass spectrometry for probing gene translation in single embryonic cells in the early frog (*Xenopus*) embryo. Frontiers Cell Development Biol 4(100):1–11
27. Lombard-Banek C, Portero EP, Onjiko RM, Nemes P (2017) New-generation mass spectrometry expands the toolbox of cell and developmental biology. Genesis 55 (1-2):1–14
28. Lombard-Banek C, Choi SB, Nemes P (2019) Single-cell proteomics in complex tissues using microprobe capillary electrophoresis mass spectrometry. In: Allbritton NL, Kovarik ML (eds) Methods in enzymology, vol 628. Academic Press, pp 263–292
29. Onjiko RM, Portero EP, Nemes P (2018) Single-cell metabolomics with capillary electrophoresis–mass spectrometry. In: Ramautar R (ed) Capillary electrophoresis–mass spectrometry for metabolomics. The Royal Society of Chemistry, pp 209–224
30. Sun L, Dubiak KM, Peuchen EH, Zhang Z, Zhu G, Huber PW, Dovichi NJ (2016) Single cell proteomics using frog (*Xenopus laevis*) blastomeres isolated from early stage embryos, which form a geometric progression in protein content. Anal Chem 88(13):6653–6657
31. Saha-Shah A, Esmaeili M, Sidoli S, Hwang H, Yang J, Klein PS, Garcia BA (2019) Single cell proteomics by data-independent acquisition to study embryonic asymmetry in *Xenopus laevis.* Anal Chem 91(14):8891–8899
32. Budnik B, Levy E, Harmange G, Slavov N (2018) SCoPE-MS: mass spectrometry of single mammalian cells quantifies proteome heterogeneity during cell differentiation. Genome Biol 19(1):1–12
33. Specht H, Emmott E, Petelski AA, Huffman G, Perlman D, Serra M, Kharchenko P, Koller A, Slavov N (2020) Single-cell proteomic and transcriptomic analysis of macrophage heterogeneity. Biorxiv:665307

34. Cahill JF, Riba J, Kertesz V (2019) Rapid, untargeted chemical profiling of single cells in their native environment. Anal Chem 91(9): 6118–6126
35. Cong Y, Liang Y, Motamedchaboki K, Huguet R, Truong T, Zhao R, Shen Y, Lopez-Ferrer D, Zhu Y, Kelly RT (2020) Improved single-cell proteome coverage using narrow-bore packed nanoLC columns and ultrasensitive mass spectrometry. Anal Chem 92(3):2665–2671
36. Li S, Plouffe BD, Belov AM, Ray S, Wang X, Murthy SK, Karger BL, Ivanov AR (2015) An integrated platform for isolation, processing, and mass spectrometry-based proteomic profiling of rare cells in whole blood. Mol Cell Proteomics 14(6):1672
37. Ivanov AR, Zang L, Karger BL (2003) Low-attomole electrospray ionization MS and MS/MS analysis of protein tryptic digests using 20-μm-i.d. polystyrene–divinylbenzene monolithic capillary columns. Anal Chem 75(20):5306–5316
38. Zhu Y, Piehowski PD, Zhao R, Chen J, Shen Y, Moore RJ, Shukla AK, Petyuk VA, Campbell-Thompson M, Mathews CE, Smith RD, Qian W-J, Kelly RT (2018) Nanodroplet processing platform for deep and quantitative proteome profiling of 10–100 mammalian cells. Nat Commun 9(1):1–10
39. Osbourn DM, Weiss DJ, Lunte CE (2000) On-line preconcentration methods for capillary electrophoresis. Electrophoresis 21(14): 2768–2779
40. Nemes P, Rubakhin SS, Aerts JT, Sweedler JV (2013) Qualitative and quantitative metabolomic investigation of single neurons by capillary electrophoresis electrospray ionization mass spectrometry. Nat Protoc 8(4):783–799
41. Drouin N, van Mever M, Zhang W, Tobolkina E, Ferre S, Servais A-C, Gou M-J, Nyssen L, Fillet M, Lageveen-Kammeijer GSM, Nouta J, Chetwynd AJ, Lynch I, Thorn JA, Meixner J, Lößner C, Taverna M, Liu S, Tran NT, Francois Y, Lechner A, Nehmé R, Al Hamoui Dit Banni G, Nasreddine R, Colas C, Lindner HH, Faserl K, Neusüß C, Nelke M, Lämmerer S, Perrin C, Bich-Muracciole C, Barbas C, Gonzálvez ÁL, Guttman A, Szigeti M, Britz-McKibbin P, Kroezen Z, Shanmuganathan M, Nemes P, Portero EP, Hankemeier T, Codesido S, González-Ruiz V, Rudaz S, Ramautar R (2020) Capillary electrophoresis-mass spectrometry at trial by metabo-ring: effective electrophoretic mobility for reproducible and robust compound annotation. Anal Chem 92(20):14103–14112
42. Ramautar R, Somsen GW, de Jong GJ (2019) CE-MS for metabolomics: developments and applications in the period 2016–2018. Electrophoresis 40(1):165–179
43. Ramautar R, Somsen GW, de Jong GJ (2017) CE–MS for metabolomics: developments and applications in the period 2014–2016. Electrophoresis 38(1):190–202
44. Liao H-W, Rubakhin SS, Philip MC, Sweedler JV (2020) Enhanced single-cell metabolomics by capillary electrophoresis electrospray ionization-mass spectrometry with field amplified sample injection. Anal Chim Acta 1118: 36–43
45. Liu J-X, Aerts JT, Rubakhin SS, Zhang X-X, Sweedler JV (2014) Analysis of endogenous nucleotides by single cell capillary electrophoresis-mass spectrometry. Analyst 139(22):5835–5842
46. Choi SB, Lombard-Banek C, Muñoz-Llancao P, Manzini MC, Nemes P (2018) Enhanced peptide detection toward single-neuron proteomics by reversed-phase fractionation capillary electrophoresis mass spectrometry. J Am Soc Mass Spectrom 29(5):913–922
47. Kawai T, Ota N, Okada K, Imasato A, Owa Y, Morita M, Tada M, Tanaka Y (2019) Ultrasensitive single cell metabolomics by capillary electrophoresis–mass spectrometry with a thin-walled tapered emitter and large-volume dual sample preconcentration. Anal Chem 91(16): 10564–10572
48. Portero EP, Nemes P (2019) Dual cationic–anionic profiling of metabolites in a single identified cell in a live *Xenopus laevis* embryo by microprobe CE-ESI-MS. Analyst 144(3): 892–900
49. Lombard-Banek C, Yu Z, Swiercz AP, Marvar PJ, Nemes P (2019) A microanalytical capillary electrophoresis mass spectrometry assay for quantifying angiotensin peptides in the brain. Anal Bioanal Chem 411(19):4661–4671
50. Choi SB, Zamarbide M, Manzini MC, Nemes P (2016) Tapered-tip capillary electrophoresis nano-electrospray ionization mass spectrometry for ultrasensitive proteomics: the mouse cortex. J Am Soc Mass Spectrom 28(4): 597–607
51. Choi SB, Muñoz-LLancao P, Manzini MC, Nemes P (2021) A data-dependent acquisition ladder for ultrasensitive (neuro)proteomics in review

52. Lombard-Banek C, Li J, Portero EP, Onjiko RM, Singer CD, Plotnick DO, Al Shabeeb RQ, Nemes P (2021) In vivo subcellular mass spectrometry enables proteo-metabolomic single-cell systems biology in a chordate embryo developing to a normally behaving tadpole (*X. laevis*). Angew Chem-Int Edit 60(23): 12852–12858
53. Choi SB, Polter AM, Nemes P (2019) Mass spectrometry-based proteomics of single dopaminergic mouse neurons identified by whole-cell electrophysiology. In: Pittsburgh conference on analytical chemistry and applied spectroscopy, Philadelphia, PA
54. Choi SB, Polter AM, Nemes P (2019) Single-cell mass spectrometry for proteomic analysis of patch-clamp electrophysiology identified dopaminergic neurons. In: United States human proteome organization, Washington DC
55. Baxi AB, Lombard-Banek C, Moody SA, Nemes P (2018) Proteomic characterization of the neural ectoderm fated cell clones in the *Xenopus laevis* embryo by high-resolution mass spectrometry. ACS Chem Neurosci 9(8): 2064–2073
56. Onjiko RM, Plotnick DO, Moody SA, Nemes P (2017) Metabolic comparison of dorsal versus ventral cells directly in the live 8-cell frog embryo by microprobe single-cell CE-ESI-MS. Anal Methods 9(34):4964–4970
57. Lombard-Banek C, Reddy S, Moody SA, Nemes P (2016) Label-free quantification of proteins in single embryonic cells with neural fate in the cleavage-stage frog (*Xenopus laevis*) embryo using capillary electrophoresis electrospray ionization high-resolution mass spectrometry (CE-ESI-HRMS). Mol Cell Proteomics 15(8):2756–2768
58. Onjiko RM, Morris SE, Moody SA, Nemes P (2016) Single-cell mass spectrometry with multi-solvent extraction identifies metabolic differences between left and right blastomeres in the 8-cell frog (*Xenopus*) embryo. Analyst 141(12):3648–3656
59. Onjiko RM, Nemes P, Moody SA (2021) Altering metabolite distribution at *Xenopus* cleavage stages affects left-right gene expression asymmetries. Genesis 59(5-6):1–8
60. Portero EP, Andrews AJ, Nemes P (2019) Probing metabolic pathways during early embryonic development using stable isotope labeling and single-cell mass spectrometry. In: ASMS conference on mass spectrometry and allied topics, Atlanta, GA
61. Portero EP, Pade L, Nemes P (2020) Single-cell mass spectrometry reveals cell-to-cell communication in the same vertebrate (*Xenopus laevis*) frog embryo. In: ASMS conference on mass spectrometry and allied topics
62. Aerts JT, Louis KR, Crandall SR, Govindaiah G, Cox CL, Sweedler JV (2014) Patch clamp electrophysiology and capillary electrophoresis-mass spectrometry metabolomics for single cell characterization. Anal Chem 86(6):3203–3208
63. Choi SB, Polter AM, Nemes P (2020) Integration of patch-clamp electrophysiology with single-cell mass spectrometry for proteomic analysis to extend the bioanalytical toolbox of neuroscience. In: ASMS Conference on Mass Spectrometry and Allied Topics
64. Moody SA (2018) Microinjection of mRNAs and oligonucleotides. Cold Spring Harb Protoc 2018(12)
65. Sive HL, Grainger RM, Harland RM (2000) Early development of *Xenopus laevis*: a laboratory manual. Cold spring harbor laboratory press. Cold Spring Harbor, NY
66. Kaech S, Banker G (2006) Culturing hippocampal neurons. Nat Protoc 1(5):2406–2415
67. Klein SL (1987) The 1st cleavage furrow demarcates the dorsal ventral axis in *Xenopus* embryos. Dev Biol 120(1):299–304
68. Polter AM, Bishop RA, Briand LA, Graziane NM, Pierce RC, Kauer JA (2014) Poststress block of kappa opioid receptors rescues long-term potentiation of inhibitory synapses and prevents reinstatement of cocaine seeking. Biol Psychiatry 76(10):785–793
69. Moody SA (1987) Fates of the blastomeres of the 32-cell-stage *Xenopus* embryo. Dev Biol 122(2):300–319
70. Moody SA (1987) Fates of the blastomeres of the 16-cell stage *Xenopus* embryo. Dev Biol 119(2):560–578
71. Liu ZC, Portero EP, Jian YR, Zhao YJ, Onjiko RM, Zeng C, Nemes P (2019) Trace, machine learning of signal images for trace-sensitive mass spectrometry: a case study from single-cell metabolomics. Anal Chem 91(9): 5768–5776
72. Xie YR, Castro DC, Bell SE, Rubakhin SS, Sweedler JV (2020) Single-cell classification using mass spectrometry through interpretable machine learning. Anal Chem 92(13): 9338–9347
73. Sun L, Zhu G, Zhang Z, Mou S, Dovichi NJ (2015) A third-generation electrokinetically pumped sheath-flow nanospray interface with

improved stability and sensitivity for automated capillary zone electrophoresis-mass spectrometry analysis of complex proteome digests. J Proteome Res 14(5):2312–2321

74. Guijas C, Montenegro-Burke JR, Domingo-Almenara X, Palermo A, Warth B, Hermann G, Koellensperger G, Huan T, Uritboonthai W, Aisporna AE, Wolan DW, Spilker ME, Benton HP, Siuzdak G (2018) METLIN: a technology platform for identifying knowns and unknowns. Anal Chem 90(5): 3156–3164

75. Wishart DS, Feunang YD, Marcu A, Guo AC, Liang K, Vázquez-Fresno R, Sajed T, Johnson D, Li C, Karu N, Sayeeda Z, Lo E, Assempour N, Berjanskii M, Singhal S, Arndt D, Liang Y, Badran H, Grant J, Serra-Cayuela A, Liu Y, Mandal R, Neveu V, Pon A, Knox C, Wilson M, Manach C, Scalbert A (2018) HMDB 4.0: the human metabolome database for 2018. Nucleic Acids Res 46(D1): D608–D617

76. Bateman A, Martin MJ, Orchard S, Magrane M, Agivetova R, Ahmad S, Alpi E, Bowler-Barnett EH, Britto R, Bursteinas B, Bye-A-Jee H, Coetzee R, Cukura A, Da Silva A, Denny P, Dogan T, Ebenezer T, Fan J, Castro LG, Garmiri P, Georghiou G, Gonzales L, Hatton-Ellis E, Hussein A, Ignatchenko A, Insana G, Ishtiaq R, Jokinen P, Joshi V, Jyothi D, Lock A, Lopez R, Luciani A, Luo J, Lussi Y, Mac-Dougall A, Madeira F, Mahmoudy M, Menchi M, Mishra A, Moulang K, Nightingale A, Oliveira CS, Pundir S, Qi GY, Raj S, Rice D, Lopez MR, Saidi R, Sampson J, Sawford T, Speretta E, Turner E, Tyagi N, Vasudev P, Volynkin V, Warner K, Watkins X, Zaru R, Zellner H, Bridge A, Poux S, Redaschi N, Aimo L, Argoud-Puy G, Auchincloss A, Axelsen K, Bansal P, Baratin D, Blatter MC, Bolleman J, Boutet E, Breuza L, Casals-Casas C, de Castro E, Echioukh KC, Coudert E, Cuche B, Doche M, Dornevil D, Estreicher A, Famiglietti ML, Feuermann M, Gasteiger E, Gehant S, Gerritsen V, Gos A, Gruaz-Gumowski N, Hinz U, Hulo C, Hyka-Nouspikel N, Jungo F, Keller G, Kerhornou A, Lara V, Le Mercier P, Lieberherr D, Lombardot T, Martin X, Masson P, Morgat A, Neto TB, Paesano S, Pedruzzi I, Pilbout S, Pourcel L, Pozzato M, Pruess M, Rivoire C, Sigrist C, Sonesson K, Stutz A, Sundaram S, Tognolli M, Verbregue L, Wu CH, Arighi CN, Arminski L, Chen CM, Chen YX, Garavelli JS, Huang HZ, Laiho K, McGarvey P, Natale DA, Ross K, Vinayaka CR, Wang QH, Wang YQ, Yeh LS, Zhang J, UniProt C (2021) UniProt: the universal protein knowledgebase in 2021. Nucleic Acids Res 49(D1):D480–D489

77. Karimi K, Fortriede JD, Lotay VS, Burns KA, Wang DZ, Fisher ME, Pells TJ, James-Zorn C, Wang Y, Ponferrada VG, Chu S, Chaturvedi P, Zorn AM, Vize PD (2018) Xenbase: a genomic, epigenomic and transcriptomic model organism database. Nucleic Acids Res 46 (D1):D861–D868

78. Zhang Y, Fonslow BR, Shan B, Baek M-C, Yates JR (2013) Protein analysis by shotgun/bottom-up proteomics. Chem Rev 113(4): 2343–2394

79. Moody SA (2000) Cell lineage analysis in *Xenopus* embryos. Methods Mol Biol 135: 331–347

80. Moody SA (2018) Lineage tracing and fate mapping in *Xenopus* embryos. Cold Spring Harb Protoc 2018(12)

81. Viczian AS, Zuber ME (2014) A simple behavioral assay for testing visual function in *Xenopus laevis*. J Vis Exp 88:e51726

82. Ting JT, Daigle TL, Chen Q, Feng GP (2014) Acute brain slice methods for adult and aging animals: application of targeted patch clamp analysis and optogenetics. In: Martina M, Taverna S (eds) Patch-clamp methods and protocols, vol 1183, 2nd edn. Methods in Molecular Biology, pp 221–242

83. Grace AA, Onn SP (1989) Morphology and electrophysiological properties of immunocytochemically identified rat dopamine neurons recorded in vitro. J Neurosci 9(10):3463–3481

84. Khokha MK, Chung C, Bustamante EL, Gaw LW, Trott KA, Yeh J, Lim N, Lin JC, Taverner N, Amaya E, Papalopulu N, Smith JC, Zorn AM, Harland RM, Grammer TC (2002) Techniques and probes for the study of *Xenopus tropicalis* development. Dev Dyn 225(4):499–510

85. Wallingford JB (2010) Preparation of fixed *Xenopus* embryos for confocal imaging. Cold Spring Harb Protoc 2010(5):1–7

Chapter 6

High-Throughput Image-Guided Microprobe Mass Spectrometric Analysis of Single Cells

Stanislav S. Rubakhin, Elena V. Romanova, and Jonathan V. Sweedler

Abstract

Characterizing the chemicals within individual cells is an important analytical capability, with mass spectrometry (MS) being one of the most chemically information-rich techniques available. There are two major approaches for characterizing single cells using MS: the assay of small-volume solutions using electrospray ionization and, the focus of this chapter, measurement of cells dispersed across a surface using matrix-assisted laser desorption/ionization (MALDI) MS. In combination with spatiometric information obtained via optical imaging tools to locate individual cells on the sampling surface, follow-up measurement via MALDI MS can profile a multitude of single cells in a high-throughput manner. Compared to conventional MS imaging of the entire sample-containing surface, preselection of accurate cell locations for targeted analysis increases selectivity and throughput as well as reduces overall data size. As a result, more informative data sets are produced in less time with fewer resources. Employing this strategy, we present a multistage protocol for optical image-guided, high-throughput single cell analysis by MALDI MS. We outline steps for the isolation and preparation of single cells prior to dispersing on a glass slide for optical imaging, automated mapping of imaged cell locations using microMS, preparing the cells for MALDI MS analysis using standards and MALDI matrices, acquisition of mass spectra from mapped individual cells, and data processing and statistical analysis. Images obtained via a range of microscopy modalities can be assessed by the microMS custom image processing software suite to determine location coordinates for cells with defined morphological and/or biochemical characteristics. Depending on the biological model and MS instrumentation used, this strategy can be further enhanced by hyphenation of different imaging modalities, e.g., electron microscopy, for multidimensional sample characterization.

Key words Single cell, MicroMS, Mass spectrometry, Pancreatic islets, Peptides, Microanalysis, High throughput, Matrix-assisted laser desorption/ionization

1 Introduction

Arguably, the cell is the most fundamental unit of life and possesses a variety of the needed mechanisms of independent function and replication. Structural similarity among cells is the basis of tissue formation and reproduction of organisms, whereas heterogeneity among similar cells is critical for organism functioning. Since its observation by Robert Hooke [1], the cell has become an

Jonathan V. Sweedler et al. (eds.), *Single Cell 'Omics of Neuronal Cells*,
Neuromethods, vol. 184, https://doi.org/10.1007/978-1-0716-2525-5_6,

important target of scientific analysis. Improvements in the sensitivity of mass spectrometry (MS) now enable single cell and subcellular analysis, allowing us to obtain unique multiplexed chemical information on the presence of inorganic compounds, metabolites, lipids, peptides, proteins, oligonucleotides, and polynucleotides in single cells. A variety of MS approaches have been adapted to single cell analysis, with secondary ionization mass spectrometry (SIMS), electrospray ionization (ESI)-MS, and matrix-assisted laser desorption/ionization (MALDI) MS being most frequently used. Each of these approaches has advantages and disadvantages, including specific requirements for sample handling and analyte extraction.

The goals of bioanalytical single cell MS are often the detection and structural and quantitative characterization of endogenous and exogenous compounds. The multiplexed nature of MS analysis permits coverage of a wide range of analyte classes, making the detection and characterization of important components of the metallome, metabolome, lipidome, peptidome, and proteome of a cell possible [2, 3]. Due to normal and pathological variabilities and heterogeneities of biological systems, often exacerbated by sample handling during preparation, experimental success depends on the analysis of a representative number of samples (replicates) to provide sufficient statistical power. As even a small sample can contain hundreds of thousands to millions of cells, such MS measurements may consume unrealistic resources and time. Therefore, optimization of data collection leads to more efficient measurements, enhanced results, and reduced data size. Though one can reduce the amount of data obtained and enhance analyses using several post-acquisition approaches, here we present a protocol that improves data acquisition time and efficiency during the sample preparation and optical imaging stages to facilitate high-throughput single cell MALDI MS analysis. An important component of this approach is the ability to filter individual cells with predetermined morphometric and biochemical characteristics and only select specific cells for follow-up MS analysis, excluding empty spaces and structures, such as debris and dust. We have optimized this protocol for the analysis of low-density populations of individual cells dispersed across a glass slide, which is a common practice in biomedical investigations and clinical applications.

1.1 Comparison with Other Approaches

Image analysis approaches have widespread application in a variety of sample characterization processes, including single cell recognition, characterization, and identification [4–7]. More broadly, optical information for object location and sorting is widely used in diverse applications, from fluorescent cell sorting [8, 9] to industrial applications of computer vision systems [10], manufacturing quality controls [11, 12], biomedical fields [13, 14], and big data-related analyses [15].

For spatial information-guided MS of single cells, two approaches are typically used: (1) operator-defined targeting of a region of interest within the sample and (2) analysis of samples deposited at predefined coordinates. The first approach is widely used in mass spectrometry imaging (MSI) where measurements are typically performed on manually selected areas having the desired morphological features, such as defined brain regions on brain slices. Measurement of systematically deposited single cells (e.g., into array patterns), including individual bacteria, has been effective with MS-based approaches, e.g., microarrays for mass spectrometry (MAMS) [16–18], cell sorting with surface deposition [19], and cell culture MS analysis [20, 21]. These cell targeting approaches have multiple benefits, including high methodological control in the sample localization process; however, they have some limitations. For example, MAMS requires specifically designed and manufactured wells (tailored to the size of the cells) that may complicate work with cellular populations of different cell sizes, such as populations of isolated hippocampal neurons having more than 100-μm-long terminals. For MSI, information is acquired on all spots in a sample, some of which are empty spaces, nonbiological structures, and cells. As result, a lower measurement throughput is achieved and larger data files are created. Moreover, because cellular debris will also be measured, the data set will require greater post-processing effort.

Also interesting are the flow-based MS approaches, which can quickly analyze hundreds to tens of thousands of individual cells delivered to the mass spectrometer in a continuous flow of solution. For example, mass cytometry utilizes fast, orderly delivery of cells and sampling of their chemical composition using an inductively coupled argon plasma (ICP) [22]. Moving away from MS, several spectroscopic methods may provide multiplexed chemical information from single cell populations including direct measurements by Raman Fourier transform (FT) spectroscopy [23], infrared spectroscopy [24], and affinity probe-based methods, such as sequential multiplex staining of cells and tissues [25, 26] targeting multiple analyte classes that is dependent on probe specificity.

1.2 Development of the Approach

Several decades of work [27–29] on the development and application of single cell MALDI MS have demonstrated its ability to aid in the discovery of new neuropeptides in selected cells. While the approach is effective when specific cells need to be assayed, throughput has been a major bottleneck, limiting MALDI MS analysis to small numbers of individual cells. To address this issue, we increased throughput by placing low-density scattered cells on a slide, locating their positions, and then using these positions to guide the laser (or ion beam for SIMS) [30–32]. Because this process was somewhat tedious, in 2017 our group developed the software package microMS [33] to automate cell finding and improve coordinate registration via a user-friendly GUI interface, and using point-based image registration to correlate fiducial markers on the microscopy image to physical coordinates on the mass spectrometer's stage. The targeted profiling steps enabled by the microMS-guided workflow include the ability to (1) determine locations/coordinates of cells or other structures dispersed across a microscope slide using an optical image; (2) select coordinates of subpopulations of detected features (e.g., cells) according to flexible, user-defined criteria, such as shape and intercellular distances; and (3) convert the cell pixel coordinates to physical stage coordinates for the measurement platform (i.e., mass spectrometer) being used. Importantly, the positional information can be combined with other analyses, such as post-MALDI MS selected cell sampling and capillary electrophoresis (CE)-MS analysis [34]. While developed for single cell analysis, with modifications the underlying microMS code is versatile and can be applied to a variety of samples, imaging modalities, and chemical analysis techniques. For more advanced users, the addition of new, off-line instrument platforms and even direct instrument control is possible. We describe the most common usage of microMS, detail additional features, and provide a starting point for its application to a range of other platforms. Preparing the biological samples and performing the MS measurements require a medium level of expertise. Our experience with training undergraduate research assistants has shown us that novice users are capable of producing repeatable results after 2–3 months of practice, and this is the audience aimed for here. A higher level of expertise or additional expert advice may be needed for tissues that are hard to dissociate or when using complex or unusual MS equipment.

1.3 Overview of the Approach

The combination of single cell isolation, single cell MS, optical imaging, optical image processing utilities, and molecular probes provides opportunities for cell chemical typing and determination of the chemical stability and heterogeneity of cell types in larger cellular populations. Though such combined measurements are not common, the key mass spectrometric and optical imaging hardware and software already exist, enabling us to link these approaches together using the integrative microMS software package.

The major methodological steps in this multistage protocol are (1) biological sample preparation, to include dissociation of individual cells via enzymatic and/or mechanical treatments, labeling with fluorescent dye, and deposition of cells onto an ITO-coated glass slide and stabilized with glycerol, followed by partial or complete glycerol removal using washing buffers; (2) acquisition of an optical image of cells under fluorescent and transmitted light; (3) processing of the acquired image with microMS to determine cell location coordinates according to user-defined criteria; (4) sample preparation for MALDI MS analysis consisting of analyte standard deposition and MALDI matrix application (after this stage, samples can be stored at low temperatures in a desiccator or in a low oxygen environment but should be acclimated to ambient temperature before MS analysis); (5) loading of the sample into a mass spectrometer and performing point-based image registration by locating fiducial markers in microMS and the same location on the MS stage; (6) loading coordinates of the targeted cells into the mass spectrometer's software; (7) acquisition of mass spectra; (8) and finally, data processing and analyses using a variety of computational tools available from different industrial and academic sources. The main goals of the analyses are the measurement of the chemistry within the cell, thus enabling chemical heterogeneity of cellular populations to be probed, and the biomarkers for cell types to be determined.

Using this approach, several hundred to a thousand cells per slide are commonly measured. Depending on the instrument used and settings selected, each slide can be analyzed in less than 1 hour, allowing at least eight slides to be processed per day. There are several options for further increasing the number of analyzed slides, including automation of sample loading/unloading as well as generation of scripts for automatic import of cell coordinates and instrument training (e.g., mass spectrometer) using preset fiducial marks, and available hardware and software. However, additional efforts should be made toward registration of fiducial marks after sample loading into a mass spectrometer. This step currently requires involvement of an operator and thus, limits measurements to operator availability.

1.4 Advantages and Disadvantages of the Approach

Analysis of intact cells provides distinct advantages in comparison with MS measurements of thin tissue sections. In most cases, tissue section samples analyzed with MS do not provide information on the chemical profile of individual cells. This may be because cells are cross-sectioned and only parts of cells are analyzed, or interference from other surrounding cells is causing difficulties in assessing cell content. In addition, even when the probe size is on the order of (or smaller than) a cell, cells are irregularly shaped and the probes are often not well-aligned with cells, so that one is not assured of sampling from an individual cell when assaying tissue sections.

In contrast, dispersing individual cells at distances typically exceeding the probe (laser) diameter ensures that different cells are not examined simultaneously. In addition, one can use wetter conditions for analyte extraction and incorporation into MALDI matrix crystals so that limited spreading of the analytes to adjacent sections is not an issue. Because of these factors, single cell MALDI MS increases data reliability by reducing single cell sample cross-contamination and decreases the likelihood that noncellular structures or multicellular aggregates are assayed. The microMS-based workflow can be adapted to a large variety of sample plates besides the standard microscopy glass slide. Fluorescent images can also be taken from nontransparent, 386-well polished MALDI sample plates and custom-designed glass or silicon sample plates [35]. Optical image-guided MS analysis can be adapted to other imaging modalities, including electron microscopy, and magnetic resonance, x-ray, and ultrasound imaging. As a last benefit of MALDI MS, after a measurement, much of the sample is left, allowing the remaining material to be characterized again. We already have demonstrated that several technologies can be combined with microMS-based analysis for single cell measurements, including CE-MS [34] and immunolabeling [36], and it is likely that others will follow.

As with many other technologies for measuring single cells, there are some limitations to the described approach, but there are also viable solutions to overcoming some of these challenges. microMS-based single cell MS analysis does not preserve information on intra-specimen localization of individual cells within a sampled tissue. To correlate single cell and tissue level data sets with cell type/state, biomarkers can be used in follow-up measurements of the same tissue utilizing MSI or immunohistochemistry, thereby addressing this issue. Another concern is that bias can be introduced when isolating cells because of differences in their mechanical stability or susceptibility to damage during dissociation. If one cell type is damaged and another is not, this would be difficult to determine based solely on the MS data. Moreover, different enzymatic treatments may be needed for isolation of astrocytes and neurons, or cells may not be stained uniformly with a specific dye. For example, we have noticed that Hoechst 33342 produces dim staining in cerebellar Purkinje neurons compared to granular cells. As a remedy, the addition of dyes for monitoring cell pyroptosis, apoptosis, or cell lysis [37], followed by optical imaging and exclusion of specific cells during MS analysis, can be introduced.

Despite the multiplexed nature of single cell MALDI MS analysis, analyte coverage is relatively low, typically with 10–200 compounds detected. This issue is due in part to low analyte amounts and also because of the relatively low dynamic range and resolving power of direct MS compared to hyphenated approaches such as CE-MS and liquid chromatography (LC)-MS. However, the introduction of ion mobility to the front end should address this drawback, and use of ultrahigh-resolution FTMS should also increase the number of analytes detected and characterized.

Despite the limitations noted above, microMS-based single cell MS is a flexible and versatile platform for analyses of cellular agglomerates and single cells that we have successfully applied to the characterization of cellular populations of individual islets of Langerhans, dorsal root ganglia, and brain regions [32, 36, 38]. Low analyte consumption techniques, such as SIMS and

MALDI MS, are hyphenated for analysis of the same cellular population and provide wider analyte coverage and additional opportunities for cell phenotyping [38]. Low analyte consumption as well as characteristic sample preparation for MALDI MS, where not all analytes are extracted and incorporated into MALDI matrix crystals, allows post-MALDI MS detection of cellular proteins using immunolabeling [36] and metabolites by CE hyphenated to ESI-MS [34]. microMS is critical for determining cell targets for more time-consuming analytical measurements such as CE ESI-MS. Typically, a single CE ESI-MS run takes 10–30 min; therefore, only a small number of cells can be analyzed. MALDI MS provides information on cell content and cell types to guide selection of single cells for a follow-up analysis by CE-MS. microMS processing of coordinates of cellular locations enables different approaches for cell content collection, including microjunction approaches as well as other analytical platforms [34].

While more standard applications of the approach are covered here, microMS can be tailored to new and diverse instruments and also offers more advanced operations, including direct instrument control. For example, many other types of single cell analysis can be hyphenated with microMS, e.g., vibrational spectroscopies such as infrared, near infrared, and Raman. Knowledge of cell coordinates allows formation of multiplatform data sets for extensive characterization of individual cells and cellular populations. Single organelle analysis is a new area for microMS-based MS [39]; however, this application is not covered in the current protocol.

1.5 Experimental Design Considerations

It is important to ensure that all components of the experimental design are compatible and sufficiently flexible for achieving the planned goals. For example, not all instrumental software packages are capable of generating optical image files that are compatible with microMS. Also, some cell labeling methods may not be equally efficient for specific cell types. Depending on the biological model, targeted analyte class, and study goals, different approaches have to be chosen for different stages of the experimental process, beginning with suitable cell isolation, labeling and storage techniques, selection of the MALDI matrix, and microscopy and MS instrumentation, and finally, choosing appropriate data analysis tools, methods, and resources.

Especially important is the quality of the cell isolation and sample preparation process to ensure accurate representation of the cell population. All components should be tested at the preliminary stage of planning. Before the actual experiment is started, multiple optimizations should be implemented and evaluated for the analytes of interest, e.g., dispersing cells at an appropriate cell density, optimizing the MALDI matrix solution and its application, and adjusting the MALDI MS parameters. All solutions and surfaces used should be tested for the presence of detectable, or not detectable but interfering, analytes and compounds. For example, polymers such as polyethylene glycol and polysiloxane are often used in molecular biology but cannot be used for this protocol. Therefore, control experiments utilizing analyte standards and well-characterized biological models should be performed.

In cases of long-term studies where samples become available over an extended time, the timing and order of their analysis should be considered. MS of biological samples is prone to batch effects; therefore, accumulation of samples and their appropriate storage must be validated in a relatively short period of time prior to analysis. To further reduce the influence of batch effects, deposition of cell populations collected from different donors on different marked areas of the same glass slide in random order is suggested.

2 Materials

2.1 Biological Materials

When developing this protocol, we used cells from the rat and human islets of Langerhans as models. Isolated human islets were provided by our collaborators at the University of Pennsylvania, with the sample collection and processing approved by the University of Pennsylvania Institutional Review Board (protocol number: 832134). Written consent was obtained from responsible parties (tissue donors or relatives of donors). The handling of human cell samples was performed in accordance with Biosafety Level 2 regulations described for the project and approved by the relevant Institutional Biosafety Committees. Adult Sprague-Dawley rats were purchased from ENVIGO (https://www.envigo.com/model/hsd-sprague-dawley-sd). Protocols for animal care and procedures were approved by an Institutional Animal Care and Use Committee and be in full compliance with local, federal, and some international guidelines for the humane care and treatment of animals.

2.2 Reagents

1. 2,5-Dihydrobenzoic acid (DHB) (MilliporeSigma, cat. no. A3539).
2. 4-(2-Hydroxyethyl)piperazine-1-ethanesulfonic acid (HEPES) (MilliporeSigma, cat. no. H3375).
3. Acetonitrile (ACN) (Thermo Fisher Scientific, cat. no. A998-4).

   !CAUTION ACN is toxic. When handling, wear gloves and use a pipetting aid.
4. Calcium chloride (MilliporeSigma, cat. no. C7902).
5. Glycerol (MilliporeSigma, cat. no. G7893).
6. Magnesium chloride hexahydrate (MilliporeSigma, cat. no. M2393).

7. Magnesium sulfate heptahydrate (MilliporeSigma, cat. no. M2773).
8. Peptide calibration standard II (angiotensin II, angiotensin I, substance P, bombesin, ACTH clip 1–17, ACTH clip 18–39, somatostatin 28, bradykinin fragment 1–7, renin substrate tetradecapeptide porcine) (Bruker Corp., part no. 222570).
9. Potassium chloride (MilliporeSigma, cat. no. P5405).
10. Potassium phosphate monobasic (MilliporeSigma, cat. no. P5655).
11. Sodium bicarbonate (MilliporeSigma, cat. no. S5761).
12. Sodium chloride (MilliporeSigma, cat. no. S5886).
13. Sodium phosphate (Thermo Fisher Scientific, cat. no. S374–500).
14. Nuclear dye Hoechst 33342 (Thermo Fisher Scientific, cat. no. 62249).
15. Trifluoroacetic acid (TFA) (MilliporeSigma, cat. no. T6508).

    !CAUTION TFA is highly corrosive. When handling, wear gloves and goggles, use a pipetting aid, and operate in a ventilated chemical hood.
16. Trypsin LE (Thermo Fisher Scientific, cat. no. 12604013).
17. Paraformaldehyde (Fisher Scientific, cat. no. 50-980-487).
18. Tris-EDTA buffer solution (Sigma-Aldrich, cat. no. 93283).
19. N-Acetyl-L-cysteine NAC (Sigma-Aldrich, cat. no. A7250).
20. Collagenase P from *Clostridium histolyticum* (Sigma-Aldrich, cat. no. 11213857001).

2.3 Laboratory Equipment

1. Diamond scriber (e.g., Electron Microscopy Sciences, cat. no. 62107-ST).
2. Microtubes, 1.5 mL (e.g., Eppendorf Safe-Lock Tubes, cat. no. 022363204).
3. 15 ml screw-cap tubes (Falcon, cat. no. 352095).
4. 50 ml screw-cap tubes (Falcon, cat. no. 352070).
5. Disposable borosilicate glass Pasteur pipets (Fisher Scientific, cat no. 13–678-20A).
6. Non-treated culture dishes D × H 60 mm × 15 mm (e.g., Corning, cat. no. CLS430589).
7. Disposable safety gloves (e.g., Microflex® 93–833, XCEED® nitrile gloves, powder-free (MilliporeSigma, cat. no. Z687448)).
8. Disposable graduated transfer pipets (e.g., Fisher Scientific, cat. no.13–711-9 AM).

9. Safety glasses (e.g., MR0111ID Mirage safety glasses (Radians, cat. no. MR0111ID).
10. Face masks (e.g., disposable PPE Latex-Free 3-Ply Face Masks, ProPac, SKU: D8012).
11. Paper absorbent wipes (e.g., Kimwipes, Kimberly-Clark Inc., cat. no. 34155).
12. Indium tin oxide (ITO)-coated microscopy glass slides (coating resistance, 70–100 ohms; dimensions, 25 × 75 × 0.8 mm) (e.g., Delta Technologies, cat. No. CG-90IN-S110).
13. Automated matrix spraying system (e.g., HTX M5 Sprayer; HTX Technologies).
14. Water bath for 34–37 °C incubations (Cole-Parmer Polystat EW-20873-45).
15. Digital multimeter – ohmmeter (e.g., HHM10-20-30-DMM-Series, www.omega.com).
16. Plastic rectangle microscope glass holders (boxes for five slides).
17. Ultra-microbalance (e.g., Cubis II, Sartorius).
18. Vacuum-sublimation apparatus (e.g., Ace complete apparatus, H × I.D. 250 mm × 152 mm, Sigma-Aldrich, cat. no. Z258717).

    !CAUTION Working with glass vessels kept under vacuum requires establishing and following specific standard operating procedure(s).
19. A Styrofoam box to hold ice
20. pH meter (e.g., FiveEasy Plus pH meter FP20-Micro-Kit, Mettler Toledo, material no. 30266940).
21. Surgical tools, including forceps, scissors, and a vessel clamp (e.g., Fine Science Tools, cat. no. 18320-11).
22. 5 mL plastic syringe with blunt end 20G–30G needle attached. Needle size depends on animal size, to correspond to the diameter of the bile duct, and should be determined experimentally. For easier operation, a fine Teflon or other plastic tubing can be attached to end of needle extending working distance and reducing effect of syringe motion on success of solution injection.
23. Guillotine for rodents (e.g., Nemi Guillotine, Braintree Scientific, cat. no. NS-80 2).
24. Interchangeable magnifier headset with LED used during surgical procedures (e.g., Fine Science Tools, cat. no. 28080-10).
25. Flexible plastic tubing (e.g., Tygon one WHATMAN Inline Filter Device, 60 psi (https://www.grainger.com/, item no. 32HK77)).

26. Two 50 mL heavy wall glass filtering flasks (e.g., https://www.usplastic.com/, item no. 76984).

2.4 Mass Spectrometry Instruments and Software

1. MALDI-TOF/TOF mass spectrometer (we use an ultrafleXtreme, Bruker Corp.)
2. Fourier transform ion cyclotron resonance (FTICR) mass spectrometer (we use a 7 T solariX, Bruker Corp.).
3. flexAnalysis software (Bruker Corp.).
4. flexControl software (Bruker Corp.).
5. flexImaging software (Bruker Corp.).

2.5 microMS Software and Hardware

1. The microMS software downloads (http://neuroproteomics.scs.illinois.edu/microMS.htm) are comprised of a compiled Windows executable, "microMS.exe" (ZIP folder), which includes a Python interpreter and all dependencies, and the Python Source Code, "microMS" (ZIP folder). The microMS User Guide can also be downloaded from this site. The operation and use of this code have been described in https://github.com/troycomi/microMS and [33].
2. Python (https://www.python.org/downloads/).
3. PyQt5 (https://www.riverbankcomputing.com/software/pyqt/download5).
4. NumPy (http://www.numpy.org/).
5. SciPy (https://www.scipy.org/).
6. Matplotlib (http://matplotlib.org/).
7. Skimage (http://scikit-image.org/docs/dev/api/skimage.html).
8. OpenSlide (http://openslide.org/) and OpenSlide Python (https://github.com/openslide/openslide-python#installation).
9. PySerial (https://pypi.python.org/pypi/pyserial).
10. The software requires a workstation running an up-to-date Windows operating system (e.g., Windows 10, 32 GB of RAM, a 1 TB hard drive), or Linux distributions (e.g., Fedora, Ubuntu).

2.6 Biochemical and Mass Spectrometry Databases

1. Lipids Maps Structure Database (http://www.lipidmaps.org/data/structure/).
2. UniProt (https://www.uniprot.org) [39].
3. Human Metabolome Database (https://hmdb.ca).
4. MassBank of North America (MoNA) (https://mona.fiehnlab.ucdavis.edu/).

5. METLIN – Scripps Research (https://metlin.scripps.edu).
6. Protein – NCBI (https://www.ncbi.nlm.nih.gov/protein).

2.7 Microscopy Equipment

1. High-performance dissection stereomicroscope (e.g., Leica MZ 7.5 with 7.9:1 zoom; Leica Microsystems Inc.).
2. Transmission light microscopy inverted microscope equipped (e.g., Axiovert 25 with AxioCam MRc camera; ZEISS Microscopy).
3. UV fluorescence microscopy upright microscope (e.g., Zeiss Axio Imager.M2 (https://www.micro-shop.zeiss.com/en/se/system/axio+imager+2-axio+imager.m2-upright+microscopes/10215/#variants) equipped with an Axiocam 503 mono; ZEISS Microscopy, cat. no. 4109664265591000).
4. Color microscope camera (e.g., AxioCam ICc 5; ZEISS Microscopy, cat. no. 4265540000000000).
5. Fluorescence illumination system (e.g., X-Cite 120Q Iris FL Light Source; Lumen Dynamics).
6. Illuminator (e.g., VIS-LED; ZEISS Microscopy, cat. no. 4230539030000000).
7. Filter set (e.g., FL Filter Set 49 DAPI, EX G365 S free; ZEISS Microscopy, cat. no. 4880499901000000).
8. UV transmitting objective (e.g., Plan-Apochromat 10×/0.45 WD = 2.0 M27; ZEISS Microscopy, cat. no. 4206409900000000, or EC Plan-Neofluar 5×/0.16 M27; ZEISS Microscopy, cat. no. 420330-9901-000).
9. Scanning stage (e.g., 130 × 85 piezo STEP (D); ZEISS Microscopy, cat. no. 4320229902000000).
10. An image acquisition and processing package (e.g., ZEN software ver. 3.1 (blue edition); ZEISS Microscopy).
11. An image processing package (e.g., Fiji, https://imagej.net/Fiji).
12. Microsoft Office package for data visualization and .txt file editing (https://www.microsoft.com/en-us/microsoft-365).

3 Methods

3.1 Reagent Preparation

1. Modified Gey's balanced salt solution (mGBSS) containing (in mM) 1.5 CaCl2, 4.9 KCl, 0.2 KH2PO4, 11 MgCl2, 0.3 MgSO4, 138 NaCl, 27.7 NaHCO3, 0.8 Na2HPO4, 25 HEPES and 10 glucose, pH 7.2.
2. Glycerol stabilization solution containing 33% (vol/vol) glycerol mixed with 67% mGBSS.

3. MALDI matrix solution containing DHB saturated in a 1:1 mixture of ACN: 0.1% (vol/vol) aqueous/TFA acid. Many additional MALDI matrices are available for investigation of small metabolites, peptides, proteins, nucleotides, and polysaccharides, and can be used in biochemical profiling of single cells.

3.2 MALDI Mass Spectrometry Calibration

Mass spectrometers should be tuned and calibrated before running high-throughput single cell analysis. Calibration of mass spectrometers, preferably, should be performed for each sample (e.g., ITO glass slide).

1. Time-of-flight (TOF)-TOF instruments: standards are placed on the same surface where samples are deposited. Even slight difference in distance between sample surface and mass spectrometer detector leads to variable measurements.
2. FTICR instruments (e.g., Bruker solariX): can be calibrated with secondary ESI ion source providing better options for calibration than MALDI ion source. Calibrants should cover the entire *m/z* range under analysis. Typically, for detection in positive mode, ion signals of peptide standards (*m/z* 700–6000) and ion signals of MALDI matrix (*m/z* 100–600) are used for instrument calibration.
3. Ion trap instruments (including FT mass spectrometers equipped with ESI and MALDI ion sources): can be calibrated via ESI source using commercially available ESI tuning mixes, or volatile salts of metals (such as sodium or lithium format).

3.3 Solutions for Human Pancreatic Islet Dissociation and Cell Labeling

Fresh trypsin LE islet dissociation media should be prepared as follows:

1. Prepare DPBS Tris-EDTA buffer: according to manufacturer's instructions, dilute 10× DPBS (no calcium, no magnesium) stock with Tris-EDTA buffer solution according to manufacturer's instructions, containing 10 mM Tris-HCl, 1 mM disodium EDTA, pH 8.0.
2. Prepare 10 mM stock solution of N-acetyl-L-cysteine (NAC) in 1× DPBS. N-Acetyl-L-cysteine is a cytoprotective agent against reactive oxygen species.
3. Add NAC into 1× DPBS Tris-EDTA buffer (1:100).
4. Dilute (1:1000) Hoechst 33342 of commercial stock (1 mg/mL) with 1× DPBS Tris-EDTA buffer NAC solution.
5. Dilute (1:10) 10× trypsin LE media with 1× DPBS Tris-EDTA buffer NAC solution. Antibiotics can be added to this solution as well to prevent bacterial growth during incubations.

3.4 Solutions for Rat Pancreatic Islet Dissociation and Cell Labeling

1. One hour before the procedure is started, prepare 10 mL (per rat) of 1.4 mg mL^{-1} collagenase P solution dissolved in mGBSS. 5 mM d-glucose and 0.05%–1% (w/v) bovine serum albumin (BSA) can be added to collagenase P solution [40].
2. Prepare 5 mL (per rat) of the glycerol stabilization solution.
3. Dilute (1:1000) Hoechst 33342 commercial stock (1 mg/mL) with the glycerol stabilization solution. Keep solution at 15 °C.

3.5 Solution Preparation

1. Ammonium acetate sample rinsing media used after cell stabilization with glycerol: dissolve 150 mM ammonium acetate (NH_4Ac) in water [41].
2. Bovine serum albumin (BSA) solution used to coat surfaces coming in touch with cells during sample preparation: dissolve 1% bovine serum in milliQ water. Can be prepared and stored at 4 °C for 1–2 weeks.

3.6 ITO Glass Slide Preparation

1. Timing 2–5 min per slide.
2. Use multimeter to check the conductivity of the ITO glass surface to verify the choice of glass side and orientation of conductive surface, which is not obvious due to transparency. Note that the use of conductive glass slides is critical for analysis with TOF mass analyzers. Measurements via ion trap mass analyzers such as FTICR systems are less reliant on substrate conductivity, which allows the use of conventional microscope glass slides for sample deposition and ultimately lowers the sample preparation cost.
3. Use a diamond scriber or laser etcher to etch 1–3-mm-wide fiducial marks (can be crosses or digits), important sample information, and markups for positioning of standards and MALDI matrix blanks on conductive side of the slide. It is important to place markups on the same surface as the samples and take care in not damaging conductive surface in a way (e.g., complete circles) that would cause some regions of the slide to lose electric conductivity with other regions.
4. Wipe slides with Kimwipes, or use clean compressed nitrogen gas stream, to remove glass dust and small shards resulting from the markup process or from formation of stress cracks on the slide during etching.
5. Store slides in properly marked containers, e.g., microscopy glass storage boxes, protecting them from dust. Inscribed or etched slides can be stored at ambient conditions indefinitely.

   !CAUTION Ensuring the glass slide surface is free from chemical contamination should be verified using a randomly chosen slide from the manufacturer's box before working on sample preparation. Additional washing steps can be implemented for removal of possible contamination.

!CAUTION Wear gloves and work on a clean, dust-free surface to avoid ITO slide contamination and fingerprints.

!CAUTION Scribing and etching leads to a decrease in glass slide mechanical stability – handle finished slides with care.

3.7 Optical Microscopy

Optical microscopy is an important part of the described procedure. Two different systems can be used – a microscope (either inverted or upright) or histology/pathology slide scanners. These instruments should be spatially and optically calibrated/tested. Evaluation of a control/reference sample using a stage micrometer (e.g., Thorlabs part no. R1L3S1P) and fluorescent microscope slides (FSK1–FSK6) is critical for reliable and reproducible measurements. Finally, the entire workflow should be tested and optimized before starting the experiment. Here is a brief description of an Axio Imager.M2 microscope setup established using the ZEN software.

1. Turn fluorescent and transmission light sources ON.
2. Set 100% light after filter cube is going to camera.
3. Select None or DIC filter for brightfield imaging, and DAPI filter for fluorescence imaging.
4. Program sequential acquisitions of brightfield and fluorescence images.
5. Find optimal illumination for both channels.
6. Update method, saving experimentally determined settings.
7. Set acquisition of Tiled images using mosaic mode. The number of tiles, their positions, and overlap are determined at this step using the Advanced Setup option.
8. Due to the typically wide areas of samples analyzed, semiautomatic adjustment of focus should be done along the slide. Therefore, establish a set of reference points across the entire sample where the optical focus is automatically adjusted. Typically, we use 9–12 evenly distributed Focus Surface areas for automatic Z-position adjustment. It is important to determine which channel – brightfield or fluorescent – provides better options for surface focusing. Depending on the objective, magnification, and specimen, different outcomes can be achieved.
9. Under visual control, adjust and save Z-positions at each reference point.
10. Due to the large size of resulting image files (2–8 GB per tiled image), preplan and establish your image storage, processing, and export pipeline. Creating an easy-to-navigate directory structure and file naming system will save time by limiting the number of times images should be loaded for evaluation.

11. Incorporate tile stitching into the pipeline to reduce file sizes and increase the precision of cell coordinate finding.
12. Correct image brightness, contrast, and in some cases, color. This correction preferably should be done before image export into the microMS-compatible bigTIFF format. For example, due to differences in brightness of stained nuclei, increase the brightness and contrast for the fluorescent channel to levels that allow capture of the maximal number of cells, which is critical despite potential loss or distortion of morphological information. Finding of all cell coordinates is a paramount goal of this procedure. Missing low-intensity stained cells will lead to their exclusion from the distance filtering process and possible signal acquisition from more than single cell.
13. Save the method for saving tiled, stitched, and optimized images into the bigTIFF format.
14. Test method performance using small-size images.

    !CAUTION Resulting images should have minimal distortions in background to avoid formation of artifacts during image acquisition. The addition of a shading correction step at this point is important for tiled images to reduce the influence of optical path artifacts onto the final image, which may interfere with determination of localization of targeted structures or introduce faux structures.

3.8 Optical and Mass Spectrometric Quality Control of Experimental Solutions and Supplies

1. Use 20× objective and inverted microscope in fluorescent and transmission light modes to check for the presence of particles, including microbes and fungi, in solutions and on experimental surfaces. Dust particles may produce fluorescent signals similar to those from biological structures. Microorganism-related compounds can be detected by MS and assigned as being from sample cells.
2. If particulate matter is found in solutions, reagents, and/or supplies, filter or replace them and incorporate additional cleanup steps to ensure working surfaces remain dust- and contaminant-free (wet wiping, blowing with pressurized gas).
3. Run quality control experiments with all supplies and solutions for sample preparation as well as analyze reagents relevant to MS measurements. Create a library of signals detected. Contaminants, such as polyethylene glycols, are present in many scientific materials and reagents, and should be controlled for.

3.9 MALDI Matrix Application

MALDI matrix should be applied using optimal conditions for analyte extraction from the sample and mixing of analytes with MALDI matrix at the most effective ratio for the analyte class under investigation [42]. An excess or deficit of matrix can lead to poor analyte ionization, lower sensitivity of detection, and/or high noise in the spectra. Matrix type, solvent composition, duration of contact between the MALDI matrix solution and sample, and rate of matrix drying affect matrix crystal formation, which in turn influences signal quality during measurements. The choice of MALDI matrix, as well as its solvents and additives, is critical for detection of a particular class of analytes. For recommendations, search the literature (e.g., a search of https://www.sigmaaldrich.com for "MALDI matrix" produced 112 entries) or instrument vendor guidelines. There are several methods for applying MALDI matrix.

1. Aerospray option 1: manual wet application using an airbrush. Multiple parameters can be adjusted for deposition of MALDI matrix onto a sample using an airbrush, including distance between airbrush nozzle and sample, spraying time, concentration of MALDI matrix, ratio of organic and aqueous solvents in this mixture, nebulizing gas flow rate, as well as physicochemical properties of solvents, e.g., volatility. Additionally, the airbrush needle/nozzle size plays a critical role in the formation of droplets of different dimensions and, respectively, different drying times. Artistic airbrushes with 0.1–0.2 mm diameter needle sizes produce the finest mist for application of MALDI matrix solution; 0.3–0.35 mm needles generate predominantly medium-sized droplets. Airbrushes with 0.4–0.5 mm needles are good for quicker application of large volumes of matrix solution.
2. Aerospray option 2: automated wet application using a commercial device. The HTX-M5 sprayer is an advantageous tool for uniform application of MALDI matrix solution onto several ITO glass slides. As an example, effective matrix deposition can be achieved with 30 mg/ml DHB in 1:1 (v:v) methanol/0.01% TFA in water applied in two passes from a nozzle positioned at a distance of 6 cm from the sample, with a nozzle velocity of 1300 mm/min, temperature of 70 °C, MALDI matrix flow rate of 100 uL/min, and matrix track spacing of 2.5 mm.
3. Sublimation (dry application): One of the best approaches for dry, thin, and relatively uniform MALDI matrix application is via sublimation [43]. This approach has demonstrated excellent performance in lipid and metabolite analysis where analytes are located on sample surfaces, e.g., on a cellular cytoplasmic membrane. However, intracellular compounds are best extracted from cells by organic solvents and internalized within the MALDI matrix crystals. The matrix sublimation/recrystallization approach was developed to enhance detectability of larger molecules, such as peptides and proteins in an *m/z* range of 1000–18,000 [44]. To determine the deposited matrix amount and surface density of the MALDI matrix,

ITO glass slides are weighed on a micro- or ultra-microbalance before and after the sublimation process. For sublimation, 200–500 mg of MALDI matrix is evenly distributed on aluminum foil placed at the bottom of the sublimation chamber; an ITO glass slide with mounted samples is affixed to a cold finger using double-sided copper tape. The assembled apparatus is placed in a heating mantle, medium vacuum (~10 mTorr) is developed inside, the cold finger is filled with iced water (~4 °C), and the entire system is acclimated for 5 min. The mantle is then heated 5–10 min to achieve the targeted thickness of MALDI matrix coating. Sublimation is stopped by removal of the heat source and slow venting of the apparatus, its disassembly, and sample removal.

4. MALDI matrix application approaches can be compared [45] and parameters should be optimized, depending on the results of preliminary measurements involving several differently treated samples.

   !CAUTION MALDI matrix application steps should be performed in a well-ventilated space, such as a chemical hood, to avoid inhalation of MALDI matrix and solvent vapors and aerosols, which may have hazardous properties.

3.10 Vacuum-Assisted Liquid Aspiration Setup

1. Liquid aspiration system built in accordance with Biosafety Level 2 regulations [46].
2. The system is connected in a chain using appropriate diameter clear, flexible plastic tubing and two 50 mL heavy wall glass filtering flasks.
3. One end of the chain is attached by the tubing to an in-house vacuum system, followed by the filter, overflow flask with punched rubber stopper with glass or plastic tube inserted, aspiration flask with punched rubber stopper with glass or plastic tube inserted, and ending with tubing of appropriate length capped with a 200 μL universal plastic pipet tip.
4. Fill 10% of both flask volumes with disinfectant liquid, e.g., freshly prepared 10% bleach solution or different chaotropic agents.

   !CAUTION Filters should be replaced at least annually.

   !CAUTION Disinfectant solution should be replaced after each daily experimental session.

   !CAUTION Depending on the appropriate biological safety regulations, the entire system, except the filter, can be autoclaved or exposed to disinfectant according to regulations for the particular hazard and disinfection approach required.

3.11 Mass Spectrometer Setup

1. Determine the laser footprint area and its position during instrument training. The laser footprint position may shift over time due to different events, including ion source cleaning

on TOF-TOF instruments. Therefore, load an ITO glass slide coated with a thin but visible layer of MALDI matrix. Completely ablate several individual spots of MALDI matrix layer with the laser.

2. Measure spot sizes and determine the morphological parameters of the laser's footprint. Note the laser footprint position in relation to the software-generated crosshair used in sample navigation, visible on the computer monitor. If laser alignment is not possible moving forward, the experimentally determined laser position should be marked on the monitor and used in instrument training.
3. Choose a method for biological data acquisition using the mass spectrometer. Manufacturer-provided default methods can be used as the starting point for sample- and analyte-specific method development. High-throughput single cell MS analysis utilizing microMS can be performed on majority of mass spectrometers equipped with sufficiently precise programmable sample or microprobe positioning. For example, we utilized two Bruker instruments [33, 38]: (1) an ultrafleXtreme mass spectrometer, set at a *m/z* window of 400–3000. The "Ultra" (~100 μm footprint) laser setting was used with 300 laser shots at 1000 Hz to generate a MALDI-time-of-flight mass spectrum of each cell; (2) a 7 T solariX FTICR mass spectrometer, operated with a *m/z* window of 150–3000; 300 scans were collected with a 4-Mega Word time domain. Each mass spectrum was acquired with 20 laser shots at 1000 Hz and 60% laser energy.
4. Using prespotted analyte standards, calibrate the mass spectrometer.
5. Optimize the chosen method. The most useful customizations include adjustment of the laser intensity, laser footprint size and geometry, number of laser shots per acquisition, detector's gain (should be adjusted with care to ensure a longer detector lifespan), *m/z* range, time of delay of ion extraction (for TOF instruments), detection polarity, as well as ion optics voltages. Preparation of high-density cell population samples for realistic tuning of instrument parameters is optimal. Comparative analysis of the parameters of several default methods followed by optimization of the best-performing one may produce faster results.

   !CAUTION To avoid mass spectrometer damage or a reduction in operational longevity, most mass spectrometer parameters should be adjusted only by specifically trained personnel or by the instrument vendor. Most MS instrument vendors provide remote diagnostics and assistance, which is a suitable for method optimization, if needed.

3.12 microMS Software Installation and Execution

microMS can be installed on either Windows or Linux computers, as follows:

3.12.1 Install microMS on a Windows Computer

1. For the base components of microMS to run properly, the following software dependencies should be downloaded and installed, with links provided at http://neuroproteomics.scs.illinois.edu/microMS.htm: PyQt5, NumPy, SciPy, Matplotlib, Skimage, PySerial, OpenSlide, and OpenSlide Python. Most dependencies of microMS are included in standard distribution packages of Python 3. We have had success using anaconda, which also includes an IDE (https://www.continuum.io/downloads.) A Python 3.X (X > = 5) version must also be installed.
2. After download, PySerial and OpenSlide require separate installations. PySerial is installed by opening Windows PowerShell and entering *pip install pyserial*. This should automatically install the most recent version.
3. OpenSlide requires additional binaries, located at http://openslide.org/download/ under Windows Binaries. Download and extract the folders, placing them in your Documents or Program files directories. The contained bin folder also must be added to the operating system path (e.g., C:\Program Files\OpenSlide\bin\; see http://www.howtogeek.com/118594/how-to-edit-your-system-path-for-easy-command-line-access/). Once added correctly, the OpenSlide Python wrapper is simply installed by entering *pip install openslide-python* in PowerShell.
4. With the dependencies installed, microMS is installed by downloading from http://neuroproteomics.scs.illinois.edu/microMS.htm and extracting the ZIP file into a designated directory (extracted size is ~560 MB). After downloading all packages and unzipping the archive, no further installation of microMS is required. The main script is run with Python microMS.py in a command prompt in the script directory or by double-clicking the *microMS.bat* file, which runs the above command and waits for a key press.
5. To run the program, double click "microMS.exe"; two windows will be opened (Fig. 1).

3.12.2 microMS Installation and Execution on Linux Computers

1. With an installation of Python 3, the required packages are installed in the terminal with *pip3 install <package>*.
2. OpenSlide is installed with *apt-get install python-openslide pip3 install openslide-python*.
3. After unzipping the microMS file, the main method is run with either Python microMS.py or ./microMS.py after executable

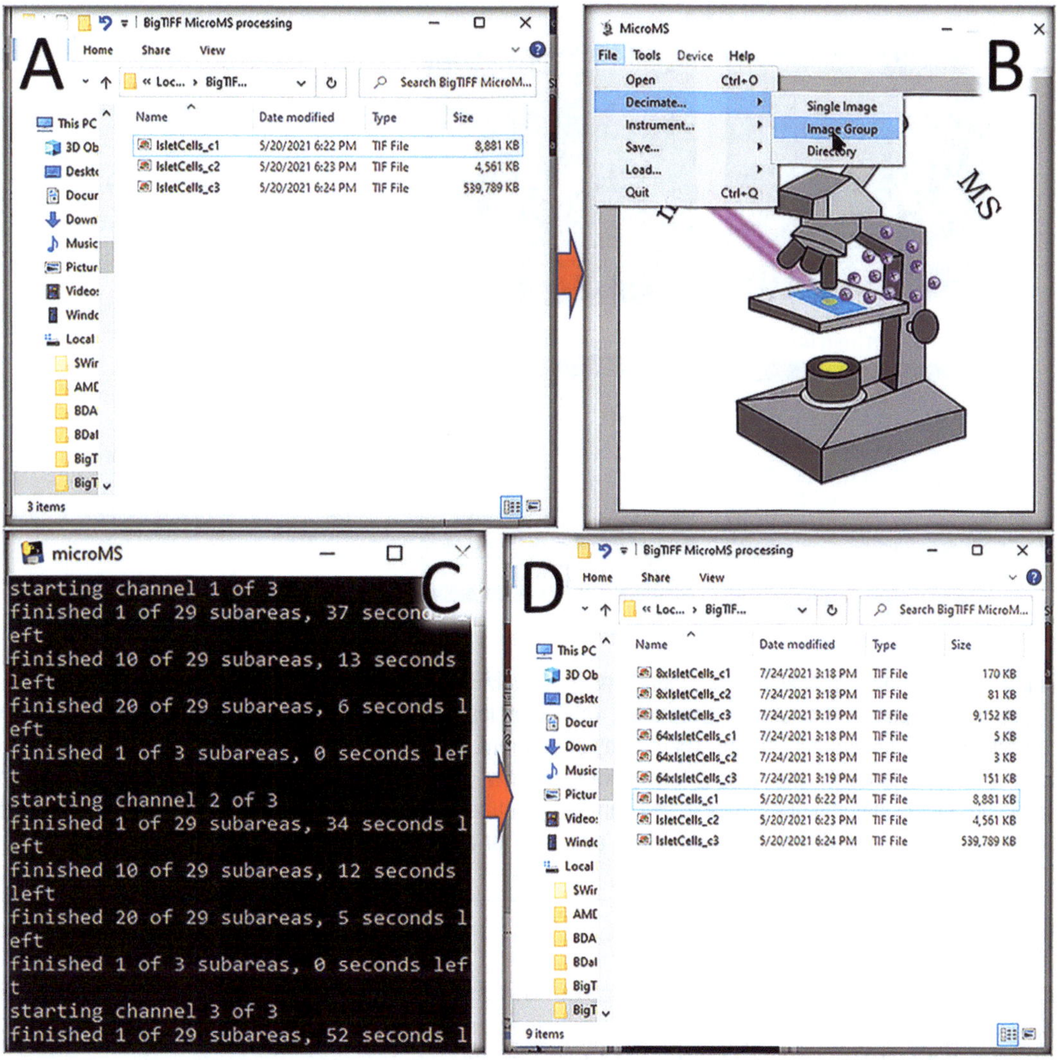

Fig. 1 BigTIFF images may be decimated to 8× and 64×, generating the file sets: "8× <image name>" and "64× <image name>," with <image name> being the name of original file (see Fig. 2). (**a**) Original bigTIFF files with names ending with optical channel abbreviations c1–c3 stored in a single directory. (**b**) Activation of bigTIFF decimation option for multichannel image processing. (**c**). bigTIFF decimation process visualized in second microMS window. (**d**) Original bigTIFF and decimated files

permission is granted (i.e., chmod +x microMS.py). Adding the microMS directory to the PATH variable will allow execution from any directory.

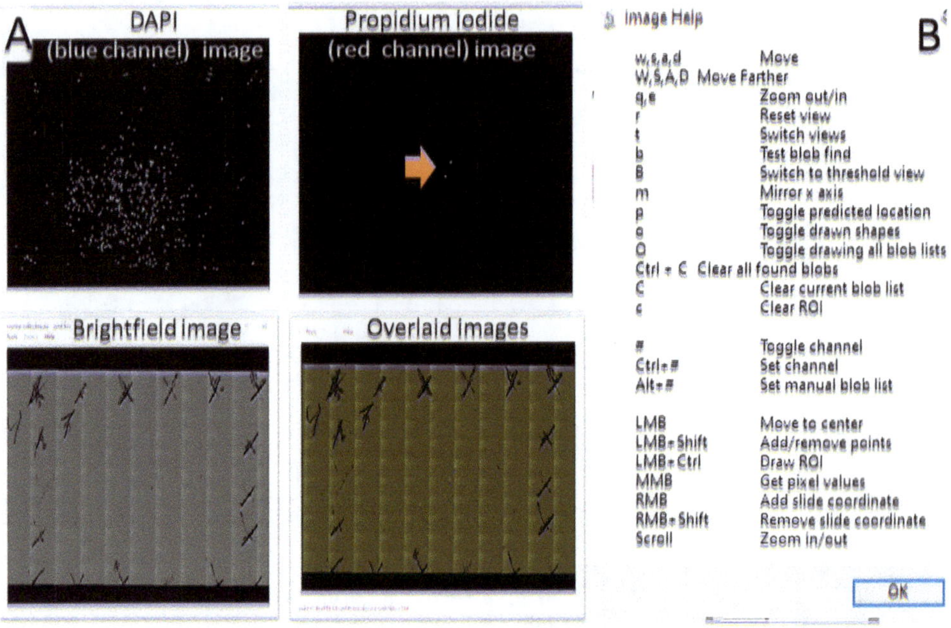

Fig. 2 Optical image processing and operations in microMS. A population of individual cells obtained from a single human pancreatic islet is examined. (**a**) Examples of optical image views available through the "switching image channels" option. The display of different image channels is controlled through the numeric keys and is cycled with the T or Z keys. Orange arrow points toward two cell-like structures exhibiting propidium iodide staining, indicative of cell damage. Such features are excluded from MALDI MS analysis. (**b**) Image help menu describing multiple options available for image processing

4 Procedures

4.1 Islet Dissociation: Human Single Cell Populations

Here we describe the steps required for single islet enzymatic treatment, dissociation, and deposition of populations of individual cells onto ITO glass slides.

1. Under stereomicroscope, manually collect 10–20 isolated human islets located in transportation media.
2. Transfer islets into a BSA-treated Petri dish filled with 4 mL of 1× DPBS Tris-EDTA buffer NAC solution and incubate for 10 min at 37 °C.
3. Transfer multiple islets into 1.5 mL tube filled with trypsin LE in 1× DPBS Tris-EDTA buffer NAC solution.
4. Incubate islets for 12 to 24 min at 37 °C.
5. Move all islets into a Petri dish coated with BSA and filled with ice-cold 1× DPBS Tris-EDTA buffer NAC solution. Keep dish on ice.

6. Move an individual islet into BSA-treated or low protein binding PCR tube filled with 50 μL of 1× DPBS Tris-EDTA buffer NAC solution.
7. Mechanically dissociate islet in a tube using a BSA-treated cut pipet tip attached to a 200 μL pipettor set to 50 μL.
8. Deposit three 500 μL droplets of 1× DPBS Tris-EDTA buffer NAC solution onto a new marked ITO glass slide. Make sure that the droplets remain separate.
9. Transfer solution with a cellular population formed during pancreatic islet dissociation into one of 1× DPBS Tris-EDTA buffer NAC solution droplets and agitate solution using careful pipettor-generated liquid motions. This is needed for more even distribution of cells on the ITO glass surface occupied by the droplet.
10. Wait for 15–20 min for cells to sediment and stick to the ITO glass slide's surface. Visual control under a microscope using a 10× objective is instrumental for determination of the best cell density and timing for the next step.
11. Fill a Pasteur pipet with ~1000 μL of the glycerol stabilization solution.
12. Replace surrounding cells 1× DPBS Tris EDTA buffer NAC solution with the glycerol stabilization solution by placing the inlet of vacuum suction system at one side of 1× DPBS Tris-EDTA buffer NAC solution droplet and adding the glycerol stabilization solution onto another side.
13. Continue to use vacuum suction system to remove the glycerol stabilization solution from the spot. Most of the solution should be efficiently removed and glycerol-stabilized cells stuck to the surface. Visible reduction in cell size can be observed due to the hyperosmotic nature of the glycerol stabilization solution. However, test measurements did not reveal the presence of endogenous peptides or lipids in the glycerol stabilization solution collected during this process. Additionally, no loss of fluorescent staining is observed. Different concentrations of up to 40% glycerol w/v are used in mammalian cell cryopreservation [47].
14. At this point, samples can be stored at 4 °C for several days, preferably in a low oxygen environment. Other storage approaches and durations should be tested experimentally for their effects on cell integrity and biochemical profiles.
15. Remove samples from storage and acclimate them to ambient conditions prior to optical imaging.
16. In most cases, some excess glycerol solution surrounding the cells remains. Glycerol solution works as stabilizer but may interfere with MALDI matrix deposition and crystallization

processes as well as influence the results of brightfield optical imaging. Therefore, remove this solution using one or two 15-s rinses with an appropriate solution, e.g., 150 mM ammonium acetate in water [41]. Remove liquid excess with a stream of nitrogen gas after each rinse. This improves the quality of extracellular solution removal.

17. If needed, add aliquots of analyte standards mixed with MALDI matrix solution onto marked spots on the ITO glass slide. These spots are important for the mass spectrometer calibration and batch mass spectra reprocessing conducted and described in Subheading 3.2.

4.2 Islet Isolation and Dissociation: Rat Single Cell Populations

Rodent islet isolation is broadly covered in the literature [40, 48, 49]. The following is a brief description of the procedure used in sample preparation for the outcomes described in Subheading 5, Anticipated Results:

1. Fill syringe with 5 mL (of the 10 mL) prepared solution containing collagenase P, BSA, and d-glucose. Preserve the remaining 5 mL of the solution.
2. Use a guillotine to decapitate Sprague-Dawley rats.
3. Surgically open the abdominal cavity.
4. Apply a clamp at the point where the common bile duct merges with the intestine, thus blocking liquid flow. Improper clamping will lead to incomplete fill of the pancreas with the digestion collagenase *P* solution.
5. Inject the pancreas with the collagenase *P* solution at the common bile duct *Y* junction, located between the liver lobes.
6. Dissect the pancreas and place into plastic tube with remaining 5 mL collagenase P solution.
7. Incubate the pancreatic tissues in a recirculating water bath for 20–30 min at 37 °C. A visually homogenous yellowish suspension should be formed. Taking several aliquots of suspension during this treatment for examination under microscope is helpful for determining the optimal duration of tissue exposure to enzyme. Incomplete digestion will result in formation of tissue pieces containing both acinar and islet cells. Over-digestion coincides with a loss of islet integrity.
8. Shake tube gently every 10 min of the incubation period.
9. Wash resulting suspension twice with ice-cold mGBSS.
10. Centrifuge the suspension for 3 min at 300 *g*.
11. Remove and discard supernatant and resuspend pellet in 10 mL of ice-cold mGBSS.
12. Place resulting suspension into a plastic Petri dish.

13. Place dish under inverted microscope and adjust illumination to achieve islet appearing as brownish structures and acinar tissue grayish or yellowish.
14. Under visual control using an inverted microscope, in within 10–20 min, manually isolate individual pancreatic islets using a micropipet. Typically, only 20–30 pancreatic islets collected from a single animal are needed for successful preparation of 10–15 single islet cell populations for microMS-based MALDI MS analysis. If longer time is needed, DNAse (3000 U/mL) should be added into the mGBSS where dissociation of pancreatic tissue is located to prevent clumping of acinar tissue pieces and islets [50].
15. Place a single islet into 20 μL glycerol stabilization solution containing Hoechst 33342.
16. Incubate for 2 h at 15 °C.
17. Triturate individual islets into single cells onto ITO-coated glass microscopy slides creating a low-density cellular population.
18. Repeat **steps 15–18**, described in Subheading 4.1.

4.3 Optical Imaging of Cellular Populations

1. Perform tiled image acquisition using a fluorescent microscope with sufficient magnification for the given cell type.
2. Each optical channel image should be preprocessed in ZEN for optimal brightness and contrast before further processing, including tile stitching and image export.
3. Images are processed using tile stitching algorithm available in ZEN.
4. Resulting files are exported using conversion to bigTIFF [51] format using ZEN software, blue edition.
5. For microMS, any optical image data processing and coordinate finding, OpenSlide-supported image file format [52, 53] is a suitable input; however, only Hamamatsu ndpi and bigTIFF images have been thoroughly tested and are accepted by default by microMS. For multichannel images, ndpi file names should end with "Brightfield" (for channel 1, brightfield) and "Triple" (for channel 2, RGB fluorescence). bigTIFF image file name prefixes should not start with "8×" or "64×." Multichannel bigTIFF image file names require the suffix "c#.tif" where "#" is a digit between 1 and 9, typically corresponding to an optical microscopy channel, e.g., brightfield at channel 1. ZEN creates a bigTIFF file for each optical channel and names resulting file with user-defined names appended by channel information, e.g., IsletCells_c1.tif and IsletCells_c2.tif for two channels.

4.4 Sample Preparation for MALDI MS Analysis

1. To achieve best results, immediately after image acquisition, coat ITO glass slides with cellular populations with MALDI matrix using the matrices, solvents, and matrix deposition methods that are optimal for the research goal. Due to the large variety of sampling approaches, study goals, and variability in the equipment used, an evaluation set of samples should be prepared and tested using several approaches as guided by published information.
2. For detection of peptides in single cells, apply 50 mg/mL DHB dissolved in 1:1 (v/v) LC-grade ethanol/water with 0.1% TFA using an automatic MALDI matrix sprayer.

   !CAUTION Fiducial marks should be free of a MALDI matrix layer because it may obstruct their visibility after loading into the mass spectrometer.
3. If quantitative analysis is planned, apply aliquots of a solution containing standards for targeted analytes on several marked spots of the MALDI matrix-coated ITO glass slide. The analytical standard solution may include MALDI matrix, or you can omit it if the solution solvents are capable of dissolving the deposited MALDI matrix layer. Choose standard deposition locations far from cell populations to avoid contamination. Also, the locations should be easy to find after sample loading into the mass spectrometer.
4. Generated samples can be stored at room temperature in a desiccator or N2 environment for 24 h. Storage at lower temperatures, e.g., 4 °C to −80 °C, helps to preserve biological samples for different MS applications [54, 55]. It is important to experimentally determine the optimal sample storage time and conditions for each sample type. Reduction of sample exposure to oxygen and excessive moisture, e.g., due to water condensation on cold specimens, can be critical for successful measurements.

4.5 Processing Optical Images with microMS: Finding Cell Coordinates and Output Filtering

Also, see the microMS manual: http://neuroproteomics.scs.illinois.edu/microMS.htm, for more details.

1. Start microMS. On startup, the main GUI window is displayed with the program icon and a second window for visualization of the operational steps and errors (Fig. 1b).
2. Opening an image in microMS can be done in several ways, depending on the image type, desired spatial resolution, and procedure goal. Images that do not need to be magnified during operation, e.g., images covering small areas, can be loaded unmodified by selecting Open in the File menu (or Ctrl + O). This will launch a file selection dialog that prompts the user to select the desired file, which will subsequently be loaded. Selecting any image from a multichannel-

data image set located in the same directory will open all channels with matching filenames, as described above. Multiple image channels (if they exist) are loaded at the same time and overlaid on top of each other without image flattening. However, by default, non-pyramidal bigTIFF images can only be zoomed out by 4×. bigTIFF images may be decimated to 8× and 64×, generating the sets of files "8 × <image name>" and "64 × <image name>" with <image name> being the name of original file (Fig. 1). This will enable further zoom levels from 1× to 1/256× (which is recommended).

(a) bigTIFF image sets are decimated by selecting the Decimate option in the File menu. Several options are available: both "Single Image" and "Image Group" options are for the selection of one bigTIFF image. However, when a single file of a multichannel-data image set (e.g., IsletCells_c1.tif) is selected with the "Image Group" option, all images (e.g., IsletCells_c2.tif) of the set will be decimated. The "Directory" option allows the selection of a single directory. All images contained in sub-directories of the selected directory will be decimated.

(b) Notice that since a single image (brightfield) was decimated, only the specified channel is displayed at higher zoom levels. After decimation is complete, the process does not need to be repeated the next time the image set is opened. Opening the single non-decimated image file through Open in the File menu will automatically load the full image stack.

3. Use the image navigation features of microMS to visually inspect loaded and displayed images. A list of mouse and keyboard controls are available through Help > Image Hotkeys (Fig. 2). Depending on the size of the image and computer performance, response time for these operations can be different. Working with image files located on network drives or USB2-connected devices may increase this time even further.

4. Perform previous step for all optical channels, switching between image features available in microMS. The same list of mouse and keyboard controls available through Help > Image hotkeys can be used. As an example, displaying different images for different optical channels is controlled through the numeric keys and is cycled with the T or Z keys (Fig. 2a).

5. Determine the sample area(s) to be processed with microMS. By default, the entire optical image area is subjected to processing. However, only a portion of the ITO glass slide is occupied by cell populations. Therefore, to reduce processing time, regions of interest (ROI) can be created on the analyzed images so that finding of cell-like structures will be restricted to

selected ROI. A single rectangular or polygonal ROI can be created by holding Ctrl + left mouse button (LMB) and dragging the cursor to form a yellow rectangle around the ROI. To create polygonal ROI, a rectangular ROI should be drawn, followed by its modification by (a) holding Ctrl + Alt, (b) placing cursor nearby one of yellow side lines, (c) dragging the cursor to a desired location leading to outlining of ROI with yellow, and (d) clicking of LMB to finalize the modification. Polygonal ROIs also can be formed by holding Ctrl + Shift and consecutively clicking LMB and dragging the cursor. The ROI can be removed by pressing the "c" key.

6. All structures detected on the images via user-defined parameter settings in microMS are called blobs (cells and other strictures). With optimal sample preparation, image acquisition, and microMS parameter settings, most blobs will represent individual cells. During the automatic blob finding process, each blob has the following parameters determined: an x-y coordinate, an effective radius, and the object's circularity. To start automatic blob finding, several blob parameters, such as minimum size (in pixels), maximum size, minimum circularity (shape), maximum circularity, threshold (lower threshold for acceptable pixel signal intensity), image channel, and color, should be imputed into Blob Find Entry window opened through Tools > Blob Options (Ctrl + B).
   (a) By default, image channel 1 corresponds to the brightfield image. The intensity thresholding is done on one of the RGB channels, allowing more specific and precise localization of the cellular features of a sample. Some of parameters can be left blank, e.g., maximum size and maximum circularity.
   (b) Activation of added parameters is done by clicking "Set Parameters." This will lead to automatic resizing of the current image to maximal magnification and markup of visible blobs detected according to the set parameters with green circles. This option represents a quick way to determine the efficiency of the selected parameters. Visualization of a different area of the image will lead to markup of visible blobs with characteristics according to the set parameters.

   !CAUTION Different strategies can be used to optimize the blob finding parameters, which are critical for determination of the coordinates for most of the individual cells. However, if the parameters are too strict, they may exclude a number of sample structures from the next steps, such as distance filtering. This may lead to mass spectra acquisition from several closely located structures due to the relatively large laser beam footprint (e.g., 100 μm). Therefore, it is important to obtain information on the localization of the maximum number of structures before distance filtering, followed by removal of size-incompatible features, e.g., large cell clusters or single nuclei released from damaged cells. This challenge can be addressed at the sample preparation stage when low-density populations of intact single cells are deposited on the ITO glass slide.

7. To find cell-like structures (i.e., blobs) (Fig. 3), use Tools > Blob Find to initiate finding according to set parameters. This process can take from several seconds to several minutes. At the end of the process, a window "Save All: Enter base filename" pops up. Enter desired file name to save information on found blob list in the .txt file located on the same directory where the original images are located.
8. Visually examine the detected blobs, which will be outlined by circles (Fig. 3b). For faster processing, only 150 blobs will be outlined in the field view. To mark all blobs, uncheck Tools > Limit Found Blobs. Alternatively, found blobs and other structures not selected as blobs but exhibiting above threshold signal intensities can be visualized as uniquely color-coded features on a dark blue background by pressing Shift + B, activating/deactivating threshold view. Pointing the cursor to a blob of interest and clicking on the middle mouse button will generate information on its area and circularity in the status bar. This information can be instrumental for optimization of the blob finding process.
9. Create an additional blob list if unsatisfactory blob finding occurs and if there is a need for comparison of different blob lists. Up to ten different blob lists can be saved and quickly compared. Alt + "the blob list number" will toggle different already populated or empty blob lists. Choose an empty blob list (0 blobs indicated in info line), change Blob Options, and repeat the blob finding process; a new blob list and corresponding file will be created and assigned chosen list number. This process will result in the appearance of a new set of marked blobs. Additionally, saved blob lists can be loaded through File > Load > Found Blobs. All created blob lists can be visualized using the key combination Shift + O. Overlapped blobs will be marked by circles of blended colors.
10. If needed, filter blob list using ROI filtering. A ROI can be created and blobs inside or outside of this ROI retained or removed. Tools > ROI Filter Retain and Tools > ROI Filter Remove are used in this procedure, which can be performed at any time after blob finding. The ROI Filter Retain operation removes all blobs outside of the specified ROI and moves the resulting list to the next open blob list. The ROI Filter Remove

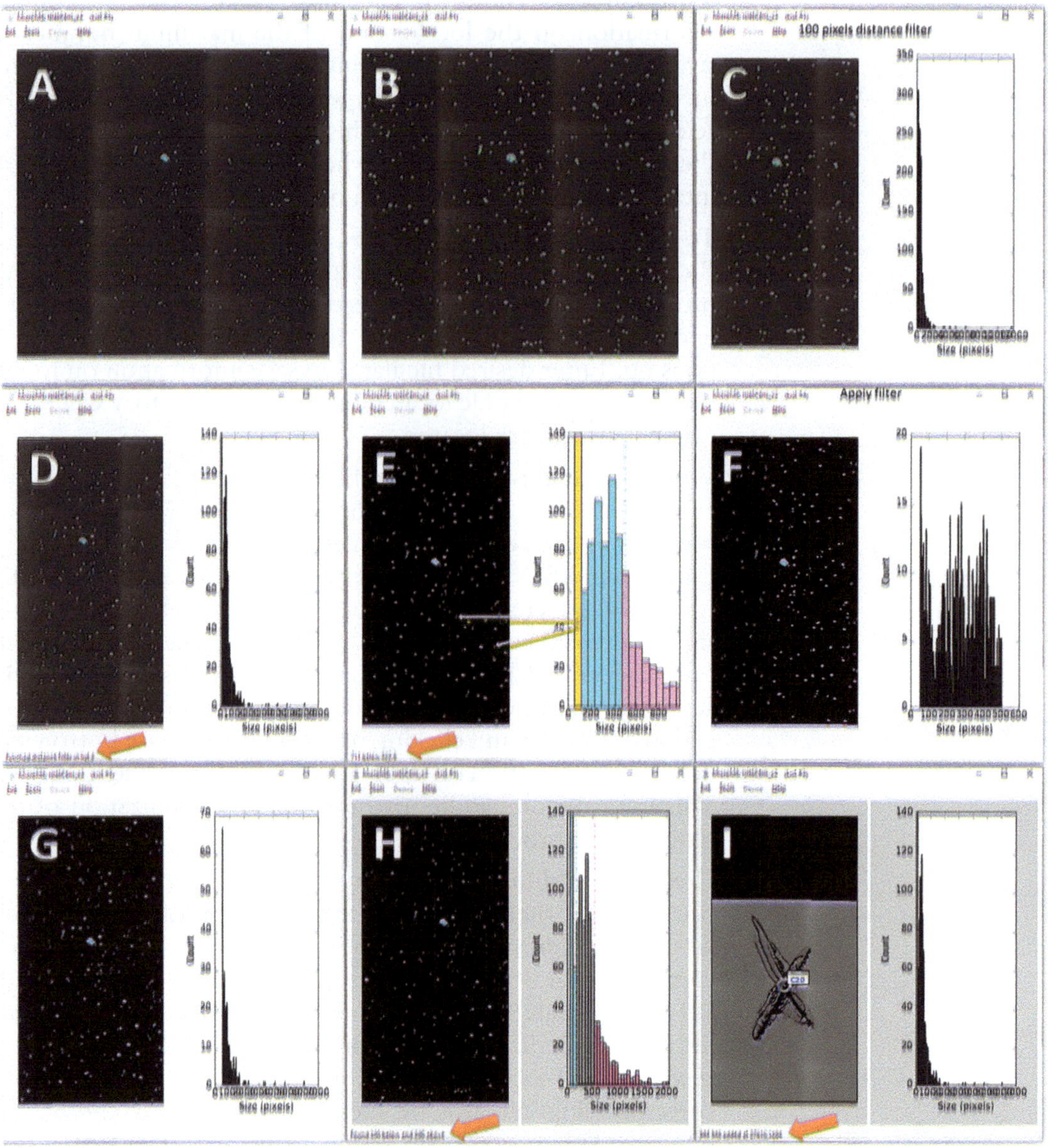

Fig. 3 Examples of major steps in blob finding and selection as well as defining of fiducial marks. (**a**) Original decimated three-channel overlay image of single human pancreatic islet cellular population. (**b**) Results of blob finding procedure with minimal restrictions performed on blue channel image. Blobs are marked with green circles. (**c**) Histogram window shows size distribution for all detected blobs. (**d**) Results of distance filtering, cutoff at 100 pixels. Blob list #2 is automatically created. Blue and orange arrows point to related displayed information. (**e**) Results of blob size filtering followed by histogram markup and filtering. Results of manual single bar markup are shown in yellow. Blue bars of the histogram relate to blue circled blobs on the image. The same is true for pink bars. Yellow lines point to two blobs corresponding to selected yellow bar. Command "Apply filter" leads to formation of two new blob lists (from **e**): (**f**) list #4 for blue bars corresponding cells and (**g**) list #5 for pink blobs. (**h**) Histogram and selected blobs for "Peak extremes" option of microMS; 200 blobs above and 200 blobs below the average are found and marked. (**i**) Example of manually added markup of a fiducial point from the brightfield image. Images of small areas of one single cell population are demonstrated. Orange arrows point to information line appearing after different calculation steps

places all blobs outside the ROI into a new list. In both cases, the original blob list is not modified and retains its number.

11. If needed, new blobs larger than 10 pixels are added to the current blob list with a Shift + left mouse button (LMB) click. Toggling of this key sequence will mark or unmark selected blob. Shift + LMB click and dragging of cursor allows the user to mark larger blobs. It is critical to manually create blobs at ITO glass slide locations where analyte standards are deposited. Mass spectra acquired from these sample areas can be used in batch mass spectra reprocessing during the data analysis phase detailed below.

12. If needed, perform filtering of blob lists using collected (during automatic blob finding) information on blob size, circularity, pairwise distance, and average signal intensity at chosen fluorescence channel. Pairwise distance filtering is a recommended first step, which allows detection and removal of blobs that are too closely located for independent analysis by MALDI MS. This and other filtering options are realized through the histogram interface of microMS. Tools > Histogram Window (Ctrl + H) opens the graphic window (Fig. 3c, d). Shortcuts for histogram navigation and function operations can be visualized through Help > Histogram Hotkeys. The histogram will be automatically updated after each operation, e.g., by switching to another blob list or executed filtering. Tools > Histogram Options leads to a pop-up window allowing a number of parameters for filtering to be selected, including image channel, blob parameter (optical channel color or morphology: size, circularity, or distance), offset (to the amount to increase (>0) or decrease (<0) the blob radius when examining the blob region), and level of optical signal intensity within the outline of the blob. To save histogram information, use File > Save > Histogram Image (.png file is generated) and File > Save > Histogram Values (tab delineated text files containing information on x-y coordinates and corresponding metric value for each blob are saved). microMS has a number of helpful functions for visual assessment of the results of histogram-based filtering, including visualization of blobs that correspond to a particular bar of the histogram using the middle mouse button (MMB) to click on a single bar (Fig. 3e). Vice versa, a LMB click on a blob will highlight the corresponding bar. Clicking capital C as well as switching to another blob list clears all histogram information. A right mouse button (RMB) click on the histogram sets the upper threshold for the filtered blob parameter (information on number of blobs above threshold is presented and responsible blobs are color marked), and LMB sets the lower threshold (information on number of blobs below threshold is presented and

responsible blobs are color marked). For automatic finding of peak extremes in a blob population, first select a range of bars in the histogram using Shift + LMB (lower cutoff) and Shift + RMB (upper cutoff) and then select Tools > Peak Extremes (Fig. 3h). A pop-up window will allow the user to set the number of blobs that will be searched for in high and low ranges selected for the blob population.

13. In instances where several filters need to be applied for even more selective defining of a targeted blob list, visually verified filtering results should be consecutively saved as new blob lists (Ctrl + A or Tools > Apply Filter) (Fig. 3f), or as new files containing blob lists (File > Save > Histogram Divisions).
14. If needed, utilize the blob patterning feature of microMS for analysis of large structures, such as multicellular pancreatic islet aggregates or even entire islets. This option allows the user to generate nine or more acquisition points per blob. Multipoint signal acquisitions can produce better results than single-point acquisition. Due to the focus of this protocol on single cell analysis, please refer to the online microMS manual for more details on this feature.
15. Save resulting blob lists using one of the following options: (a) File > Save > Current Blobs (only current blob list is saved), (b) File > Save >Histogram Divisions (saves information on histogram filtered low- and high-intensity populations as separate files), and (c) File > Save > All Lists of Blobs (all blob lists are saved in separate files). Editable text files will be created. Loading of saved blob lists using File > Load > Found Blobs will overwrite current information, which should be saved in advance if important.

4.6 High-Throughput Single Cell MALDI MS: microMS and Mass Spectrometer Information Co-registration

1. Take sample coated with MALDI matrix from storage and, if needed, acclimate to room temperature, preferably under vacuum.
2. Load acclimated specimens into a MALDI TOF-TOF mass spectrometer.
3. Using spots with calibrants, blank areas, and a test area with cells deposited, create and optimize instrumental methods appropriate for the experimental conditions and research goals (Fig. 4).
4. To co-register the optical imaging information and the mass spectrometer coordinates, run both microMS and the mass spectrometer operating software (e.g., Bruker flexControl).
5. Ablate a single spot on the MALDI matrix-covered area of the sample to visually determine the spot's location on a video output from the mass spectrometer. Mark this position for easy finding during coordinate system alignment. The laser

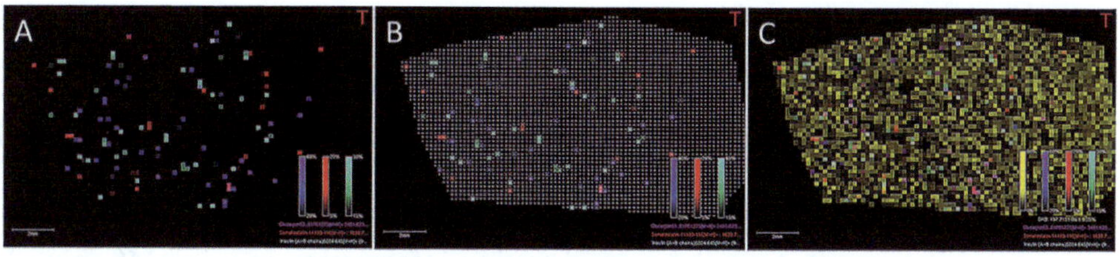

Fig. 4 Mass spectrometry imaging-based testing of a sample area prepared for microMS-assisted analysis. Cells are from a single human pancreatic islet. (**a**) Molecular image of distributions of three major pancreatic islet hormones – glucagon (pink), somatostatin (red), and insulin (green). (**b**) The same image as in (**a**) with the positions of laser sample probe shown. (**c**) The same as in (**a**) with distribution of MALDI matrix signal (yellow) added

beam footprint position may change over time due to mechanical misalignment that may occur during regular manual ion source cleanup. Even a 50 μm shift may lead to significant alteration of data output.

6. Load processed optical images for the sample under investigation into microMS. Only filtered blobs that have been targeted for analysis should be circled on the microMS displayed image.
7. Choose the analysis instrument in microMS by selecting File > Instrument. Several options are available by default, including the Bruker ultrafleXtreme MALDI TOF-TOF and solariX FTICR instruments.
8. Co-register the fiducial markers on the optical image displayed in microMS and the same physical markers visible on the video output from the mass spectrometer. Under visual control, align the sample plate position and the first fiducial mark on the mass spectrometer's computer monitor where the laser beam appears during acquisition. Find the same fiducial mark on the image displayed in microMS and click RMB (Fig. 3i). A pop-up window will appear asking the user to "Enter plate coordinate"; the x-y coordinates of the sample plate aligned in the mass spectrometer should be entered, e.g., 9494 3493, where the first number is the x-coordinate and the second is the y-coordinate, separated by a space. The actual sample plate coordinates can be found in several places within the instrument's software. For example, for the ultrafleXtreme, flexControl displays this information in the Sample control window. Copy and paste the displayed values into the pop-up window in microMS and click OK. A blue circle marked with the letter and number corresponding to the default plate spot numbering for a Bruker MTP glass slide II adapter will appear. Move to another fiducial mark in microMS, find the corresponding fiducial on the mass spectrometer, and repeat the process. More fiducial marks used for each ITO glass slide provides enhanced coordinate alignment. At times an individual fiducial

appears inconsistent with others; this is marked with a red circle to indicate the fiducial with the worst alignment compared to the other fiducial marks. If it appears there is an error in a specific mark, this can be deleted by pressing RMB + Shift. It is recommended to proceed to completion with the remaining fiducial marks.

9. Save the co-registration information in microMS using File > Save > Registration; a .msreg (mass spectral registration) file will be created containing information on the instrument name and fiducial set characteristics. The saved co-registration can be loaded through File > Load > Registration.
10. Save the fiducial positions using File > Save > Fiducial Positions, producing an instrument-specific .xeo file with information on fiducials. This file can be loaded into a mass spectrometer to determine the accuracy of the co-registration process.
11. Save the information on found and filtered, currently displayed blob list using File > Save > Instrument Positions forming an . xeo file. File is instrument-specific. During the save process, three prompts will appear. The first allows the user to specify file name and location. The second allows the user to set the maximum number of blob positions that will be saved. The third provides a choice between two options related to the order in which blobs will be analyzed with MS. microMS incorporates a simplified version of the traveling salesperson path (TSP) optimization. The TSP approach determines the minimum stage movements required to target all selected locations. Thus, choosing the TSP option significantly reduces the sample plate movement time needed for analysis of all blobs. However, there is a correlation between computer performance, number of blobs, and time needed to calculate the optimal path using the TSP algorithm. Considering that the steps of this protocol, starting with co-registration of fiducial marks, require mass spectrometer engagement, TSP calculations for very large blob lists may require more than the 3 min allocated for this operation. Therefore, finding the optimal number of blobs for such operation can be determined for your specific computer system. Splitting of large blob lists into several smaller ones by identifying ROI is one approach to reduce the sizes of blob lists without compromising the number of blobs analyzed. We have found that TSP is very useful for most measurements.
12. It is a good idea to repeat the measurement of a small portion each sample (Fig. 5). This test allows the user to determine if the MALDI matrix and spatially co-localized analytes are completely desorbed at each targeted cell. Incomplete use of

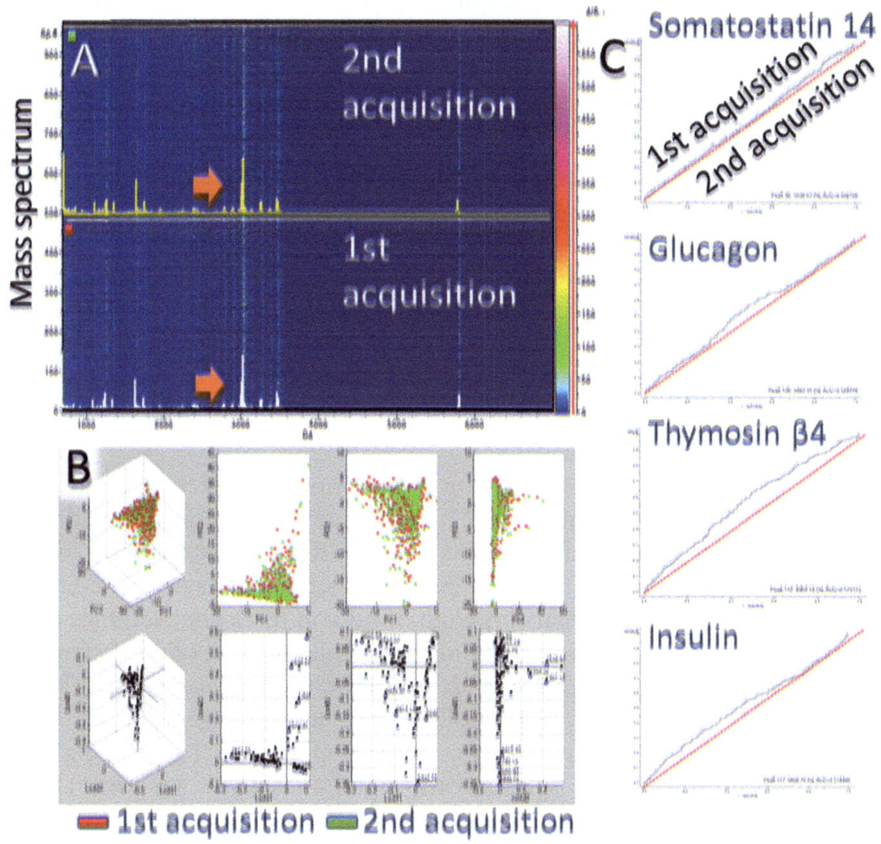

Fig. 5 Test for completeness of MALDI matrix ablation during microMS-assisted high-throughput single cell MALDI MS analysis. The same human pancreatic islet cells were analyzed twice. (**a**) Gel view of mass spectra acquired twice from 498 cells. Arrows point to average mass spectra for all cells acquired during corresponding set of acquisitions. The mass spectra are overplayed with gel views. (**b**) Results of PCA of acquired data sets. (**c**) Receiver operator characteristic curves for selected analytes detected in repeated measurements of the same sample

sample/matrix in a single run suggests that one is not producing the best possible signals. However, it also creates an opportunity to reanalyze the same cells using different instrument settings, including the use of MS2-based analyte sequencing.

4.7 High-Throughput Single Cell MALDI MS: Data Acquisition

Here we describe the procedures for a MALDI TOF-TOF mass spectrometer (we used the Bruker ultrafleXtreme mass spectrometer and software); similar procedures can be used for FTICR mass spectrometers (e.g., the Bruker solariX).

1. Copy microMS-generated fiducial positions and the instrument registration .xeo file into Data\Methods\GeometryFiles.
2. In flexControl, load the Method intended for analysis of microMS-defined blobs.

3. Optimize acquisition parameters and calibrate the mass spectrometer using the Method and analyte standards deposited on ITO glass slides.
4. Using flexControl, create an AutoXecute Method through the Method: Edit option in the AutoXecute window. Use the calibrated and optimized method during AutoXecute Method creation. The number of laser shots per acquisition and laser intensity used for an automatic run should be specified in the AutoXecute Method. Importantly, if batch reprocessing of acquired mass spectra are planned for the data analysis stage (discussed below), also include and specify standard calibration spots and flexAnalysis methods. These parameters are defined during the Run Editor operation described in the following steps.
5. Create a new run by selecting Run: New option in AutoXecute window. The AutoXecute Run Editor window will pop up. Choose File > New Run Wizard (Ctrl + Shift + N) to open the New Run: Target geometry page. Mark Start with wizard at this page. Find and mark the .xeo file. Hit Next; the laser power tuning page will appear. Follow instructions on the page and click Next. Calibration and sample spot selection page will appear. Follow instructions by pointing, holding, and dragging cursor from top left to bottom right corner of empty space on the right side of the window. This will select all microMS-determined blobs listed in the .xeo file. Hit Next. The Run Parameter page will appear where the AutoXecute Method, Data Destination: Data Directory, and Sample Name should be specified. Keep lengths of both to a reasonable minimum if peak list export into Excel spreadsheets is planned (described below in Data Analysis and Processing sections). More options are listed and can be used during the AutoXecute run. Importantly, flexAnalysis MS and MS/MS methods should be included in the AutoXecute run at this point to allow some types of future batch mass spectra reprocessing, if needed. (Consult the flexControl manual or Help for instructions.) Hit Next. A summary of the New Run Finish page will appear. Click Finish. This will close the New Run Wizard.
6. In the remaining open AutoXecute Run Editor, click, hold, and drag cursor, starting from upper left corner of unmarked empty space located on the left upper side of the window. Releasing the cursor at the right bottom corner will again mark all microMS-determined blobs, which were invisible before this selection step. Hitting OK opens the prompt for naming and saving the AutoXSequence file. Saving the file results in closure of the AutoXecute Run Editor and opens the window for

Messages. This usually informs the user about Factory teaching points; these cannot be loaded in case of custom teaching done through co-registration of image file in microMS and the video output of mass spectrometer. The Messages window can be closed.

7. Hit Start Automatic Run in the AutoXecute window. Automatic acquisition of data from all specified blobs will start. As a result, the data directory specified during the New Run Wizard procedures will be populated with folders containing individual mass spectra.

4.8 Data Processing and Analysis Using ClinProTools

ClinProTools is a user-friendly program with an integrated set of tools for biomarker discovery and clinical proteomics search. It supports MS data post-processing and statistical profiling workflow to detect biomarker patterns indicative of specific cell types. We find the program help files and user manual informative regarding the available analysis functions and guidelines on the choice of statistics. Included below is an example of a workflow for a two-group experimental comparison.

!CAUTION Mass accuracy of MS measurements depends on many factors [56–58]. Therefore, before analysis of each data set, determine if recalibration of the mass spectra is needed. The entire data set or just representative mass spectra should be loaded into ClinProTools (or flexAnalysis as described in the following section) and precision and accuracy of measurements determined. Both programs allow mass spectrum recalibration.

1. Split the entire data set into several new folders according to the desired comparison traits. For example, we split 485 mass spectra acquired from one pancreatic islet population into two folders.
2. Set ClinProTools Data Preparation > Settings Spectra Preparation parameters.
3. Set Data Preparation > Settings Peak Calculation parameters.
4. Sequentially load two sets of mass spectra from the two different folders into ClinProTools.
5. Using File > Save Class Paths, create an XLM file with information on the location of all individual mass spectra loaded for analysis.
6. Run Statistical Analysis > PCA. Confirm Scaling Method: Level. New window with Score plots and Loading plots will appear, and automatically selected peaks will be marked in the mass spectra window. There are two main outputs of the analysis. One is related to finding of characteristic profiles for

different mass spectral groups corresponding to different cellular populations. Loading plots allow the user to determine specific signals that contribute to profile differences among cellular phenotypes. The other is the ability to export calculated information on signal parameters into a table format, as described below.

7. Save results in an Excel-compatible XML3 format using File > Peak List Export option. This option also allows the user to export information on all selected peaks in several other formats. The XML3 format provides information on the class and spectra paths as well as peak area and intensity as two separate tables. Importantly, ClinProTools generates information on intensity and area for all mass spectra, including ones that have no peaks in the automatically selected *m/z* region. As a result, the generated tables have no missing values, which are typical for many types of mass spectrometric measurement outputs. This allows straightforward statistical analysis, including sorting of mass spectra lists by biomarker information and reloading of mass spectra sorted by groups into ClinProTools for more detailed statistical analyses.
8. To open the XML3 file in Excel, a style sheet reference for ClinProtPeakList.xsl should be found in C:\BDAL\ClinProTools_3_0\Files and copied to the same directory where the XML3 file with results is located. When opening the file, a pop-up with warning message about file format and extension will appear and should be ignored by clicking "Yes."
9. In most cases, signal intensity is one of the most common parameters in the statistical analysis of MALDI MS-acquired data. Therefore, choose Intensity table and perform data sorting according to chosen biomarkers.
10. Evaluate individual mass spectra to determine the minimal signal intensity for a particular biomarker and split the table into two sections. Perform data sorting for another biomarker in the remaining data set and repeat the steps above. As a result, the entire data set will be divided into a number of classes according to signal intensities of the chosen biomarkers.
11. Using the Excel CONCATENATE function, transform the resulting table into a single column table with information structured as follows:

```
<ClinProtSpectraImport Version="0.0">
 <Class Name="Class A">
  <Element Path="C:\A\0_M16_1SLin\fid"/>
  <Element Path="C:\A\0_M17_1SLin\fid"/>
  <Element Path="C:\A\0_M18_1SLin\fid"/>
 </Class>
 <Class Name="Class B">
  <Element Path="C:\B\0_M19_1SLin\fid"/>
  <Element Path="C:\B\0_M20_1SLin\fid"/>
 </Class>
</ClinProtSpectraImport>
```

where "Class A" can be replaced with custom name for the data class and information on data path, e.g., "Y:\microMS \Islet cells\1\0_x_17134y_16568\1\1SRef" enclosed in parenthesis.

12. Save resulting table in Extensible Markup Language (XML) file using Excel. Alternatively, the single column table can be copied into the free software Notepad++ (https://notepad-plus-plus.org/) for additional checking, formatting, and then saving as an XML file.
13. Open newly created XML file in ClinProTools through File > Open Spectral Import XML.
14. Perform new set of calculations, including PCA, pLSA, and unsupervised clustering.
15. Export results through Reports option.

4.9 Data Processing and Analysis Using flexAnalysis

One drawback of using ClinProTools for a large number of mass spectra is the need for significant data reduction, e.g., by a factor of 30–50. This leads to a noticeable reduction in mass resolution and the accuracy of *m/z* values determined for each of detected compounds. Therefore, complementary analysis of data sets can be performed in flexAnalysis.

1. Open flexAnalysis and load representative mass spectra for all predetermined classes.
2. Load mass spectra of analyte standards acquired during the same set of measurements.

3. Determine if mass spectra require recalibration. Sometimes mass spectra acquired at different ITO slide locations require recalibration, which can be done manually in flexAnalysis. Unfortunately, no batch recalibration option is available unless calibration mass spectra are included in the AutoXecute run sequence.
4. Using Method > Edit Processing Method (Active Settings), optimize Mass list and Processing parameters or use default settings.
5. Perform Mass List > Find operation resulting in peak finding, markup, and population of Mass List window with selected set of parameters. A variety of individual signal parameters can be chosen, including *m/z*, time-of-flight, signal intensity, signal-to-noise ratio, quality factor, resolution, signal area, relative intensity, full width at half maximum (FWHM), and chi-square value.
6. The Export\Mass List to Excel command creates an individual spreadsheet in a new Excel file for each reprocessed mass spectrum. Spreadsheets are automatically named with the directory name containing the mass spectrum. Due to restrictions on the length of spreadsheet names, some directory names are truncated. The process of spreadsheet creation will stop when a second identical name is generated. Therefore, it is strongly recommended to keep the length of data set names to a reasonable minimum. Additionally, with the availability of proper data converters allowing access to proprietary data formats, mass spectra generated by instruments of different vendors can be converted into general formats such as mzXML, which are readable by third-party software, e.g., MathWorks Mass Spectrometry and Bioanalytics toolbox [59, 60]. Bruker's CompassXport is one such converter.
7. Use the Process Menu > Batch process macro for reprocessing of multiple mass spectra. Mass spectra recalibration, baseline correction, smoothing, normalization, and peak picking are among the recommended reprocessing steps [61]. For large sets of mass spectra, a data reduction procedure can be implemented [62]. Mass spectra in a chosen folder, or those related to particular AutoXecute run, can be selected for reprocessing. Importantly, flexAnalysis MS and MS/MS methods should be included when creating the AutoXecute run described above. A variety of reprocessing functions can be performed in batch mode, including mass spectra smoothing, peak picking, and peak list export. flexAnalysis methods are editable through the Method > Edit Script option. Multiple processes can be included in a single flexAnalysis method.

8. Correlate ClinProTools- and flexAnalysis-generated peak lists and use the most accurate *m/z* values, which typically are generated by flexAnalysis.

4.10 Ways to Assign the Detected Signals

Typically, single cells contain insufficient material to allow structural characterization of multiple analytes using MALDI MS/MS. Correlation to CE-ESI MS/MS or LC-ESI MS/MS data on identical tissue or bulk cell suspensions often provides more information on the chemical identity of biological constituents. However, there is often a partial overlap between ions detected by MALDI MS and ESI MS, making direct comparison of data sets less straightforward. There are several approaches available to help address this challenge. One option is a combination of single cell MALDI MS data, published information on analyte presence in cells and related tissues, and identifications from the LC-MS/MS data.

Alternatively, putative assignment of detected peptides can be based on mass match to predicted peptides from known proteins available in different databases such as UniProt [63, 64] and the NCBI Protein Database [65]. In silico peptides should be calculated not only as typical monoprotonated $[M + H]^+$ ions but also frequently detected sodiated $[M + Na]^+$ and potassiated $[M + K]^+$ molecular ions typically detected by MALDI MS operating in positive polarity mode. Steps for using UniProt to create a list of putative ions for human islets of Langerhans, and NeuroPred, a useful tool for prediction of prohormone/secreted protein post-translational processing (http://stagbeetle.animal.uiuc.edu/cgi-bin/neuropred.py) [66], are provided below.

4.10.1 Using UniProt

1. Go to https://www.uniprot.org/ and initiate search using keywords related to annotation type and organism, e.g., annotation:(type:peptide) organism:human. (Our search resulted in 454 entries with known endogenous peptides.)
2. To visualize information and links on peptides, use the "Columns" option to add new non-default columns such as Peptide, Transit Peptide, Modified residue, Post-translational modification, and Disulfide bond.
3. Mark and copy entire area containing the search results and paste it into an Excel spreadsheet. This information can be also saved into a variety of formats (including Excel) via the "Download" option. However, links and particularly useful formatting in sub-tables will be lost.
4. Unmerge all columns in Excel and expand information on Entry, Entry Name, Gene Name, and Protein Name specific to each peptide to newly formed cells in corresponding columns.

5. Align in the same rows, information on Modified Residues and peptides that have them. New columns may need to be added because multiple modified residues are often present in the same peptide.
6. Add an additional column for peptide sequences, e.g., "Peptide Sequence," which can be found by clicking on corresponding links in the "Position(s)column. A new web browser BLAST window will open, presenting information on the peptide sequence.
7. Copy and paste this information into new "Peptide Sequence" column.
8. When the "Peptide Sequence" column is populated, add new column "Peptide Monoisotopic Mass."
9. Download the NIST Mass and Fragment Calculator from https://www.nist.gov/services-resources/software/nist-mass-and-fragment-calculator-software.
10. Copy and paste entire "Peptide Sequence" column into Sequence window of the Calculator. Select "Monoisotopic mass of peptide sequence" and click "Formula." A table containing the Peptide formula, its monoisotopic mass, and sequence will appear in lower window.
11. Copy and incorporate this table into the Excel spreadsheet.
12. Create a list of putative ions by adding to or subtracting from peptide monoisotopic mass values associated with PTMs specified in the table, as well as proton, sodium, and potassium adducts. Information on monoisotopic masses for these entries can be found at several locations including https://www.uniprot.org/docs/ptmlist and https://fiehnlab.ucdavis.edu/staff/kind/metabolomics/ms-adduct-calculator/. This procedure will create at least three *m/z* entries for each peptide. Pay attention to peptides that have two chains linked by disulfide bonds, e.g., insulin. Only separate chains are listed in the Peptide tables. Therefore, their combined monoisotopic masses should be calculated and corresponding PTM and ion-forming modifications calculated.
13. Create a combined table of putative and measured peptides. Sort all entries and find close to identical *m/z* values.
14. If all ions cannot be identified using UniProt, test the information on identified peptides using the peptide mass fingerprinting [66–68] approach [64, 65, 67] for protein identification. Several open-source modeling tools are available on the ProteinProspector portal https://prospector.ucsf.edu.

4.10.2 Using NeuroPred

The basic steps are outlined below, with detailed instructions available on the NeuroPred site. There is no batch processing option in NeuroPred, and predictions can be made for individual prohormones.

1. Use the FASTA sequence format to paste the protein of interest into The Sequence Submission Box of NeuroPred http://stagbeetle.animal.uiuc.edu/neuropred/neuropredinput.html#sequenceformat).
2. Fill the prediction parameters box as desired (include signal peptide length, PTMs of interest, etc.). Importantly, select a relevant animal model for your search before submitting the search. In our case, we selected the Mammalian prediction model.
3. NeuroPred will generate a list of theoretically possible peptides resulting from endogenous enzymatic prohormone processing according to known cleavage rules observed in mammals. Note that resulting list of putative processed peptides needs additional sorting upon export in Excel to remove peptides with untrimmed basic residues and products of missed cleavage (if this option was enabled during setup).
4. Combine all tables, perform additional data sorting, and perform additional search.
5. Ambiguous mass matches to predicted peptides can be validated by modeling the isotopic cluster of predicted sequence against experimentally observed peptide isotopic envelop pattern. In flexAnalysis, this can be achieved by using the SmartFormula function.

5 Anticipated Results

microMS-supported high-throughput single cell MALDI MS provides individual profiles on populations of thousands of cells. Because of the nature of MALDI MS and the low amounts of material within a cell, only the most abundant analytes are detected, while assessment of large populations enables statistical comparisons and insights into cellular heterogeneity. For example, within the populations of individual cells from rodent pancreatic islets, peptide profiles representing the expected cell types (α, β, δ, and PP cells) containing the canonical islet peptides are detected (Fig. 6a) [32]. The approach described here allows detection of more than 300 molecular futures in the 1000–7000 molecular mass range among individual cells. With optimal sample preparation and tuning of the MALDI TOF-TOF mass spectrometer, even a single cell sample can generate mass spectra of higher molecular weight peptides at >8 kDa with a mass resolution and accuracy from sub

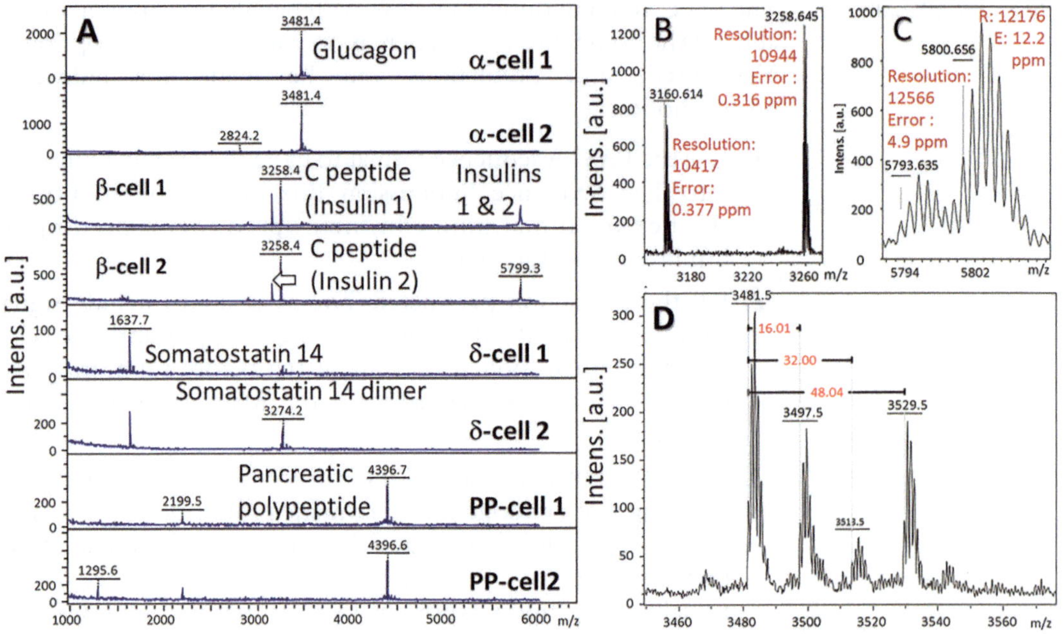

Fig. 6 Results of microMS-assisted MALDI MS analysis of a single rat pancreatic islet cell population. (**a**) Examples of mass spectra acquired from eight cells representing main cellular types in the pancreatic islets. (**b**) Mass spectrum demonstrates insulin 1C and insulin 2C peptide signals with mass resolution and error indicated. (**c**) Mass spectrum demonstrating insulin 1 and insulin 2 signals with mass resolution and error indicated. (**d**) Mass spectrum demonstrating putative multiple oxidation of glucagon. Intens. – relative intensity of signals. (a.u.) – arbitrary units. "Error" in ppm is a deviation between the measured mass and the theoretical mass of the selected peak. "Error" is calculated using "SmartFormula Manually" option in flexAnalysis. "Resolution" is mass resolution calculated as part of "Mass list" parameters in flexAnalysis

ppm to single digits (Fig. 6b, c). However, the relative intensity of signals varies between cells. A number of rare observations have included putative multiple oxidation of glucagon (Fig. 6d) and dimerization of somatostatin 14 (Fig. 6a). Both biological and experimental reasons for signal variability should be investigated further.

As mentioned above, image-guided single cell MALDI MS analysis is capable of providing information on the chemical heterogeneity of cellular populations. Proper data processing and analysis are critical for this evaluation. Dispersing a cellular population on an ITO glass surface leads to loss of information on tissue localization, e.g., known localization of β-cells in the islet core surrounded by inhomogeneous layers of α, δ, and PP cells [69]. Therefore, detected chemical biomarkers are becoming major guiding factors for cell-type determination. Importantly, other approaches for cell-type determination before or after MALDI MS analysis can be used in conjunction with microMS, including immunostaining [36]. Such biomarkers can be predicted using published information or direct data set analysis. As an example, a data set acquired

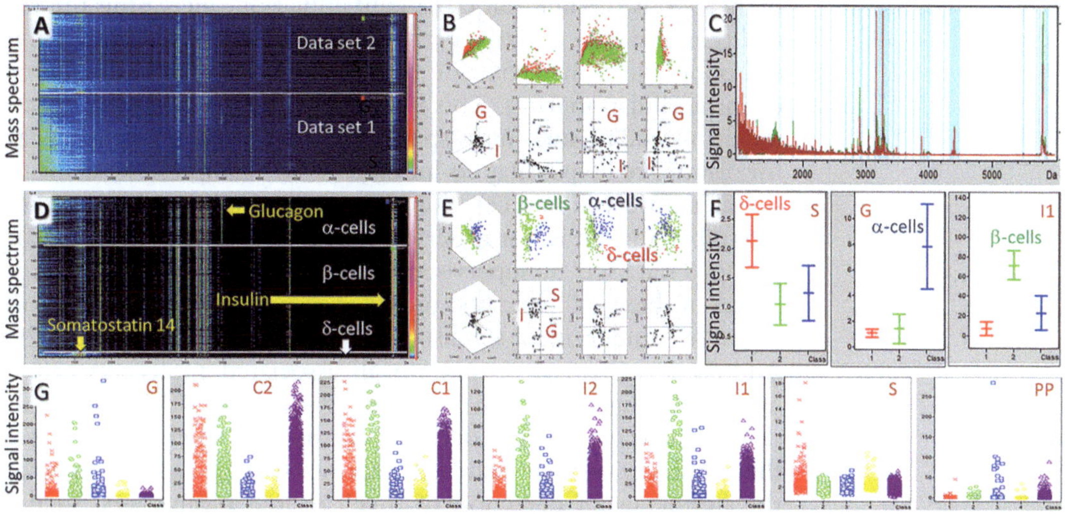

Fig. 7 Chemical cellular heterogeneity of rat pancreatic islets via microMS-assisted MALDI MS analysis. (**a**) Gel view of mass spectra acquired from cellular population of a single pancreatic islet divided into two data sets. (**b**) PCA of the data sets with loading plots demonstrating co-localization of signals related to different cell types, e.g., insulin (I)- and glucagon (G)-expressing cells. (**c**) Average mass spectrum for both data sets with signals automatically picked by ClinProTools (blue lines). (**d**) Gel view of filtered and grouped by listed above biomarkers and somatostatin 14 (S) entire set of mass spectra. (**e**) Results of PCA of three data sets formed by biomarker-guided filtering and grouping of entire set of mass spectra. (**f**) Histograms of relative intensities of somatostatin 14 (class 1) insulins 1 and 2 (class 2), and glucagon (class 3) signals detected in cells of different types. (**g**) Distributions of relative signal intensities at *m/z* areas of somatostatin 14 (S), glucagon (G), insulin 1 C-peptide (C1), insulin 2 C-peptide (C2), pancreatic polypeptide (PP), insulin 1 (I1), and insulin 2 (I2). These distributions were determined during ClinProTools analysis of mass spectra of cellular populations obtained from five individual rat pancreatic islets (marked on x-axis as class 1–5)

from a cellular population of a single rat pancreatic islet was split into two and evaluated using unsupervised PCA (Fig. 7a–c). Related loading plots revealed compounds that affected the diversity of cellular populations most with Load2 vs. Load3 exhibited two major influences – signals for glucagon and insulin – which were accompanied by other related compounds (Fig. 7b). As expected, score plots did show minor differences between the two populations. Such differences can be important to determine sample inhomogeneity related to experimental or biological factors. Influences of experimental factors, such as gradients in the MALDI matrix layer thickness or localized sample contamination, have to be evaluated to exclude their detrimental effects on measurement results. However, this approach needs further development in terms of cell-to-cell reproducibility (which may be biology-, sample preparation-, or MS measurement-related). Nevertheless, the molecular signatures enable classification of the entire data set into groups related to known cell types (Fig. 7d). This step helps not only to group cells but also to remove cells that exhibit higher levels of chemical cross-contamination, increasing the reliability of the data analysis. Perhaps dissociation of islets into single cell populations may coincide with loss of cell integrity and release of intracellular analytes. This is noticed when mixed mass spectra are acquired. In such mass spectra, characteristic molecular peaks for distinct cell types may co-localize. As an example, a weak insulin signal is observed in α cells (orders of magnitude weaker than in β cells), but insulin is not observed in δ cells (Fig. 7d). Reanalyzing grouped data sets by supervised PCA demonstrated expected separation of three cell types and provided more information on association of unidentified signals with known cell types (Fig. 7e). Additional statistical output on all associated molecules can be generated (Fig. 7f). Extension of image-guided MALDI MS analysis to cellular populations of different pancreatic islets demonstrates the variability of their cellular composition (Fig. 7g). As expected, the frequency of detection of β cell-related biomarkers, such as insulins 1 and 2, as well as their C peptides, is highest in all five examined islets. However, only one islet demonstrated cells with strong somatostatin 14 signals, and another islet exhibited a large number of insulin-producing cells but a small number of glucagon-expressing cells. Ratios of relative intensities for insulin 1/insulin 1C peptide and insulin 2/insulin 2C peptide are similar for each of the examined pancreatic islets, which is indicative of good repeatability of measurements. These results corroborate well with previous findings on the cellular heterogeneity of pancreatic islets [69, 70]. We expect similar results are possible on many other cell types.

Acknowledgments

The authors gratefully acknowledge support from the American Diabetes Association Pathway award number 1-18-VSN-19, the National Institute on Drug Abuse award number P30 DA018310, and the National Institute of Mental Health, award number 1 U01 MH109062. microMS was designed and programmed by Dr. Troy J. Comi. Dr. Elizabeth K. Neumann and Dr. Thanh D. Do made significant contributions in the development and application of microMS. Dr. Ali Naji and Dr. Chengyang Liu (Division of Transplantation, Department of Surgery, Hospital of the University of Pennsylvania, Philadelphia, PA) provided live isolated human islets for this project. We thank Dan Castro for valuable comments.

References

1. Hooke R (1665) Micrographia or, some physiological descriptions of minute bodies made by magnifying glasses: with observations and inquiries thereupon. J. Martyn and J. Allestry, London
2. Comi TJ, Do TD, Rubakhin SS et al (2017) Categorizing cells on the basis of their chemical profiles: progress in single-cell mass Spectrometry. J Am Chem Soc 139(11):3920–3929. https://doi.org/10.1021/jacs.6b12822
3. Neumann EK, Do TD, Comi TJ et al (2019) Exploring the fundamental structures of life: non-targeted, chemical analysis of single cells and subcellular structures. Angew Chem Int Ed Engl 58(28):9348–9364. https://doi.org/10.1002/anie.201811951
4. Dey S (2018) Hands-on image processing with python: expert techniques for advanced image analysis and effective interpretation of image data. Packt Publishing Ltd, Birmingham
5. Prasad S, Chanussot J (eds) (2020) Hyperspectral image analysis: advances in machine learning and signal processing. Springer International Publishing, Switzerland AG. https://doi.org/10.1007/978-3-030-38617-7
6. Umbaugh SE (2017) Digital image processing and analysis: applications with MATLAB and CVIPtools. CRC Press, Boca Raton
7. Klaus DT (2017) Guide to medical image analysis: methods and algorithms. Springer, London. https://doi.org/10.1007/978-1-4471-7320-5
8. Fu AY, Spence C, Scherer A et al (1999) A microfabricated fluorescence-activated cell sorter. Nat Biotechnol 17(11):1109–1111. https://doi.org/10.1038/15095
9. Hawley TS, Hawley RG (eds) (2017) Flow cytometry protocols, 4th edn. Springer, New York
10. Blasco J, Cubero S, Gomez-Sanchis J et al (2009) Development of a machine for the automatic sorting of pomegranate (Punica granatum) arils based on computer vision. J Food Eng 90(1):27–34. https://doi.org/10.1016/j.jfoodeng.2008.05.035
11. Cheng Y, Jafari MA (2008) Vision-based online process control in manufacturing applications. IEEE Trans Autom Sci Eng 5(1):140–153. https://doi.org/10.1109/Tase.2007.912058
12. Huang SH, Pan YC (2015) Automated visual inspection in the semiconductor industry: a survey. Comput Ind 66:1–10. https://doi.org/10.1016/j.compind.2014.10.006
13. Jardini AL, Larosa MA, Maciel R et al (2014) Cranial reconstruction: 3D biomodel and custom-built implant created using additive manufacturing. J Cranio Maxill Surg 42(8):1877–1884. https://doi.org/10.1016/j.jcms.2014.07.006
14. Jannin P, Fitzpatrick JM, Hawkes DJ et al (2002) Validation of medical image processing in image-guided therapy. Ieee Trans Med Imaging 21(12):1445–1449. https://doi.org/10.1109/Tmi.2002.806568
15. Zhang QC, Yang LT, Chen ZK et al (2018) A survey on deep learning for big data. Inform Fusion 42:146–157. https://doi.org/10.1016/j.inffus.2017.10.006
16. Urban PL, Jefimovs K, Amantonico A et al (2010) High-density micro-arrays for mass spectrometry. Lab Chip 10(23):3206–3209. https://doi.org/10.1039/c0lc00211a
17. Ibanez AJ, Svatos A (2020) Applications of MicroArrays for Mass Spectrometry (MAMS) in single-cell metabolomics. Methods Mol Biol 2064:73–88. https://doi.org/10.1007/978-1-4939-9831-9_6
18. Yang T, Gao D, Jin F et al (2016) Surface-printed microdot array chips coupled with matrix-assisted laser desorption/ionization mass spectrometry for high-throughput single-cell patterning and phospholipid analysis. Rapid Commun Mass Spectrom 30(Suppl 1):73–79. https://doi.org/10.1002/rcm.7628
19. Zimmermann M, Escrig S, Hubschmann T et al (2015) Phenotypic heterogeneity in metabolic traits among single cells of a rare bacterial species in its natural environment quantified with a combination of flow cell sorting and NanoSIMS. Front Microbiol 6:243. https://doi.org/10.3389/fmicb.2015.00243
20. Zimmerman TA, Rubakhin SS, Sweedler JV (2011) MALDI mass spectrometry imaging of neuronal cell cultures. J Am Soc Mass Spectrom 22(5):828–836. https://doi.org/10.1007/s13361-011-0111-2
21. Schober Y, Guenther S, Spengler B et al (2012) Single cell matrix-assisted laser desorption/ionization mass spectrometry imaging. Anal Chem 84(15):6293–6297. https://doi.org/10.1021/ac301337h
22. Spitzer MH, Nolan GP (2016) Mass cytometry: single cells, many features. Cell 165(4):780–791
23. Wu H, Volponi JV, Oliver AE et al (2011) In vivo lipidomics using single-cell Raman spectroscopy. Proc Natl Acad Sci U S A 108(9):3809–3814

24. Baker MJ, Trevisan J, Bassan P et al (2014) Using Fourier transform IR spectroscopy to analyze biological materials. Nat Protoc 9(8): 1771
25. Bolognesi MM, Manzoni M, Scalia CR et al (2017) Multiplex staining by sequential immunostaining and antibody removal on routine tissue sections. J Histochem Cytochem 65(8): 431–444. https://doi.org/10.1369/0022155417719419
26. Dixon AR, Bathany C, Tsuei M et al (2015) Recent developments in multiplexing techniques for immunohistochemistry. Expert Rev Mol Diagn 15(9):1171–1186. https://doi.org/10.1586/14737159.2015.1069182
27. Lanni EJ, Rubakhin SS, Sweedler JV (2012) Mass spectrometry imaging and profiling of single cells. J Proteome 75(16):5036–5051. https://doi.org/10.1016/j.jprot.2012.03.017
28. Li L, Garden RW, Sweedler JV (2000) Single-cell MALDI: a new tool for direct peptide profiling. Trends Biotechnol 18(4):151–160. https://doi.org/10.1016/s0167-7799(00)01427-x
29. Chen R, Li L (2010) Mass spectral imaging and profiling of neuropeptides at the organ and cellular domains. Anal Bioanal Chem 397(8): 3185–3193. https://doi.org/10.1007/s00216-010-3723-7
30. Ong TH, Kissick DJ, Jansson ET et al (2015) Classification of large cellular populations and discovery of rare cells using single cell matrix-assisted laser desorption/ionization time-of-flight mass Spectrometry. Anal Chem 87(14): 7036–7042. https://doi.org/10.1021/acs.analchem.5b01557
31. Do TD, Comi TJ, Dunham SJ et al (2017) Single cell profiling using ionic liquid matrix-enhanced secondary ion mass Spectrometry for neuronal cell type differentiation. Anal Chem 89(5):3078–3086. https://doi.org/10.1021/acs.analchem.6b04819
32. Jansson ET, Comi TJ, Rubakhin SS et al (2016) Single cell peptide heterogeneity of rat islets of Langerhans. ACS Chem Biol 11(9): 2588–2595. https://doi.org/10.1021/acschembio.6b00602
33. Comi TJ, Neumann EK, Do TD et al (2017) microMS: a python platform for image-guided mass Spectrometry profiling. J Am Soc Mass Spectrom 28(9):1919–1928. https://doi.org/10.1007/s13361-017-1704-1
34. Comi TJ, Makurath MA, Philip MC et al (2017) MALDI MS guided liquid microjunction extraction for capillary electrophoresis-electrospray ionization MS analysis of single pancreatic islet cells. Anal Chem 89(14): 7765–7772. https://doi.org/10.1021/acs.analchem.7b01782
35. Choe K, Xue P, Zhao H et al (2021) macroMS: image-guided analysis of random objects by matrix-assisted laser desorption/ionization time-of-flight mass spectrometry. J Am Soc Mass Spectrom 32(5):1180–1188. https://doi.org/10.1021/jasms.1c00013
36. Neumann EK, Comi TJ, Rubakhin SS et al (2019) Lipid heterogeneity between astrocytes and neurons revealed by single-cell MALDI-MS combined with Immunocytochemical classification. Angew Chem Int Ed Engl 58(18): 5910–5914. https://doi.org/10.1002/anie.201812892
37. DiPeso L, Ji DX, Vance RE et al (2017) Cell death and cell lysis are separable events during pyroptosis. Cell Death Discov 3. https://doi.org/10.1038/cddiscovery.2017.70
38. Do TD, Ellis JF, Neumann EK et al (2018) Optically guided single cell mass Spectrometry of rat dorsal root ganglia to profile lipids, peptides and proteins. ChemPhysChem 19(10): 1180–1191. https://doi.org/10.1002/cphc.201701364
39. Castro DC, Xie YR, Rubakhin SS et al (2021) Image-guided MALDI mass Spectrometry for high-throughput single-organelle characterization. Nat Methods 18:1233–1238
40. Villarreal D, Pradhan G, Wu CS et al (2019) A simple high efficiency protocol for pancreatic islet isolation from mice. J Vis Exp 150. https://doi.org/10.3791/57048
41. Berman ES, Fortson SL, Checchi KD et al (2008) Preparation of single cells for imaging/profiling mass spectrometry. J Am Soc Mass Spectrom 19(8):1230–1236. https://doi.org/10.1016/j.jasms.2008.05.006
42. Batoy SMAB, Akhmetova E, Miladinovic S et al (2008) Developments in MALDI mass spectrometry: the quest for the perfect matrix. Appl Spectrosc Rev 43(6):485–550. https://doi.org/10.1080/05704920802108198
43. Hankin JA, Barkley RM, Murphy RC (2007) Sublimation as a method of matrix application for mass spectrometric imaging. J Am Soc Mass Spectrom 18(9):1646–1652. https://doi.org/10.1016/j.jasms.2007.06.010
44. Yang J, Caprioli RM (2011) Matrix sublimation/recrystallization for imaging proteins by mass spectrometry at high spatial resolution. Anal Chem 83(14):5728–5734. https://doi.org/10.1021/ac200998a
45. Gemperline E, Rawson S, Li L (2014) Optimization and comparison of multiple MALDI matrix application methods for small molecule mass spectrometric imaging. Anal Chem

86(20):10030–10035. https://doi.org/10.1021/ac5028534

46. Meechan PJ, Potts J (eds) (2020) Biosafety in microbiological and biomedical laboratories, 6th edn. Center for Disease Control and Prevention; National Institutes of Health, https://www.cdc.gov/labs/BMBL.html
47. Capicciotti CJ, Kurach JD, Turner TR et al (2015) Small molecule ice recrystallization inhibitors enable freezing of human red blood cells with reduced glycerol concentrations. Sci Rep 5:9692. https://doi.org/10.1038/srep09692
48. Carter JD, Dula SB, Corbin KL et al (2009) A practical guide to rodent islet isolation and assessment. Biol Proced Online 11:3–31. https://doi.org/10.1007/s12575-009-9021-0
49. Corbin KL, West HL, Brodsky S et al (2021) A practical guide to rodent islet isolation and assessment revisited. Biol Proced Online 23(1):7. https://doi.org/10.1186/s12575-021-00143-x
50. Neuman JC, Truchan NA, Joseph JW et al (2014) A method for mouse pancreatic islet isolation and intracellular cAMP determination. J Vis Exp 88:e50374. https://doi.org/10.3791/50374
51. bigTIFF (2021) http://bigtiff.org/
52. OpenSlide (2021) https://openslide.org/
53. Goode A, Gilbert B, Harkes J et al (2013) OpenSlide: a vendor-neutral software foundation for digital pathology. J Pathol Inform 4:27. https://doi.org/10.4103/2153-3539.119005
54. Bugovsky S, Winkler W, Balika W et al (2015) Long time storage (archiving) of peptide, protein and tryptic digest samples on disposable nano-coated polymer targets for MALDI MS. EuPA Open Proteom 8:48–54. https://doi.org/10.1016/j.euprot.2015.07.013
55. Romanova EV, Rubakhin SS, Sweedler JV (2008) One-step sampling, extraction, and storage protocol for peptidomics using dihydroxybenzoic acid. Anal Chem 80(9):3379–3386. https://doi.org/10.1021/ac7026047
56. Malys BJ, Piotrowski ML, Owens KG (2018) Diagnosing and correcting mass accuracy and signal intensity error due to initial ion position variations in a MALDI TOFMS. J Am Soc Mass Spectrom 29(2):422–434. https://doi.org/10.1007/s13361-017-1845-2
57. Christian NP, Arnold RJ, Reilly JP (2000) Improved calibration of time-of-flight mass spectra by simplex optimization of electrostatic ion calculations. Anal Chem 72(14):3327–3337. https://doi.org/10.1021/ac991500h
58. Barry JA, Robichaud G, Muddiman DC (2013) Mass recalibration of FT-ICR mass spectrometry imaging data using the average frequency shift of ambient ions. J Am Soc Mass Spectrom 24(7):1137–1145. https://doi.org/10.1007/s13361-013-0659-0
59. Mass spectrometry data format (2021) https://en.wikipedia.org/wiki/Mass_spectrometry_data_format
60. Mass Spectrometry and Bioanalytics (2021) https://www.mathworks.com/help/bioinfo/mass-spectrometry-and-bioanalytics.html
61. Norris JL, Cornett DS, Mobley JA et al (2007) Processing MALDI mass spectra to improve mass spectral direct tissue analysis. Int J Mass Spectrom 260(2–3):212–221. https://doi.org/10.1016/j.ijms.2006.10.005
62. McDonnell LA, van Remoortere A, de Velde N et al (2010) Imaging mass spectrometry data reduction: automated feature identification and extraction. J Am Soc Mass Spectrom 21(12):1969–1978. https://doi.org/10.1016/j.jasms.2010.08.008
63. UniProt (2021) https://www.uniprot.org/
64. Consortium TU (2020) UniProt: the universal protein knowledgebase in 2021. Nucleic Acids Res 49(D1):D480–D489. https://doi.org/10.1093/nar/gkaa1100
65. NCBI > Proteins > Protein Database (2021). https://www.ncbi.nlm.nih.gov/protein
66. Yates JR 3rd, Speicher S, Griffin PR et al (1993) Peptide mass maps: a highly informative approach to protein identification. Anal Biochem 214(2):397–408. https://doi.org/10.1006/abio.1993.1514
67. Henzel WJ, Billeci TM, Stults JT et al (1993) Identifying proteins from two-dimensional gels by molecular mass searching of peptide fragments in protein sequence databases. Proc Natl Acad Sci U S A 90(11):5011–5015. https://doi.org/10.1073/pnas.90.11.5011
68. Pappin DJ, Hojrup P, Bleasby AJ (1993) Rapid identification of proteins by peptide-mass fingerprinting. Curr Biol 3(6):327–332. https://doi.org/10.1016/0960-9822(93)90195-t
69. Cabrera O, Berman DM, Kenyon NS et al (2006) The unique cytoarchitecture of human pancreatic islets has implications for islet cell function. Proc Natl Acad Sci U S A 103(7):2334–2339. https://doi.org/10.1073/pnas.0510790103
70. Pisania A, Weir GC, O'Neil JJ et al (2010) Quantitative analysis of cell composition and purity of human pancreatic islet preparations. Lab Investig 90(11):1661–1675. https://doi.org/10.1038/labinvest.2010.124

Chapter 7

Single-Cell Resolution Immunoblotting

Louise L. Hansen, Gabriela Lomeli, Julea Vlassakis, and Amy E. Herr

Abstract

Tremendous advances in genomic and transcriptomic tools with single-cell resolution have uncovered the roles of cell-to-cell heterogeneity in biological systems. Yet, protein analysis tools for single cells are limited. Analysis of endogenous, unmodified protein targets in single cells uses either mass spectrometry or immunoassays. To bring single-cell sensitivity and proteoform specificity, microfluidic immunoblots permit the rapid electrophoretic separation of proteins on the basis of size (single-cell western blot) or isoelectric point (single-cell isoelectric focusing). The specificity permitted by combining biomolecular separations and immunoaffinity-based target detection is attractive for interrogation of proteoform targets. During the last decade, innovation in sample preparation, hydrogel formulations, immunoprobing, and analysis has broadened the pertinence of these tools. In this chapter, we describe the general principles, protocols, and considerations of single-cell western blotting and single-cell isoelectric focusing with the aim of informing the reader so as to adopt and adapt these cytometry tools to their own lines of biological inquiry.

Key words Electrophoresis, scWB, scIEF, Soft lithography, In situ, Open microfluidics, Hydrogels, Polyacrylamide

1 Introduction

For both intercellular and intracellular communication, proteins act as dynamic functional components and signaling molecules. The expression and state (e.g., unmodified, modified, interacting, etc.) of protein species define the fundamentals of cellular phenotype and function. Characteristic gene expression profiles are associated with specific neuron types and functions and underlie numerous diseases [1]. Highly differentiated neuronal cell types have a specific cell morphology, cellular localization within the brain, and expression of functionally important proteins. Accordingly, there are several hundred distinct neuronal cell types [2]. During the neuronal differentiation process, cells undergo extensive remodeling of the neuron proteome (the complete repertoire of proteins expressed at a given time, for a given set of conditions). Consider, for example, the differentiation of rat hippocampal neurons.

Jonathan V. Sweedler et al. (eds.), *Single Cell 'Omics of Neuronal Cells*,
Neuromethods, vol. 184, https://doi.org/10.1007/978-1-0716-2525-5_7,

Measured in culture, nearly 40% of the 4,354 protein targets evaluated showed a two-fold expression level change, by high-resolution liquid chromatography-tandem mass spectrometry [3]. Drastic proteomic changes mark development, as well as neurological diseases including Parkinson's disease, amyotrophic lateral sclerosis, and Alzheimer's disease [4–7]. Yet, the regulatory pathways involved in neuronal development and disrupted in neurological disease lack comprehensive proteomic characterization [8].

Heterogeneity is inherent in cellular processes. The flow of genetic information suggests we can predict protein expression from RNA or DNA. However, an increasing number of single-cell and population studies show poor correlation between some gene products and protein expression at the single-cell level [9–12]. The discrepancy between mRNA expression and protein abundance is partially attributable to variable protein translation and degradation rates [13]. Further, the prevalence of neuronal proteoforms, or distinct versions of a protein that arise from alternative splicing or post-translational modification, contributes to the complexity of linking protein abundance with RNA or DNA measurements [14]. For example, the nervous system exhibits a particularly high usage of alternative splicing to generate transcript isoforms during neurogenesis, resulting in the generation of a multitude of neuronal properties encoded by a limited number of genes [15, 16]. Direct measurement of protein target levels is essential to understand the role of specific biomolecules in cell biology, as well as the protein expression changes that drive behavior in healthy and aberrant cells. Bulk measurements mask cell-to-cell heterogeneity, including at the molecular level. Deeper understanding of the composition and dynamics of the proteome at the single-cell level will further our understanding of the inner workings of neuronal cells during differentiation, neurogenesis, and disease development.

Consequently, precision tools designed to selectively measure protein abundance, form, and location in single cells are needed. For complex biological questions and high-resolution phenotyping, highly selective methods open the door to measuring not just which proteins are important but which forms (proteoforms) have functional impact. Protein cytometry tools are designed to detect specific proteoforms in each individual cell, with sufficient throughput to capture intrapopulation variation. For a given protein species, these tools then report on differences in chemical modification, molecular mass, interactions, conformation, and a myriad of other physicochemical parameters among the population of expressed members of that species. Single-cell proteomics is made possible by powerful analytical tools, primarily mass spectrometry and immunoassays including immunofluorescence, proximity ligation assays (PLA), immunocytochemistry (ICC), and mass and flow cytometry (Table 1).

Table 1
Benchmarking of single-cell resolution protein assays

Assay	Multiplexing	Throughput	Selectivity
Immunofluorescence	5 (with conventional filter sets)	100–1,000 s	Immunoaffinity
Mass cytometry	~37	1,000 cells/s	Immunoaffinity
Proximity ligation assay	~22 targets	~100 s	Immunoaffinity
Flow cytometry	~17 targets	10,000 minimum (~20,000 for FACS)	Immunoaffinity
Single-cell immunoblotting	~11 targets (with stripping and reprobing)	1,000 s	Immunoaffinity and protein sizing (scWB) or pI (scIEF)

Proteomic science is driven by mass spectrometry. With bottom-up mass spectrometry, proteins in a complex mixture are digested into small peptide fragments prior to detection by mass spectrometry. Bottom-up mass spectrometry identifies thousands of different proteins and post-translational modifications in a single sample. However, the protein-to-peptide digestion masks molecular stoichiometry [17, 18]. Alternatively, top-down mass spectrometry analyzes intact proteins to detect proteoform expression levels (thus reporting molecular stoichiometry). Yet, top-down mass spectrometry offers detection sensitivity and throughput insufficient for single-cell analyses [19–21].

Immunoassays are the de facto standard for measuring protein targets at the single-cell level. Immunoassays detect a priori identified targets by using an antibody probe designed to recognize and bind to the target epitope. After probing with fluorescently-labeled immunoaffinity reagents (antibody probes), cells are evaluated by a fluorescence-readout technique such as microscopy. ICC maintains single-cell information and is able to assess cell-to-cell heterogeneity. However, the use of antibody probes predisposes ICC to cross-reactivity effects, high background, and limited multiplexing capabilities [22–24]. Flow cytometry improves upon the throughput of ICC, with the trade-off of requiring a large number of cells for analysis. While powerful, the selectivity of immunoassays is fundamentally limited by the availability and specificity of antibody probes, as well as by fixation artifacts [25]. In considering the vast constellation of proteoforms known to exist, the development and application of an antibody probe specific to each proteoform serves as a major limiting factor.

To enhance the specificity of immunoassays, a first stage – typically an electrophoretic separation – is prepended to a second-stage immunoassay. This type of two-stage assay is called an immunoblot. Instead of simply reporting reactivity of protein target with antibody probes (as in an immunoassay), the immunoblot reports two characteristics: the physicochemical characteristic measured by the separation and the reactivity. The capacity to report two characteristics of each protein target increases assay specificity. A widely used type of immunoblot, known as a western blot, concatenates protein sizing using electrophoresis with immunoprobing. The size-based separation physically resolves proteins of different sizes, then reports reactivity with antibody probes, even when using a pan-specific antibody probe. The bulk western blot technique is one of the most widely used analytical tools. Over the last decade, new microscale tools for biological research have pushed the limits of bulk systems on sensitivity and speed by the adaptation of microfabrication techniques from the semiconductor industry [26].

In the electrophoresis stage, the isoelectric point and molecular mass are two powerful physicochemical properties upon which to resolve protein targets in immunoblots. Here, we focus on the implementation of microscale immunoblots by presenting protocol considerations for single-cell western blot (scWB) and single-cell isoelectric focusing (scIEF) platforms and look forward by including discussion of emerging innovations.

2 Working Principles of the Single-Cell Immunoblots

Microanalytical tools open new avenues for biological and biochemical inquiry [26]. With the miniaturization of bulk methods, analytical methods with greater sensitivity and throughput have become possible. An early example of scIEF highlights the utility and role of single-cell electrophoretic assays within neurobiology. After the initial discovery of tubulin proteoforms from bulk brain tissue lysate, it was unclear whether these diverse tubulins were attributed to cell-to-cell heterogeneity or proteoform heterogeneity at the single-cell level [27]. With a version of scIEF in conjunction with radioactive labeling, multiple tubulin proteoforms were found to exist in single neuron cells [27, 28].

Microfluidic assays like the scWB and scIEF precisely control mass and fluid handling for efficient, reproducible, and timely analysis of biological targets, all while maintaining high local analyte concentrations. For electrophoresis, miniaturization enhances performance. scWB and scIEF modalities eliminate the need to pool cells to achieve detectable protein concentration, allowing single-cell resolution for the detection of intrapopulation protein expression level variations. Cell handling losses and target dilution are mitigated by using microfluidic features to match the reaction volume to the volume of an intact cell to maintain intracellular concentration levels even after cell lysis [29]. Minimizing cell lysis and electrophoresis times decreases the dilutive effects of diffusion, facilitating the detection of low-abundance molecules from single cells. Careful control of sample injection (i.e., using the small injector diameter of the microwell) and reducing electrophoresis run times (i.e., durations of seconds, not hours) lower overall separation dispersion, making separation of protein targets with small molecular mass differences possible [30]. Lastly, microfluidic integration readily supports sample multiplexing, with hundreds to thousands of single-cell separations completed in just minutes.

The microfluidic scWB platform follows the same general steps as the bulk western blot to separate proteins based on their size (molecular weight) within a polyacrylamide gel (PAG) (Fig. 1). In short, the methodology includes (1) single-cell isolation in PAG microwells, (2) rapid lysis, (3) electrophoresis, (4) protein immobilization within the PAG by UV capture, and (5) protein detection via immunoprobing. The major discrepancies between the conventional benchtop western blot assay and the scWB lie in the sample types and blotting methods. In the bulk methods, pooled cell lysate is introduced into the respective wells. In comparison, a single cell is introduced into each microwell prior to lysis in both the scIEF and scWB described here. Also, conventional western blots rely on the transfer of proteins from the separation gel to the blotting membrane prior to antibody-based detection. The transfer step is bypassed in the microfluidic devices by covalently binding the proteins within the hydrogel to maintain spatial resolution (UV capture), which permits the direct incubation of the hydrogel with the antibody probes to detect the target antigens.

The scWB method is suitable for analysis of proteins in the 15–90 kDa range with an 8%T gel and has an estimated limit of detection at 27,000 molecules using purified EGFP (equivalent to 45 zmol), both features that make the device appropriate for detecting a range of protein targets [9, 31, 32]. Resolution of various molecular mass targets can be achieved by optimizing the molecular sieving matrix to the sample of interest (e.g., pore-gradient gels or altering the %T of a uniform gel) [29, 33]. The smallest reported molecular mass difference resolved with size-based separation in a single-cell western blot is 4 kDa, which corresponds to ~37 amino acids or 11% molecular mass difference [30]. With longer separation lengths and tuning of PAG pore size, resolution of protein species ranging from 36 to 289 kDa is possible [34]. The planar "open microfluidic" form factor ensures compatibility of immunoprobing readout with commonly available fluorescence-readout technology (i.e., laser-based microarray scanners, fluorescence microscopy).

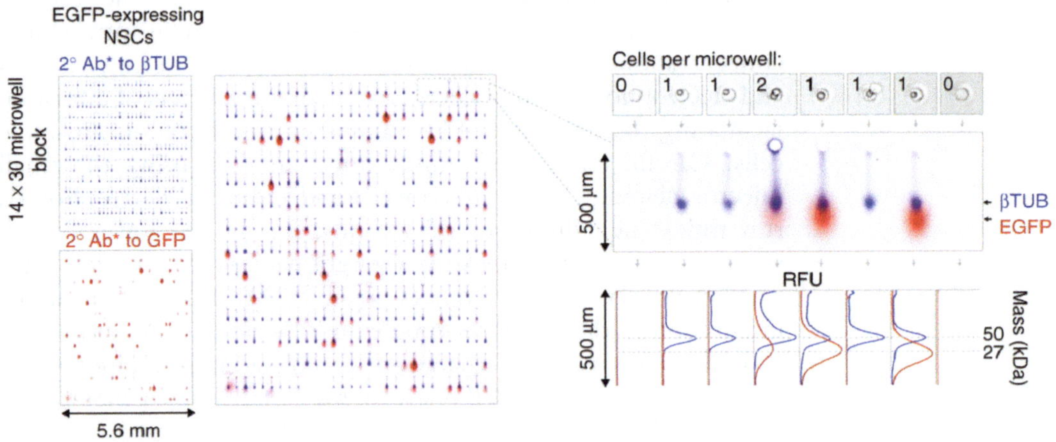

Fig. 1 Single-cell western blotting brings microarray-like throughput to high-specificity protein identification for single cells. (Left) False-color fluorescence micrographs of the scWB PAG array after antibody probing against β-tubulin (AF647 secondary antibody, blue) and EGFP (AF555 secondary antibody, red) in 420 EGFP-expressing neuronal stem cells (NSCs). (Right) Bright-field imaging determines the number of cells per microwell. The inset shows representative micrographs and protein intensity profiles for β-tubulin and EGFP (RFU, relative fluorescence units). Reprinted with permission from Springer Nature, Nature Methods Ref. [32], Copyright 2014

scIEF provides a complementary separation mechanism to resolve protein targets based on differences in the isoelectric point, pI (i.e., the pH at which a protein target has no net charge) [35]. scIEF immunoblots are particularly powerful for specific detection of chemical and post-translational modifications that only subtly alter the size of a protein (making detection by western blot difficult, if not impossible) [36]. Bulk IEF has a long-standing history as both an independent tool (preparative, as well as analytical) and in combination with orthogonal separation mechanisms. In 2D gel electrophoresis, pI-based IEF is followed by size-based western blotting [35, 37]. scIEF employs a two-layer open microfluidic device to achieve separation of proteins from single-cell lysate (Fig. 2a). The main functions of the lid layer are (1) lysis buffer delivery and (2) pH gradient generation. The functions of the bottom layer are (1) single-cell encapsulation and (2) protein immobilization. The lid layer consists of a three-domain PAG with acidic (anolyte) and basic (catholyte) regions flanking a central focusing region, which contains the lysis detergents and pH gradient-forming carrier ampholytes. The bottom layer is a single-cell immunoblot microdevice. To bring the exquisite proteoform selectivity of the bulk method to the single-cell level, scIEF was developed to complement the scWB, and following a similar procedure: (1) single-cell isolation in PAG microwells, (2) addition of the lid gel to initiate cell lysis and introduce carrier ampholytes, (3) electric field is applied for isoelectric focusing, (4) protein

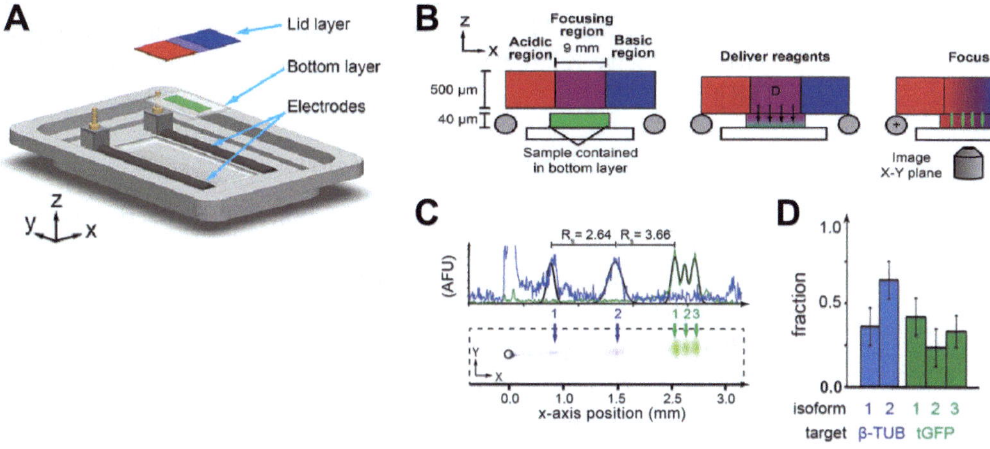

Fig. 2 scIEF can resolve proteoforms that are not distinguishable with other methods, including scWB. (**a**) Schematic of an open IEF microdevice within a custom 3D-printed electrophoresis chamber with graphite electrodes embedded in the surface. A lid PAG layer is used to deliver the pH gradient. (**b**) Cross-section schematic of the delivery and focusing of sample within the bottom layer. (**c**) False-color fluorescence micrograph and trace show detection of two β-tubulin and three TurboGFP isoforms. (**d**) Relative isoform fractions ($n_{b\text{-TUB}} = 3$, $n_{tGFP} = 8$). (**a**) and (**b**) reproduced from 2018 with permission from John Wiley and Sons, Electrophoresis [45]. (**c**) and (**d**) reproduced from 2016 with permission from John Wiley and Sons, Angewandte Chemie International Edition [36]

immobilization within the PAG by UV capture, and (5) protein detection via immunoprobing (Fig. 2b). scIEF can resolve proteoforms differing by a single charge unit and has an approximate lower limit of detection of 42,000 molecules [36]. Tentori et al. demonstrated scIEF separation of three endogenous protein targets (lamin A/C, β-tubulin, and recombinantly expressed TurboGFP) from a glioblastoma cell line [36] (Fig. 2c, d). Two β-tubulin and three TurboGFP proteoforms were detected with denaturing lysis buffer, which are not resolvable with scWB. As a member of the versatile single-cell immunoblotting family, scIEF has the potential to be an important protein analysis tool for a wide array of biological questions.

2.1 Separating Protein Targets by Size (scWB)

The governing principles of scWB are analogous to those of bulk western blots. Electrophoresis is the migration of charged particles under an applied electric field. Electrophoretic separations are conducted in a wide range of separation media, including free solution and sieving matrix. Given the pore size of polyacrylamide gel and the characteristic dimension of most proteins, polyacrylamide gel electrophoresis (PAGE) is well suited to resolve protein targets. During PAGE, protein species electromigrate through a porous polyacrylamide hydrogel matrix under the applied electric field. The porous matrix differentially impedes protein electromigration on the basis of size and structure of each protein molecule.

In free solution, particles migrate with a velocity proportional to zeta potential. However, in a sieving matrix, particles migrate with a velocity related to particle size, as described by Ferguson [38]:

$$\log\left(\mu_{\mathrm{EP,gel}}\right) = \log\left(\mu_{\mathrm{EP,fs}}\right) - K_{\mathrm{r}}T \tag{1}$$

where $\mu_{\mathrm{EP,\ gel}}$ is the electrophoretic mobility of the particle through the gel; $\mu_{\mathrm{EP,\ fs}}$ is the electrophoretic mobility in free solution; K_r is the empirically determined retardation coefficient that describes how much the sieving matrix reduces the electrophoretic mobility of the particle; and T is the percentage (w/v) of total monomer in solution (acrylamide and bis).

Protein sizing is PAGE analysis of denatured, reduced proteins coated with the anionic detergent sodium dodecyl sulfate (SDS). Also called SDS-PAGE, protein sizing relies on the uniform mass-to-charge ratio coating of SDS that linearizes proteins and gives all proteins the same free solution electrophoretic mobility ($\mu_{\mathrm{EP,\ fs}}$), making K_r dependent only on protein length [39, 40]. In this way, the electrophoretic mobility of protein targets in SDS-PAGE is related to the molecular mass of the proteins with a log-linear relationship. In comparison, non-denaturing conditions (e.g., native PAGE) result in electromigration proportional to the zeta potential ($\mu_{\mathrm{EP,\ fs}}$) and the retardation coefficient [41, 42]. The quaternary structures of proteins affect retardation in native PAGE, making protein identification by migration distance more confounding than when using a protein sizing mechanism.

Consequently, the choice of first-stage electrophoretic assay tunes the immunoblot performance to answer the question at hand. SDS-PAGE is the most common type of gel electrophoresis used in western blotting and scWB. The electrophoretic separation of denatured proteins by molecular mass is particularly well suited for interrogation of truncated protein isoforms that can make antibody-based distinction of different isoforms challenging without the added selectivity of a size-based separation [43].

2.2 Separating Protein Targets by Isoelectric Point (scIEF)

scIEF miniaturizes the fundamental principles governing bulk IEF. The pI of a given protein is dictated by both the amino acid sequence and any post-translational modifications [35, 44]. Proteins rich in amino acids with acidic side chains will have a lower pI than proteins with a higher proportion of amino acids with basic side chains. To accomplish a charge-based protein separation, a stable pH gradient is established. The pH boundary conditions define the pH gradient (slope, range) and, thus, determine which proteins will be resolved. For example, a pH 4–7 gradient would be unsuitable for analysis of proteins that are more acidic (pI < 4) or more basic (pI > 7). Although alternative methods exist for creating a pH gradient [35], scIEF utilizes a series of commercially available

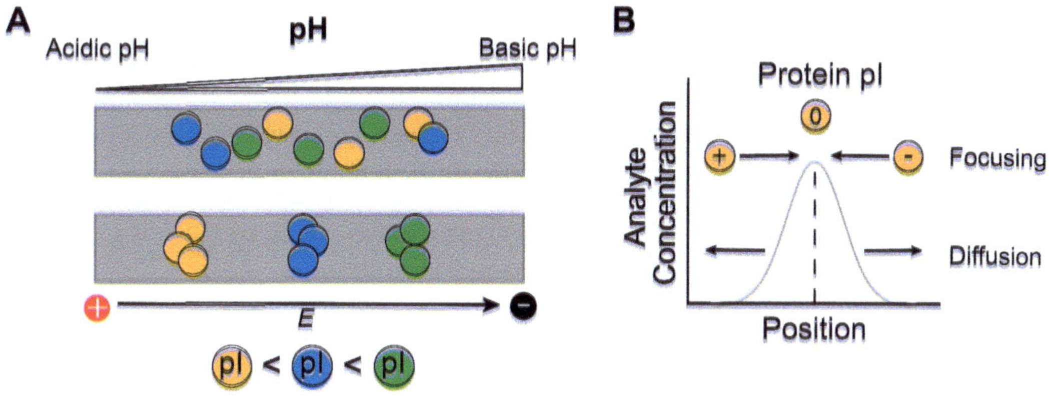

Fig. 3 Isoelectric focusing uses equilibrium electrophoresis in the presences of a stable pH gradient to report isoelectric point, not molecular mass, of each target. (**a**) Schematic of protein migration in IEF. A heterogeneous protein sample is loaded into a gel with a pH gradient. When the voltage is applied, the proteins migrate to the location at which they have no net charge (pI = pH). (**b**) Schematic representation of protein peak diffusion and restorative focusing during electrophoresis. Species are focused into bands as proteins gain a positive charge at pH < pI and a negative charge at pH > pI and move toward the cathode and anode, respectively

carrier ampholytes that stack in the presence of an applied electric potential to create a pH gradient [36]. When proteins electromigrate into the pH gradient region from a boundary, the biomolecules traverse the pH gradient according to protein charge which is impacted by dynamic protonation and deprotonation (Fig. 3a). Migration halts when proteins reach a region in the gradient where a net-zero charge is achieved (where pH = pI). While the pH gradient is stable, the focused proteins will be in dynamic equilibrium in their focused position, as diffusion causes proteins to migrate away from pH = pI, and electrophoretic force returns the proteins back to pH = pI (Fig. 3b) [35]. In scIEF, this focusing effect keeps proteins from single-cell lysate highly concentrated even across several millimeters of separation lane [36].

In the original scIEF design, a single pH gradient region limited the throughput capability of the assay to single-row 50 microwells, resulting in data from ~10 cells per chip [36]. To increase throughput, a first of its kind multi-domain hydrogel with back-to-back pH gradient regions was developed [45]. Although not yet implemented with single cells, arrayed pH gradients facilitate an increase in sample size for the scIEF platform. With each additional pH gradient, on additional lane of microwells can be assayed with simultaneous isoelectric focusing. To fit more pH gradients in the same area, the separation length is reduced, which in turn lowers the resolving power, creating a trade-off between separation performance and throughput [45].

Accordingly, one area for improvement in scIEF is throughput. scIEF has a throughput suitable for rare-cell analysis, but not well suited to analysis of large populations of cells. A scIEF chip houses ~50 microwells, which reduces the number of analyzed cells to ~10 after cell settling and quality control losses. While 6–8 min is sufficient for protein focusing, the pH gradient remains stable for at least 20 min [45, 46]. However, due to the sample preparation and run times, the lid gel must be made fresh for each experimental run. The run time per microdevice is ~2.5 h. Active cell settling, longer separation lanes, and the use of arrayed IEF lids are potential avenues for improvement in throughput.

3 Method Overview

The following sections outline the major steps of scIEF and scWB (Fig. 4). The key steps are summarized as follows: (1) immunoblot hydrogel fabrication, (2) sample preparation and introduction of cells, (3) polyacrylamide gel electrophoresis (PAGE), (4) immunoprobing, and (5) data analysis. Each step presents a series of design considerations and points of optimization for proper protein separation.

3.1 Immunoblot Hydrogel Fabrication

Single-cell immunoblot microdevices are "open microfluidic" devices, composed of a thin polyacrylamide gel (PAG) layer affixed to a glass slide (or other planar support). Anywhere between tens (for scIEF) to thousands of microwells (for scWB) decorate the PAG surface, enabling the isolation of individual cells for subsequent analysis. The microwell pattern and geometry are crafted from a reusable mold which has been functionalized to promote PAG grafting to the support and detachment from the mold after gel precursor polymerization. The chemical and physical properties of the PAG are highly tunable, which supports the extended applications of single-cell immunoblot assays.

3.1.1 Mold Fabrication

To fabricate PAG microdevices dotted with microwells, the mold consists of photoresist micropillars on a wafer. To fabricate the mold, standard soft lithography protocols are followed for the generation of SU-8 epoxy-based negative photoresist patterns on silicon substrates. The thickness of the SU-8 corresponds to the height of the micropillar and therefore the depth of the microwell in the final device. The cell size (e.g., diameter) informs selection of the SU-8 layer thickness. The photomask design will dictate the microwell shape, diameter, and spacing. The microwell diameter must be close to the average cell diameter to prevent multiple cells per well. The starting well diameter for standard scWB and scIEF is 32 μm. Single-cell electrophoretic assays have been performed with circular microwells to match the spherical shape of cells in suspension. A study of circular, triangular, and rectangular microwell geometries revealed that, for the analysis of most mammalian

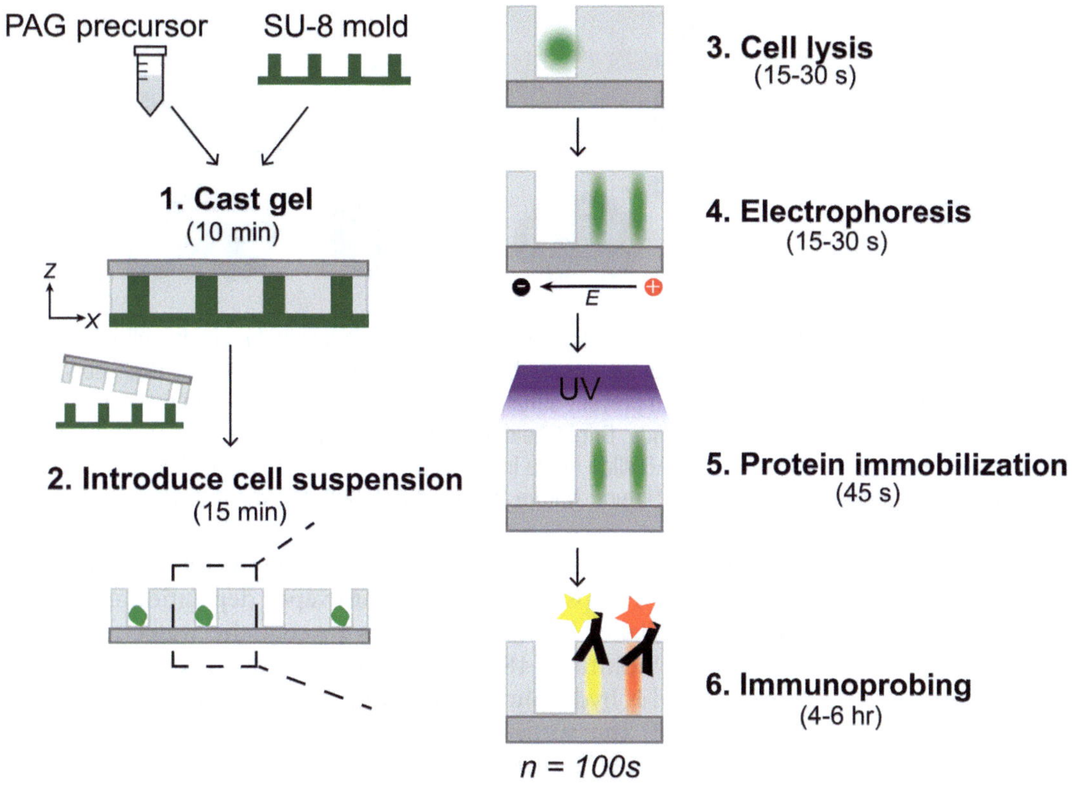

Fig. 4 Single-cell immunoblotting is a multistep workflow that adapts bulk protocols to an open microfluidic device. For example, the scWB assay is completed in steps: (1) With an SU-8 patterned mold and gel precursor, a polyacrylamide gel with microwells is cast onto a substrate. The gel acts as both the molecular sieving matrix and the blotting scaffold. (2) A cell suspension is introduced for gravitational settling. (3) The introduction of the bifunctional lysis/electrophoresis buffer initiates cell lysis. (4) Electrophoresis separates proteins. (5) Proteins are immobilized within the gel matrix with UV-mediated capture. (6) Fluorescent-conjugated antibodies detect the protein species based on immunoaffinity. A laser-based microarray scanner reads out the protein signal from 100 s of cell simultaneously. scIEF follows the same general outline with a modification to step 3, where a lid gel is used to introduce buffers and pH gradient-forming carrier ampholytes

cells, microwell geometry does not affect the resulting protein separation [47]. Only in the case of the largest mammalian cell was separation performance improved with rectangular microwell.

The microwell distribution pattern across the PAG depends on the final application of the microdevice (Fig. 5a). Microwell spacing in the direction orthogonal to the electric field (l_y) is constrained by the spreading and crossover of adjacent protein peaks as a result of lateral diffusion. For scWB, the spacing along the separation axis (l_x) depends on the migration distance of protein targets during electrophoresis. For scIEF, the number of arrayed microwells depends on the number of pH gradient regions.

The final step in mold fabrication is to increase the hydrophobicity of the SU-8 mold for clean demolding of the PAG. The most

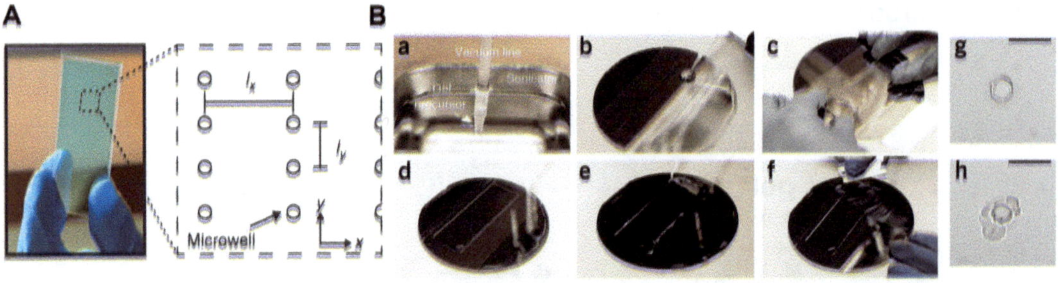

Fig. 5 Fabrication of PAG device for scWB and scIEF with patterned microwells for single-cell isolation. (**a**) Photograph and schematic of the PAG slide. Increasing the separation axis (l_x) increases the PAGE separation, and *y*-axis pitch (l_y) should be optimized to prevent diffusive overlap. (**b**) The step-by-step process of casting PAG onto an SU-8 mold with a silanized slide. Bright-field micrographs show a well-formed (*g*) and deformed (*h*) microwell. Scale bars, 100 μm. The PAG slide in the image is a standard microscope slide (25 × 75 mm). Reprinted with permission from Springer Nature: Nature Protocols Ref. [29], Copyright 2016

common strategy to attain this hydrophobicity for single-cell immunoblot chips is silanization with vapor deposition of dichlorodimethylsilane. Another option is to coat the wafer mold with Gel Slick™ solution [48]. We recommend a quick test to ensure the wafer mold is hydrophobic by testing whether water droplets bead and roll off the surface. If the wafer mold is not hydrophobic, silanize again with dichlorodimethylsilane or reapply Gel Slick™.

3.1.2 PAG Supporting Substrate

The PAG support must be functionalized to graft to the PAG gel. For glass slides, silanization with 3-(trimethoxysilyl)propyl methacrylate leads to the self-assembly of acrylate-terminated monolayers, which can covalently attach to the PAG [49]. Briefly, glass slides are first incubated with a silane solution composed of a 2:3:5 ratio mixture of 3-(trimethoxysilyl)propyl methacrylate, glacial acetic acid, and deionized water for 30 min [29, 50]. Then, slides are washed in methanol and deionized water and dried in a nitrogen stream. Silanized glass slides should not be stored for more than 2 months prior to use [29]. However, the PAG support does not have to be a glass slide. Single-cell immunoblotting chips can also be made with GelBond™, a commercially available plastic film that readily supports PAG [48]. One advantage of using GelBond™ is that the film can be easily cut. Laser cutting of PAG microwells on GelBond™ was used to excise settled nuclei for dual mRNA and protein expression analysis from single embryos [48].

3.1.3 PAG Polymerization

A standard PAG precursor contains the following basic components: the primary monomer (acrylamide), the crosslinker (e.g., bis-acrylamide), and the polymerization initiators [51]. While the PAG is composed primarily of acrylamide monomers, additional properties can be achieved by incorporating copolymers [52]. A key additive for single-cell immunoblotting assays is the inclusion of a photo-capture molecule to immobilize proteins in the 3D PAG, replacing the transfer and blotting of proteins to a separate membrane in conventional western blotting. A range of photo-capture molecules are available, including N-[3-[(3-benzoylphenyl)-formamido]propyl] methacrylamide (BPMAC), tetrazole [53], and diazirine [54, 55]. Conventionally, scWB and scIEF use BPMAC for UV capture of proteins within the PAG after electrophoresis. BPMAC is added to the gel precursor and is polymerized into the PAG backbone through its methacrylamide reactive group. Once the gel precursor components are combined, the solution is degassed (as oxygen quenches the polymerization reaction) and sandwiched between the supporting substrate and wafer mold (Fig. 5b). Polymerization occurs in 15–20 min via a free-radical reaction. Free-radical polymerization can be initiated using several different approaches, most commonly chemically or photo-initiated. While chemical polymerization (initiated by ammonium persulfate) is most common with scIEF and scWB, photopolymerization by the inclusion of UV-activated initiators such as VA-086 is advantageous for gradient gel polymerization through a photomask [56]. Once polymerization is complete, the substrate-bound PAG can be removed from the wafer mold: The single-cell immunoblot chip is ready for use. The most common failure mode during fabrication of single-cell immunoblot chips is gel sticking during demolding. Gels with larger pores (e.g., 6%T) are more likely to stick to the mold than more robust gels with smaller pores (e.g., 8% T). The first step to troubleshooting gel sticking is to make sure one is adding the right amount of initiator to the PAG precursor and mixing the precursor thoroughly. If insufficient initiator is added, the gel will not polymerize fully. Too much initiator results in premature polymer-chain termination and therefore less robust gelation [51].

PAG pore size is an important variable in electrophoretic assays, impacting separation resolution, limit of detection, and immunoprobing efficiency. In general, a smaller pore-size (higher total acrylamide concentration) PAG improves separation resolution [31], but small gel pores decrease immunoprobing efficiency due to size exclusion of antibody probes from the PAG [57]. Like conventional bulk westerns, single-cell protein sizing can be performed in PAG of either uniform pore size or gradient pore size. For uniform pore-size PAG, the final acrylamide and crosslinker concentration is adjusted to the desired concentration, and chemical polymerization is typically used. While more laborious to make, gradient pore-size PAG offers greater separation resolution over a large range of protein masses compared to uniform gels. Multiple back-to-back pore gradients can be created in PAG with photo-initiated polymerization through a grayscale mask [29]. The grayscale mask modulates the UV dose for photopolymerization, leading to faster polymerization and therefore smaller pores in lighter regions and vice versa. The inclusion of an acid-labile cross-linker in the PAG allows the gradient-pore PAG to be reverted to a uniform PAG for even immunoprobing [33].

In scIEF, the PAG serves as an anti-convective and immobilization medium. The pore size does not play as important a role in separation resolution as in the size-based separations of scWB. For this reason, larger pore-size or lower-density gels are preferred to optimize the antibody immunoprobing step. In addition to diluting the acrylamide monomer to a lower concentration (6%T), the inclusion of a porogen, such as PEG, can further increase the pore size of PAG for IEF applications [58]. Inclusion of 2% PEG for the creation of laterally aggregated PAGs led to a threefold increase in antibody probe partitioning in the PAG and a twofold reduction in background signal. While porogen gels have not been applied to scIEF yet, these highly porous PAGs present a significant materials advancement in immunoprobed IEF.

3.2 Sample Preparation and Introduction of Cells

Various cell types have been evaluated with the scWB, scIEF, and similar technology. Often, immortalized cell lines are used with these platforms. Sinkala et al. implemented the scWB protocol for the analysis of rare-cell circulating tumor cells [59]. More complex sample types are also possible, such as primary tissue from human breast tumor or mouse smooth muscle [43, 60]. Lastly, complex multicell samples such as embryos and blastomeres are possible with the appropriate adjustment to microarray features [48].

The choice of an open microfluidic design strategy is driven, in part, by the ease with which cells can be introduced and isolated in microwells. A cell suspension solution is introduced to the microwell array, and cells settle into the microwells by gravity. The conventional scWB calls for ~2×10^5 cells in 200 μL (~1×10^6 cells per mL) per microdevice [29]. After 10 min of cell settling, the microdevice is washed with PBS to limit the number of cells adhered to the PAG surface that would add background and noise to the assay. The cell distribution can result in zero, one, or multiple cells in each microwell, yielding a distribution skewed from the expected Poisson distribution [29]. With single-cell western blotting, the microwell occupancy is ideally unity. Microwell occupancy can be correlated to housekeeping protein expression level by scWB [32]. Previous work reported 40–50% single-cell occupancy when working with neural stem cells [32]. Poor cell settling of negatively buoyant cells results from multiple factors, including (1) deformed microwells, (2) poorly optimized microwell parameters, (3) inadequate starting cell concentration, and (4) cell clumping. Optimized microwell dimensions and spacing are crucial for proper assay function. If cell settling issue arises, we recommend confirming the integrity and size of the microwells and inspecting the microdevice under a light microscope to assess cell clumping (that may be mitigated with a cell strainer). Often, issues arising during sample introductions are mitigated with proper PAG fabrication and accurate mold dimensions.

To reduce cell loss that occurs when using passive sedimentation to seat a subpopulation of cells in microwells, alternate sample preparation protocols adopt "active" methods for cell isolation in microwells. Kim et al. implemented lab-on-a-disc technology in conjunction with the scWB platform to handle rare sample types [61]. Centrifugal cell preparation minimized cell loss (<30%) while achieving >70% single-cell occupancy microwells. Alternatively, dielectrophoresis (DEP) has been implemented with other microwell-based microfluidic designs for cell manipulation [62, 63]. In cases where cells can be pre-labeled with magnetic microparticles, magnetophoresis is an option to manipulate cells [64, 65].

Not all biological inquiries are compatible with suspended cells, such as in cases where detachment of cells from culture surfaces prior to analysis confounds the biological evaluation. Cell detachment is known to perturb membrane proteins, cytoskeletal proteins, and signaling proteins, and the long handling times can scramble measurement of dynamic cellular processes [66, 67]. Furthermore, assays performed on cultured cells preserve the extracellular matrix context. For scWB, decorating the surface-accessible PAG of the microwell with fibronectin or other adhesive ligands integrates cell culture with the analytical tool, setting the stage for western blotting of single, adherent cells without the need to detach cells prior to analysis [68]. The PAG is functionalized through fibronectin-embedded polymerization that generates a fibronectin layer around the exposed surface of the microwell. HeLa cells cultured overnight on the in situ scWB platform achieved adherent phenotypes. The cell-laden microwells are dosed with stimulants prior to proceeding with the scWB workflow. Zhang et al. implemented in situ scWB to evaluate the phosphorylation of MAP kinases (ERK1/2, p38) under hypertonic conditions [68]. Without on-chip culture, unintended phosphorylation from trypsinization and sample preparation steps obscure the true phosphorylation state of the cells.

For direct lysis and in situ IEF of cultured single and clustered cells, the series of discrete microwells stippled in the PAG layer on the chip has been replaced with a microcontact-printed strip of fibronectin patterned on the surface of a planar PAG [46]. Instead of settling cells in microwells, cells adhere and grow on the fibronectin-patterned PAG, where receptor proteins are recovered (Fig. 6). Lysis, focusing, and photo-capture proceed as per the canonical scIEF assay. Protein targets that would otherwise be undetectable or masked in scIEF, such as the cell receptor proteins CD44 and β-integrin, can be captured with in situ IEF. Su et al. measured cytoskeletal-associated proteins (lamin A/C and β-tubulin) from cultured, adherent BJ fibroblast cells seeded on a PAG without microwells [46].

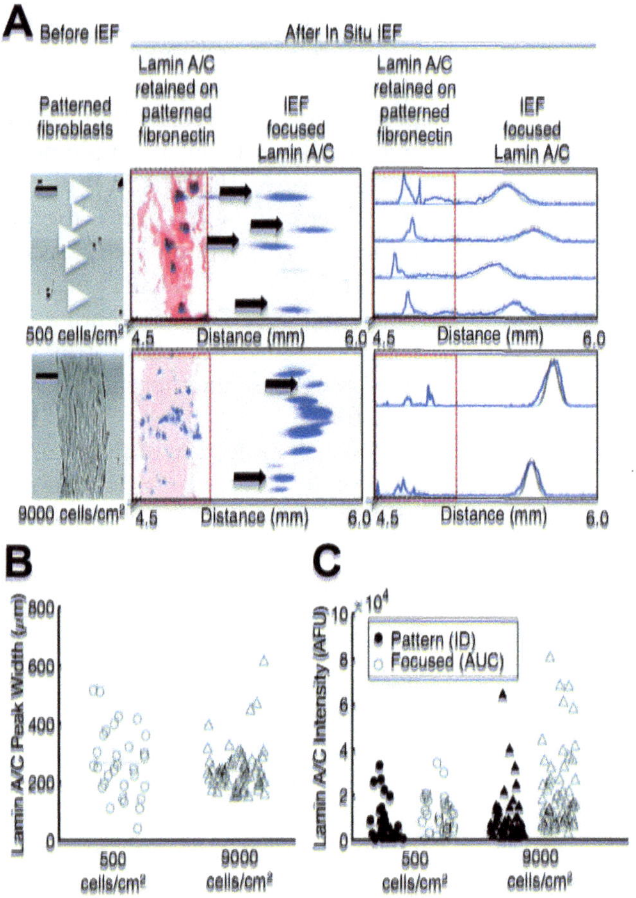

Fig. 6 In situ scIEF separates and detects proteins from single and clustered cells without the need for cell detachment prior to analysis. (**a**) Bright-field images of single cells and ~80% confluent cells prior to in situ IEF (left) and false-color inverted fluorescence micrographs of lamin A/C that is retained on the fibronectin pattern and focused into a Gaussian peak (middle). The protein peaks are segmented, and fluorescence intensity profiles are generated (right). The black line indicates Gaussian fit. (**b**) Plot of protein peak width as a function of patterned cell density. Separation performance, as measured by lamin A/C peak width, does not significantly change, indicating that protein abundance is not saturating the system in the case of higher seeding density. (**c**) Plot of amount of protein (lamin A/C) that is retained on the fibronectin pattern (ID) and focused (AUC). The bars represent the mean. Scale bars, 200 μm. AFU, arbitrary fluorescence units; ID, integrated density; AUC, area under the curve; pI, isoelectric point. Reproduced from Ref. [46], Lab on a Chip 2017, with permissions from the Royal Society of Chemistry

3.3 Polyacrylamide Gel Electrophoresis (PAGE)

3.3.1 In-Well Chemical Lysis and Protein Solubilization

Immediately after cell isolation in microwells, in-well chemical lysis and protein solubilization is initiated by the rapid introduction of a bifunctional lysis/electrophoresis buffer. In the canonical scWB, buffer is delivered rapidly for the synchronous initiation of cell lysis to the entire array via the open microfluidic device format. Numerical modeling of the convective buffer addition for a 30 μm deep microwell revealed recirculating flow in the top 2/3 of the microwell, generating a regime with Peclet number >1.0 [29]. In IEF, the lysis/electrophoresis buffer is introduced by the lid gel that also enforces the pH gradient. This lid gel is applied onto the PAG, and the reactive species diffuse from one layer to another.

Optimization of protein solubilization conditions is necessary. Proteins must be extracted from the biological sample and kept in solution during the separation process, a challenge due to the cell membranes and the crowded biomolecular environment. The conventional scWB protocol calls for a dual-function lysis and electrophoresis buffer composed of 0.5% Tris-glycine, 0.5% (w/v) sodium dodecyl sulfate (SDS), 0.25% (w/v) sodium deoxycholate (Na-DOC), and 0.1% (v/v) Triton X-100 [29]. Ionic detergents, such as SDS and Na-DOC, solubilize the cell membrane and denature proteins, while nonionic detergents and lysis duration enhance membrane solubilization [69]. Protein solubilization is quantified by imaging the cell lysis process for cell lines engineered to express fluorescent proteins. In addition, if the protein peaks appear non-Gaussian after antibody-based detection, optimization of protein solubility can ensure more predictable PAGE behavior. The concentration of detergents can be increased above the critical micelle concentration, and denaturants such as urea can be added to aid denaturation. Hardy, lysis-resistant cell types, such as circulating tumor cells, may require increased detergent concentrations, lysis time, and lysis temperature for complete lysis [59]. For example, a buffer composed of 1% SDS and 0.25% Na-DOC at 55 °C yielded complete solubilization of circulating tumor cells by visual inspection of GAPDH injection into the immunoblot gel.

In scIEF, the options for the lysis and solubilization buffer are limited to species that will not alter the pI of the protein, such as zwitterionic and nonionic detergents and chaotropic agents such as urea and thiourea [70]. The buffer composition can have a drastic impact on which proteoform species are detected in scIEF. For example, scIEF under denaturing conditions (3.6% (w/v) CHAPS, 1% (v/v) Triton X-100, 0.0125% (w/v) digitonin, 7 M urea, 2 M thiourea) yielded distinct peaks for two β-tubulin and three TurboGFP proteoforms, while native conditions (excluding urea and thiourea) yielded a single peak for both β-tubulin and TurboGFP [36].

A multistep lysis protocol provides another parameter by which to achieve additional molecular specificity. With differential detergent fractionation, subcellular western blotting permits the

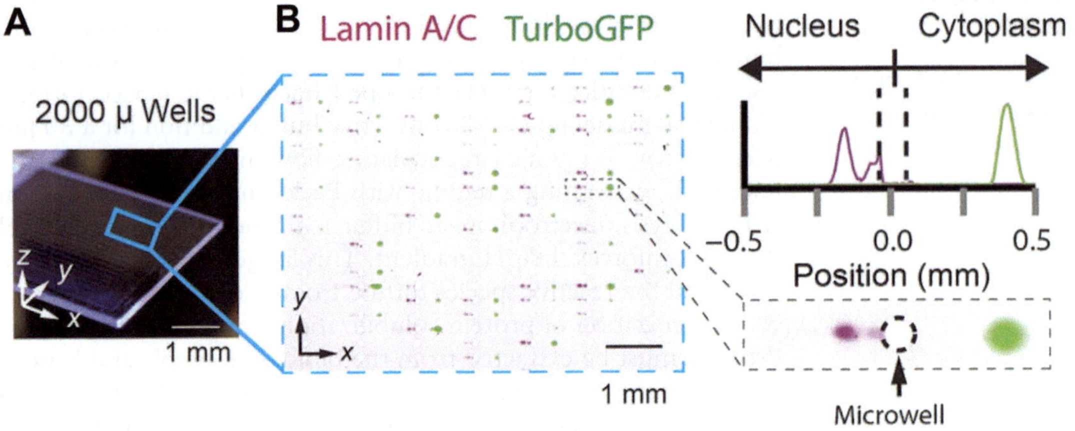

Fig. 7 Microfluidic subcellular scWB reports protein localization to the cytoplasmic or nuclear compartment of single cells. (**a**) Photograph of the PAG microdevice with microwell arrays where the microwell is situated at the center of the separation lane. (**b**) Fluorescence micrographs of the subcellular western blot array (U373 cells, $N = 56$) for lamin A/C (magenta) and TurboGFP (green). The intensity plot shows lamin A/C in the nuclear fraction and TurboGFP in the cytoplasmic fraction after a two-step, bidirectional PAGE. Reproduced from Ref. [71], Microsystems & Nanoengineering 2017, with permissions from Springer Nature

sequential lysis of the outer cell membrane and then the nucleus by dosing each cell with a series of distinct, optimized lysis buffers (Fig. 7) [71]. The lysis solutions are introduced via a PAG "lid" similar to the chemical patterning systems described earlier for the scIEF system. Separate top lids are impregnated with the various buffers, which can be sequentially overlaid with the open PA device for diffusive delivery of detergents. The application of a sequence of optimized lysis buffers permits assignment of subcellular location to each protein target (nuclear vs. cytoplasmic). One example is nuclear translation of NF-κB, which translocates to the nucleus after activation [71]. The first lid gel selectively lyses the cell membrane but maintains the nuclear membrane, which is disrupted by the second lid gel. A two-step, bidirectional PAGE is performed after each lid gel is applied to separate the lysate from each step in a district region of the separation axis.

3.3.2 Electrophoresis and Protein Immobilization

To initiate electrophoresis (EP) after cell lysis, an electric field (40–50 V/cm for scWB and 200 V/cm for scIEF) is applied across the PAG. Proteins are electrophoresed through the microwell walls and into the PAG [32]. Electrophoresis is carried out in custom electrophoresis chambers with electrodes (platinum wires or graphite bars) that directly interface with the bifunctional lysis/electrophoresis buffer or lid gel in the case of scIEF. Optimization of the protein solubilization and electrophoresis process permits the separation of protein targets of close molecular weight. Kang et al.

resolved the truncated HER2 isoform (~90–115 kDa) from the full-length protein (185 kDa) in HER2-positive breast tumors without isoform-specific antibodies with scWB [43].

The upper limit on the applied electric field strength is determined by Joule heating and power supply capacity. Microfluidic devices benefit from favorable heat transfer characteristics that allow for high electric fields (100–1000 V/cm) [45]. Joule (or resistive) heating is expected when high-conductivity buffers are utilized. Extensive Joule heating in the lysis/electrophoresis buffer increases diffusivity and diffusive losses of the proteins out of the microwell, resulting in lower separation performance [30]. Autothermal runaway can be mitigated by introducing a buffer exchange to an electrophoresis buffer with lower Joule heating than the lysis buffer. This increases separation resolution enough to resolve endogenous GAPDH (36 kDa) and PS6 (32 kDa) with a 4 kDa molecular mass difference. For example, exchanging from a high-conductivity bifunctional buffer to a low-conductivity electrophoresis-only buffer between cell lysis and electrophoresis results in a 1.6-fold improvement in separation resolution.

After 15–45 s (scWB) or 6–8 min (scIEF) electrophoresis, photoactivated protein immobilization within the PAG prevents diffusional losses of the protein from the gel into the bulk buffer solution, keeping the local concentration of protein sufficiently high for detection. The PAG becomes the immobilization scaffold for subsequent in-gel immunoassay. Immobilization is done by 45 s UV capture of the proteins by photoactive benzophenone species embedded in the polyacrylamide during chip fabrication. In general, immobilization relies on a UV-initiated hydrogen abstraction reaction to covalently bind proteins to the photoreactive species.

3.4 Immunoprobing

3.4.1 Detection Sensitivity of In-Gel Immunoblotting

In-gel immunoblots are performed using primary antibodies and fluorescently labeled secondary antibodies which diffuse directly into the gel to bind the immobilized protein target of interest. Typically, the gel is sandwiched with immunoprobe solution on a glass substrate. One advantage to in-gel immunoprobing is elimination of a separate electro-transfer step used in conventional immunoblotting. Electro-transfer is rapid, but results in unequal transfer efficiencies. As an alternate to probing on a binding membrane, an "in-gel" immobilization approach can be used. In this approach, separated protein targets are immobilized (fixed) within the PAG and antibody probe is introduced into the gel to bind to the immobilized protein target (i.e., diffusion introduction into the ~40 μm gel thickness in ~100 s [72] and binding for ~2 h for primary probing and ~1 h for secondary probing [29]).

While useful, in-gel immunoprobing sets fundamental limits on assay analytical sensitivity because the immunoprobes distribute unequally between the antibody probing solution and gel phases due to thermodynamic partitioning. At equilibrium, the in-gel antibody concentration, $[Ab^*]_{gel}$, is reduced by a factor of the partition coefficient, $K_{partition}$, from the antibody concentration in the bulk solution, $[Ab^*]_0$:

$$[Ab^*]_{gel} = K_{partition} \times [Ab^*]_0 \tag{2}$$

With an optimized imaging setup for determining in-gel partition coefficients, we found that the $K_{partition}$ for antibody in a 6%T gel was 0.097 ± 0.007 (from $n = 4$ UV-activated single-cell western blot gels) [57]. Thus, less than 10% of the solution antibody concentration is available to bind to protein within the gel. The low partition coefficient for antibody in polyacrylamide gel is critical to assay performance as the maximum immunocomplex, $[C]_{max}$, depends on the in-gel antibody concentration:

$$[C]_{max} = \frac{k_{on}[Ab^*]_{gel}[Ag]_0}{k_{on}[Ab^*]_{gel} + k_{off}} \tag{3}$$

where k_{on} ($M^{-1}\ s^{-1}$) and k_{off} (s^{-1}) are the association and dissociation rate constants, respectively, and $[Ag]_0$ is the concentration of protein within the gel. For an in-gel protein concentration just below the median-expressed protein copy number for the mammalian proteome (~0.75 nM) [9], detected with low- to moderate-affinity antibodies (K_D ~10^{-6} to 10^{-7} M), the $[C]_{max}$ falls below the single-cell western blot limit of detection (determined to be ~27,000 molecules of protein [29]). Low-abundance protein can be detected with either a high-affinity antibody or with a greater concentration of a low- to moderate-affinity antibody.

Given the dependence of immunoassay signal on in-gel antibody (probe) concentration, we have investigated methods for (1) increasing probe loading into the gel and (2) increasing uniformity of antibody probe delivery to the gel. Using antibody probe solution to rehydrate a dehydrated single-cell immunoblot gel helps overcome thermodynamic partitioning (Fig. 8) [72]. The dehydrated gel swells in a volume of antibody solution closely matched to the water volume fraction of the gel to drive all the immunoprobe into the gel at equilibrium. We found ~4–11-fold increase in the in-gel probe levels using this approach. Increased probe loading leads to a ~five-fold increase in signal-to-noise ratio (SNR) of scWB protein peaks, indicating enhanced sensitivity for dehydrated gel immunoprobing. However, nonuniform distribution of probe in dehydrated gels was observed. In cases where the partition coefficient is nonuniform across the gel, such as in gradient gels, the concomitant nonuniform probe loading can be overcome by pore expansion. Incorporation of an acid-labile crosslinker (diacrylamide ketal (DK)) into the gel precursor solution results in a hydrogel having both bis-acrylamide and DK crosslinks [33]. The DK crosslinks can be removed by incubation of the gels in 1% HCl prior to immunoprobing. Thus, the pore size can be homogenized

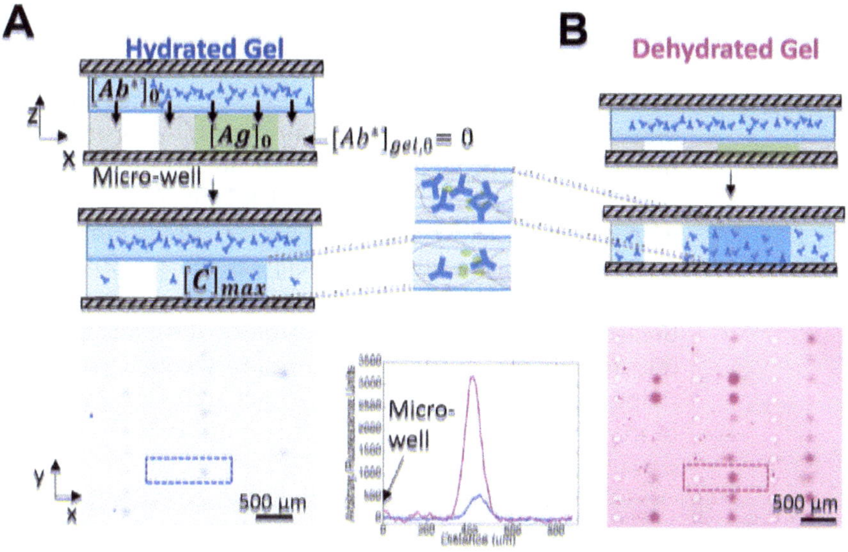

Fig. 8 Increased immunoprobe delivery increases immunoassay signal in dehydrated gels compared to hydrated gels. Schematic representation of hydrated (**a**) and dehydrated (**b**) gel immunoprobing and false-color fluorescence micrographs and representative intensity profiles of immunoprobed GFP. Reprinted with permission from Ref. [72]. Copyright 2015, American Chemical Society

after the separation and photoimmobilizing steps and before immunoprobing. Decrosslinking the gels resulted in uniform detection of protein along a ~5–9%T gradient gel, whereas lack of decrosslinking decreased immunoassay signal over tenfold along the gradient. Lastly, nonuniform immunoprobe introduction in standard hydrated gel immunoprobing can be mitigated with stirring of the antibody probe solution during incubation [73].

3.4.2 Multiplexing in Single-Cell Immunoblotting

Spectral multiplexing of the fluorescent antibody probe conjugates plays an important role in target multiplexing from a single sample. Spectral overlap of fluorescence signal limits the target multiplexing. As a result, immunoassays detect ~4–5 proteins, dependent on the fluorophores, filter set, and excitation source of the imaging system [74]. To overcome this target-multiplexing limitation, several groups have introduced chemical quenching or bleaching to probe the sample with additional antibodies in the same fluorescence channels. With these approaches, up to 61 proteins were detected with photobleaching [75], or ~60 proteins via cyclic dye oxidation [76].

A main single-cell western blot target-multiplexing strategy is a chemical stripping and reprobing approach akin to the process used in conventional western blotting. Stripping removes the immunoprobes at the conclusion of the single-cell immunoblot. Gels are immersed in a stripping buffer containing ~2% SDS and 0.8% beta-mercaptoethanol and incubated at 55 °C for at least 30 min

[29, 32]. The SDS detergent denatures while beta-mercaptoethanol reduces disulfide bonds of the immunocomplexes. After the stripping buffer is rinsed from the gel, gels are reprobed for detection with a new set of immunoprobes against different protein targets. Kang et al. profiled protein expression of ten different protein targets within PI3K/AKT/mTOR and MAPK signaling pathways in HER2-positive breast cancer cells with stripping and reprobing (Fig. 9) [43]. Furthermore, up to eight unique intracellular protein targets were measured directly from single circulating tumor cells from patients with advanced breast cancer, showing the applicability of the stripping and reprobing strategy for multiplexing in primary patient samples [59]. The stripping strategy is conducive to gel archiving for later reprobing in iterative biological hypothesis testing, even up to ~7 months of storage [31].

A main limitation of stripping and reprobing is loss of immunoassay signal with increasing numbers of stripping and reprobing cycles. Upon 12 cycles of stripping and reprobing, a >50% loss in immunoassay signal is observed [32]. By monitoring fluorescently labeled recombinant protein immobilized in 30–40 μm thick polyacrylamide gel across 29 cycles of stripping and reprobing, Gopal et al. found that protein loss is a dominant mechanism of signal loss [77]. The greatest protein loss out of the gel occurs within the first five cycles of stripping and reprobing. Importantly, the percentage of protein lost decreased for proteins with a larger hydrodynamic radius. Thus, one should consider both the expected abundance of the proteins of interest and their hydrodynamic radius when determining which proteins to immunoprobe in earlier versus later stripping and reprobing rounds [9, 77].

3.5 Data Analysis Considerations for Single-Cell Immunoblots

3.5.1 Goals of Quantitative Single-Cell Immunoblot Analysis

Two key types of information are generated by single-cell immunoblots: (1) protein amount and (2) protein location (based on mass in western blotting or charge in isoelectric focusing). Protein amounts (here relative, not absolute) can indicate whether a given protein is expressed heterogeneously across a cell population and how levels compare to other protein targets. Meanwhile, protein location along the separation axis of the immunoblot is critical in proteoform or off-target antibody binding identification. For separation of a protein and its proteoform(s), we quantify the separation resolution (R_s) to determine the degree of separation. The R_s is given by:

$$R_s = \frac{\Delta x}{\frac{1}{2}(4\sigma_1 + 4\sigma_2)} \tag{4}$$

where Δx is the center-to-center distance between protein peaks and σ_1 and σ_2 are the peak widths of the two proteins to be resolved [78]. A separation is considered baseline resolved when $R_s = 1.0$,

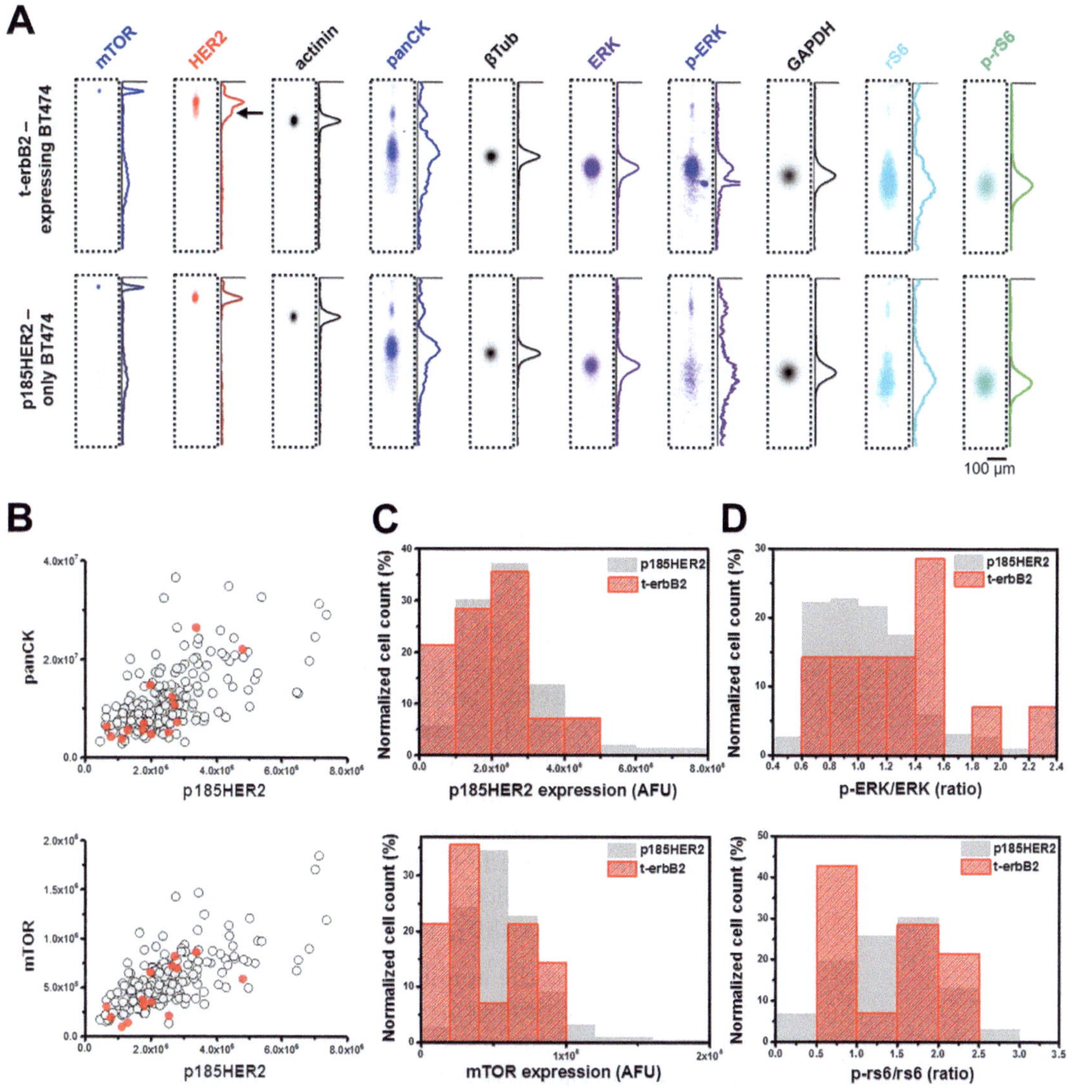

Fig. 9 Multiplexed detection of PI3K/AKT/mTOR and MAPK signaling proteins in HER2-positive breast cancer using single-cell western blotting. (**a**) False-color fluorescence micrographs and intensity profiles of ten proteins from BT474 cells that express truncated HER2 (t-erbB2) or the p185 full-length HER2 protein (p185HER2). (**b**) Scatter plot of panCK or mTOR protein versus p185HER2 levels calculated from peak area under the curve. The open circle symbols represent single BT474s expressing p185HER2 only, and the filled circles were cells expressing t-erbB2. (**c**) Histograms of p185HER2 and mTOR expression in cells expressing p185HER2 and t-erbB2 normalized to cell count. (**d**) Histograms of p-ERK/ERK and p-rs6/rs6 in the p185HER2 and t-erbB2 expressing cells normalized to cell count. Reproduced from Ref. [43], npj Precision Oncology 2018, with permissions from Springer Nature

or when Gaussian-shaped protein peaks of the same magnitude overlap by only ~2.3%. Thus, as R_s of two protein peaks approaches 1.0, there is increased confidence in discrimination of a given proteoform or off-target antibody binding peak.

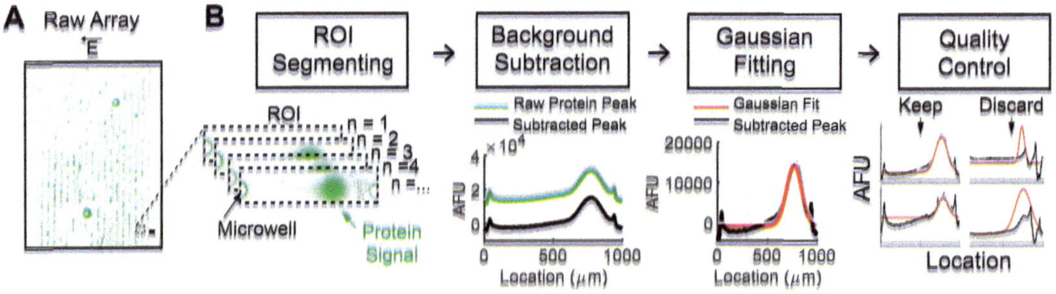

Fig. 10 Analysis pipeline for single-cell immunoblot micrographs. (**a**) False-color fluorescence micrograph of the raw array of a single-cell western blot for GFP. (**b**) Analysis steps include (i) ROI segmentation, (ii) background subtraction, (iii) Gaussian fitting, and (iv) user-driven quality control. (Reproduced from Vlassakis *et al.* Proceedings of MicroTAS 2020)

Quantification of single-cell immunoblots merges requirements of protein microarray [79] and conventional western blot [80] or electropherogram analysis (Fig. 10). First, protein signal must be distinguished from background signal (e.g., from antibody adsorption to the gel) and noise that may be variable across the array [72, 73] or gradient separation lane [33]. Thus, background subtraction, or baseline correction as it is often referred to in separation science, needs to effectively remove background without differentially impacting measured protein amount [80]. The resulting peak after background subtraction is considered detectable if the signal-to-noise ratio (SNR) is greater than 3 (assuming random Gaussian noise). Second, if the peak is detectable, the protein amount may be quantified by Gaussian fitting for area-under-the-curve (AUC) analysis [80]. We use a Gaussian function: $f(x) = ae^{-\left(\frac{x-\mu}{\sigma}\right)^2}$, where a is the amplitude, x is the position, σ is the peak width, and μ is the peak center location, because protein diffusion during electrophoresis results in Gaussian-shaped protein peaks [78]. Protein amount is determined by summing the AUC of either the Gaussian function or of the background-subtracted intensity profile over $\mu \pm 2\sigma$ (the 95% confidence interval for a Gaussian distribution) [81]. Non-Gaussian protein peaks are observed when protein is incompletely solubilized with SDS, and thus peak shape reveals the efficacy of sample preparation [29]. Finally, the Gaussian fit-determined peak center location and width determine separation resolution with another protein or approximate molecular mass if using a sizing ladder [82] or isoelectric point using pH markers [36].

3.5.2 Single-Cell Immunoblot Analysis Pipeline

Single-cell immunoblot analysis is automated in the Summit algorithm, which is comprised of four functions and available online (https://github.com/herrlabucb/summit). The algorithm segments the array into regions of interest, performs background subtraction on peak intensity profiles, fits the protein peaks to a Gaussian, and displays a GUI for user-directed quality control. All parameters that impact quantification (e.g., the Gaussian fit parameters and AUC) are stored within a field-accessible data structure. During quality control, the user inspects the Gaussian fits and intensity profiles with optional input parameters for goodness of fit (i.e., R^2 value) and SNR. Empirically, $R^2 > 0.7$ yields fits with sufficiently stringent peak boundary identification. The user then manually selects peaks to be discarded if punctate noise (e.g., antibody aggregates) or artifacts such as gel scratches are present. The high-parameter proteomic data extracted from this analysis pipeline can then be analyzed with a multitude of single-cell statistical techniques depending on the experimental hypothesis.

Acknowledgments

We would like to thank Herr Lab members, past and present, for contributing to much of the innovation presented in this chapter.

References

1. Zeng H, Sanes JR (2017) Neuronal cell-type classification: challenges, opportunities and the path forward. Nat Rev Neurosci 9:530–546
2. Kitchen RR, Rozowsky JS, Gerstein MB, Nairn AC (2014) Decoding neuroproteomics: integrating the genome, translatome and functional anatomy. Nat Neurosci 17(11): 1491–1499
3. Frese CK, Mikhaylova M, Stucchi R et al (2017) Quantitative map of proteome dynamics during neuronal differentiation. Cell Rep 18(6):1527–1542
4. Donev R, Kolev M, Millet B, Thome J (2009) Neuronal death in Alzheimer's disease and therapeutic opportunities. J Cell Mol Med 13(11–12):4329–4348
5. Trojanowski JQ, Goedert M, Iwatsubo T, Lee VMY (1998) Fatal attractions: abnormal protein aggregation and neuron death in Parkinson's disease and Lewy body dementia. Cell Death Differ 5:832–837
6. Gorman AM (2008) Neuronal cell death in neurodegenerative diseases: recurring themes around protein handling. J Cell Mol Med 12(6A):2263–2280
7. Fischer LR, Glass JD (2007) Axonal degeneration in motor neuron disease. Neurodegener Dis 4(6):431–442
8. Wilson RS, Nairn AC (2018) Cell-type-specific proteomics: a neuroscience perspective. Proteomes 6(4):51
9. Li JJ, Bickel PJ, Biggin MD (2014) System wide analyses have underestimated protein abundances and the importance of transcription in mammals. PeerJ 2:e270
10. Schwanhäusser B, Busse D, Li N et al (2011) Global quantification of mammalian gene expression control. Nature 473:337–342
11. Newman JRS, Ghaemmaghami S, Ihmels J et al (2006) Single-cell proteomic analysis of S. cerevisiae reveals the architecture of biological noise. Nature 441:840–846
12. Chen G, Gharib TG, Huang CC et al (2002) Discordant protein and mRNA expression in lung adenocarcinomas. Mol Cell Proteomics 1(4):304–313
13. De Sousa AR, Penalva LO, Marcotte EM, Vogel C (2009) Global signatures of protein and mRNA expression levels. Mol BioSyst 5: 1512–1526
14. Nery TGM, Silva EM, Tavares R, Passetti F (2018) The challenge to search for new nervous system disease biomarker candidates: the opportunity to use the proteogenomics approach. J Mol Neurosci 64:150–164
15. Furlanis E, Traunmüller L, Fucile G, Scheiffele P (2019) Landscape of ribosome-engaged transcript isoforms reveals extensive neuronal-cell-class-specific alternative splicing programs. Nat Neurosci 22:1709–1717
16. Raj B, Blencowe BJ (2015) Alternative splicing in the mammalian nervous system: recent

insights into mechanisms and functional roles. Neuron 87(1):14–27

17. Nesvizhskii AI, Aebersold R (2005) Interpretation of shotgun proteomic data. Mol Cell Proteomics 4(10):1419–1440
18. Gingras A-C, Aebersold R, Raught B (2005) Advances in protein complex analysis using mass spectrometry. J Physiol 563:11–21
19. Link AJ, Eng J, Schieltz DM et al (1999) Direct analysis of protein complexes using mass spectrometry. Nat Biotechnol 17: 676–682
20. Lesur A, Domon B (2015) Advances in high-resolution accurate mass spectrometry application to targeted proteomics. Proteomics 15(5–6):880–890
21. Yates JR, Ruse CI, Nakorchevsky A (2009) Proteomics by mass spectrometry: approaches, advances, and applications. Annu Rev Biomed Eng 11:49–79
22. Berod A, Hartman BK, Pujol JF (1981) Importance of fixation in immunohistochemistry: use of formaldehyde solutions at variable pH for the localization of tyrosine hydroxylase. J Histochem Cytochem 29:844–850
23. Voskuil JLA (2014) Commercial antibodies and their validation. F1000Res 3:232
24. Ivell R, Teerds K, Hoffman GE (2014) Proper application of antibodies for immunohistochemical detection: antibody crimes and how to prevent them. Endocrinology 155(3): 676–687
25. Teves SS, An L, Hansen AS et al (2016) A dynamic mode of mitotic bookmarking by transcription factors. Elife 5:e22280
26. Duncombe TA, Tentori AM, Herr AE (2015) Microfluidics: reframing biological enquiry. Nat Rev Mol Cell Biol 16:554–567
27. Gozes I, Ivashko-Pachima Y, Kapitansky O et al (2019) Single-cell analysis of cytoskeleton dynamics: from isoelectric focusing to live cell imaging and RNA-seq. J Neurosci Methods 323:119–124
28. Gozes I, Sweadner KJ (1981) Multiple tubulin forms are expressed by a single neurone. Nature 294:477–480
29. Kang C-C, Yamauchi KA, Vlassakis J et al (2016) Single cell-resolution western blotting. Nat Protoc 11:1508–1530
30. Vlassakis J, Herr AE (2017) Joule heating-induced dispersion in open microfluidic electrophoretic cytometry. Anal Chem 89: 12787–12796
31. Kang C-C, Lin J-MG XZ et al (2014) Single-cell western blotting after whole-cell imaging to assess cancer chemotherapeutic response. Anal Chem 86:10429–10436
32. Hughes AJ, Spelke DP, Xu Z et al (2014) Single-cell western blotting. Nat Methods 11: 455–464
33. Duncombe TA, Kang C-C, Maity S et al (2016) Hydrogel pore-size modulation for enhanced single-cell western blotting. Adv Mater 28:327–334
34. Tan KY, Herr AE (2020) Ferguson analysis of protein electromigration during single-cell electrophoresis in an open microfluidic device. Analyst 145:3732–3741
35. Righetti PG (1983) Isoelectric focusing: theory, methodology and applications. Elsevier, Amsterdam
36. Tentori AM, Yamauchi KA, Herr AE (2016) Detection of isoforms differing by a single-charge unit in individual cells. Angew Chem Int Ed 55:12431–12435
37. Friedman DB, Hoving S, Westermeier R (2009) Chapter 30 Isoelectric focusing and two-dimensional gel electrophoresis. Methods Enzymol 463:515–540
38. Ferguson KA (1964) Starch-gel electrophoresis—application to the classification of pituitary proteins and polypeptides. Metabolism 13: 985–1002
39. Burnette WN (1981) "Western blotting": electrophoretic transfer of proteins from sodium dodecyl sulfate-polyacrylamide gels to unmodified nitrocellulose and radiographic detection with antibody and radioiodinated protein a. Anal Biochem 122:195–203
40. Bischoff KM, Shi L, Kennelly PJ (1998) The detection of enzyme activity following sodium dodecyl sulfate- polyacrylamide gel electrophoresis. Anal Biochem 260:1–17
41. Blattler DP, Reithel FJ (1970) Molecular weight determinations and the influence of gel density, protein charges and protein shape in polyacrylamide gel electrophoresis. J Chromatogr A 46:286–292
42. Schaägger H, Cramer WA, von Jagow G (1994) Analysis of molecular masses and oligomeric states of protein complexes by blue native electrophoresis and isolation of membrane protein complexes by two-dimensional native electrophoresis. Anal Biochem 217: 220–230
43. Kang C-C, Ward TM, Bockhorn J et al (2018) Electrophoretic cytopathology resolves ERBB2 forms with single-cell resolution. npj Precis Oncol 2:1–10
44. Kahle J, Wätzig H (2018) Determination of protein charge variants with (imaged) capillary

isoelectric focusing and capillary zone electrophoresis. Electrophoresis 39:2492–2511
45. Yamauchi KA, Tentori AM, Herr AE (2018) Arrayed isoelectric focusing using photopatterned multi-domain hydrogels. Electrophoresis 39:1040–1047
46. Su EJ, Herr AE (2017) Electrophoretic cytometry of adherent cells. Lab Chip 17: 4312–4323
47. Pan Q, Herr AE (2018) Geometry-induced injection dispersion in single-cell protein electrophoresis. Anal Chim Acta 1000:214–222
48. Rosàs-Canyelles E, Modzelewski AJ, Geldert A et al (2020) Assessing heterogeneity among single embryos and single blastomeres using open microfluidic design. Sci Adv 6:eaay1751
49. He M, Herr AE (2009) Microfluidic polyacrylamide gel electrophoresis with in situ immunoblotting for native protein analysis. Anal Chem 81:8177–8184
50. Hughes AJ, Herr AE (2010) Quantitative enzyme activity determination with zeptomole sensitivity by microfluidic gradient-gel zymography. Anal Chem 82:3803–3811
51. Menter P, Drive AN (2000) Acrylamide polymerization – a practical approach. Bio-Rad Lab, Hercules
52. Neira HD, Jeeawoody S, Herr AE (2020) Reversible functionalization of clickable polyacrylamide gels with protein and graft copolymers. Adv Funct Mater 30:2005010
53. Zhang T, Li S, Warden AR et al (2020) A photoclick hydrogel for enhanced single-cell immunoblotting. Adv Funct Mater 30:1–8
54. Dormán G, Prestwich GD (1994) Benzophenone Photophores in biochemistry. Biochemistry 33:5661–5673
55. Blawas AS, Reichert WM (1998) Protein patterning. Biomaterials 19:595–609
56. Occhetta P, Visone R, Russo L et al (2015) VA-086 methacrylate gelatine photopolymerizable hydrogels: a parametric study for highly biocompatible 3D cell embedding. J Biomed Mater Res Part A 103:2109–2117
57. Su A, Smith BE, Herr AE (2020) *In situ* measurement of thermodynamic partitioning in open hydrogels. Anal Chem 92:875–883
58. Jeeawoody S, Yamauchi KA, Su A, Herr AE (2020) Laterally aggregated polyacrylamide gels for immunoprobed isoelectric focusing. Anal Chem 92:3180–3188
59. Sinkala E, Sollier-Christen E, Renier C et al (2017) Profiling protein expression in circulating tumour cells using microfluidic western blotting. Nat Commun 8:1–12
60. Rosàs-Canyelles E, Dai T, Li S, Herr AE (2018) Mouse-to-mouse variation in maturation heterogeneity of smooth muscle cells. Lab Chip 18:1875–1883
61. Kim JJ, Sinkala E, Herr AE (2017) High-selectivity cytology via lab-on-a-disc western blotting of individual cells. Lab Chip 17: 855–863
62. Kim SH, Yamamoto T, Fourmy D, Fujii T (2011) Electroactive microwell arrays for highly efficient single-cell trapping and analysis. Small 7:3239–3247
63. Thomas RS, Morgan H, Green NG (2009) Negative DEP traps for single cell immobilisation. Lab Chip 9:1534–1540
64. Plouffe BD, Murthy SK, Lewis LH (2015) Fundamentals and application of magnetic particles in cell isolation and enrichment: a review. Reports Prog Phys 78:016601
65. Pamme N, Wilhelm C (2006) Continuous sorting of magnetic cells via on-chip free-flow magnetophoresis. Lab Chip 6:974–980
66. Witz IP (2008) Tumor-microenvironment interactions: dangerous liaisons. Adv Cancer Res 100:203–229
67. Paschos KA, Canovas D, Bird NC (2009) Enzymatic function of multiple origins regulates the progression of colorectal cancer and the development of metastases. Hippokratia 13:23–31
68. Zhang Y, Naguro I, Herr AE (2019) In situ single-cell Western blot on adherent cell culture. Angew Chem 131:14067–14072
69. Shukolyukov SA (2012) Digitonin and sodium dodecylsulfate-solubilized frog rho-dopsin: behavior under native and denaturing polyacrylamide gel electrophoresis. Adv Biol Chem 02:84–91
70. Garfin DE (2003) Gel electrophoresis of proteins. In: Davey J, Lord M (eds) Essential cell biology, volume 1: cell structure, a practical approach. Oxford University Press, Oxford, pp 197–268
71. Yamauchi KA, Herr AE (2017) Subcellular western blotting of single cells. Microsystems Nanoeng 3:16079
72. Vlassakis J, Herr AE (2015) Effect of polymer hydration state on in-gel immunoassays. Anal Chem 87:11030–11038
73. Geldert A, Huang H, Herr AE (2020) Probe-target hybridization depends on spatial uniformity of initial concentration condition across large-format chips. Sci Rep 10:1–12
74. Stack EC, Wang C, Roman KA, Hoyt CC (2014) Multiplexed immunohistochemistry,

imaging, and quantitation: a review, with an assessment of Tyramide signal amplification, multispectral imaging and multiplex analysis. Methods 70:46–58
75. Gerdes MJ, Sevinsky CJ, Sood A et al (2013) Highly multiplexed single-cell analysis of formalin-fixed, paraffin-embedded cancer tissue. Proc Natl Acad Sci U S A 110: 11982–11987
76. Lin J, Izar B, Wang S et al (2018) Highly multiplexed immunofluorescence imaging of human tissues and tumors using t-CyCIF and conventional optical microscopes. eLife 7:1–46
77. Gopal A, Herr AE (2019) Multiplexed in-gel microfluidic immunoassays: characterizing protein target loss during reprobing of benzophenone-modified hydrogels. Sci Rep 9: 15389
78. Giddings JC (1991) Unified separation science. John Wiley & Sons, New York, pp 97–101
79. Ju Z, Liu W, Roebuck PL et al (2015) Development of a robust classifier for quality control of reverse-phase protein arrays. Bioinformatics 31:912–918
80. Gassmann M, Grenacher B, Rohde B, Vogel J (2009) Quantifying Western blots: pitfalls of densitometry. Electrophoresis 30:1845–1855
81. Glantz S (2005) Primer of biostatistics. McGraw-Hill Medical, New York
82. Kim JJ, Chan PPY, Vlassakis J et al (2018) Microparticle delivery of protein markers for single-cell western blotting from microwells. Small 14(48):e1802865

Chapter 8

Quantifying Enzyme Activity and Gene Expression Within Single Cells Using a Multiplexed Capillary Electrophoresis Platform

Matthew M. Anttila, Brae V. Petersen, Luke A. Gallion, Manibarathi Vaithiyanathan, and Nancy L. Allbritton

Abstract

To date, numerous chemical cytometry platforms have been developed to study a wide range of biologically relevant analytes in single cells by employing creative combinations of chemical sensors with separation methods and detection instrumentation. However, despite many impressive developments over the past decade, the scope of chemical cytometry methods has remained limited, in part due to a reliance on the use of chemical sensors for analyte detection. Traditionally, these chemical sensors have been restricted to detecting a single analyte or a closely related subset of analytes. Additionally, the use of multiple sensors simultaneously can quickly increase the hardware and experimental complexity, thereby restricting the technology to a handful of academic laboratories. In contrast, single-cell sequencing technologies have seen widespread adoption over the past decade and facilitate measurements of thousands of unique genes simultaneously, yielding a rich data set describing the cellular genotype or RNA expression pattern. Furthermore, sequencing methods and hardware have grown increasingly more robust and cost-effective over time. While these developments are impressive, methods relying on nucleic acid sequencing alone (genotype and RNA expression) provide a limited view into the biochemical processes and overall cellular phenotype. Therefore, to improve upon current single-cell sequencing methods, our lab has developed a combined platform that integrates capillary electrophoresis-based chemical cytometry measurements with the power of single-cell sequencing. In this manuscript, we describe the background and theory behind our multiplexed CE-based, single-cell analysis platform and provide detailed instruction as to its construction and operation when performing simultaneous measurements of enzyme activity and gene expression in single cells.

Key words Single-cell analysis, Capillary electrophoresis, RNA-seq, Kinase, Chemical cytometry

1 Introduction

1.1 Challenges Facing Multiplexed Single-Cell Analysis

To perform accurate measurements on single cells, multiple analytical challenges must be addressed. Single mammalian cells are small in size (~100–1000 μm^3) [1] and extremely low in volume (~0.1–10 pL) [2]. Each cell possesses a vast number of different

Jonathan V. Sweedler et al. (eds.), *Single Cell 'Omics of Neuronal Cells*,
Neuromethods, vol. 184, https://doi.org/10.1007/978-1-0716-2525-5_8,

molecular species so that most species are present in only minute quantities (as little as 10^{-21} moles) [3]. These attributes make multiplexed analysis of multiple cellular features and assay of large numbers of cells particularly challenging and demand that all single-cell analytical methods be systematically validated. Difficulty in understanding the biological meaning of data acquired from single cells occurs due to the heterogeneity among "seemingly" identical cell populations [4]. These differences are driven by stochastic gene expression, cell state variance, nutrient availability, random noise within critical biochemical pathways, and other attributes [5, 6]. Such variability within cell populations mitigates the utility of bulk population measurements, which report an attribute averaged across all cells. This signal averaging has the effect of obfuscating subtle nuances in signaling that correspond to subpopulations whose asynchronous behavior can potentially indicate a nascent etiology [7]. Furthermore, cellular heterogeneity becomes increasingly complex among cells that are not genetically identical, for example, tumors in which genetic alterations accumulate within cells, serving to further increase the complexity of the sample and subsequent data analysis [8, 9].

Methods to elucidate the diverse array of cellular types within a population continue to show great promise in many areas of biology and medicine. For example, single-cell sequencing methods were used to deconvolute the complex cellular genotypes of glioblastoma tumors (GBM), which have shown to be consistently variable across patient cohorts [10]. Indeed, single-cell RNA-seq (scRNA-seq) data suggest that most GBMs consist of at least four unique cell types: neural-progenitor-like, oligodendrocyte-progenitor-like, astrocyte-like, and mesenchymal-like [10]. Importantly, each subpopulation can be further divided into smaller subpopulations based on factors such as cell-cycle stage or local tumor microenvironment (TME) [11]. The TME furthers the cellular heterogeneity of the tumor due to differences in extracellular matrix and gradients of signaling factors, nutrients, and oxygen, ultimately leading to the formation of rare subpopulations [12]. Increasingly, single-cell assays are enabling researchers and clinicians to characterize many properties of single cells including their metastatic potential, drug resistance, and growth rate among other biochemical processes [6, 10, 13, 14].

1.2 Introduction to Chemical Cytometry

The term cytometry is used to describe a suite of analytical methods that measure a variety of characteristics of single cells [3]. Cytometric methods can currently be broken down into five main subcategories: (1) enumeration cytometry, where cells are counted and clustered into subpopulations using morphological differences [15]; (2) imaging cytometry, where intracellular and surface markers are characterized using fluorescent dyes and/or antibodies (Ab); (3) flow cytometry, which utilizes light scatter, fluorescent stains, and antibodies (Ab) to characterize cell populations based on granularity, intracellular or cell surface markers, and metabolites [16]; (4) mass cytometry, which employs Abs labeled with a lanthanoid and/or actinoid metal ion to permit the simultaneous characterization of over 40 cellular markers [17]; and (5) chemical cytometry, which uses instrumentation from analytical chemistry to measure the chemical contents of cells [3, 18]. Chemical cytometry has been performed utilizing matrix-assisted laser desorption/ionization (MALDI) mass spectrometry, Raman spectroscopy, and electrochemistry [18–21]. Other forms of chemical cytometry employ a chemical sensor to facilitate the detection and quantitation of a chemical at the level of a single cell. To date, an impressive array of chemical sensors have been developed for use in chemical cytometry including sensors for gases, ions, amino acids, peptides, proteins, sugars, lipids, and polynucleotides [18].

Interrogating properties of single cells with chemical cytometry is challenging due to the small volume of cells, the extremely low quantities of analytes within single cells, and the very large number of other potentially interfering molecules, as described above in Subheading 1.1 [3]. As such, chemical cytometric assays must be sufficiently sensitive to interrogate the cellular feature of interest while overcoming noise that may arise from the large numbers of other molecules within the intracellular milieu. Therefore, fluorescence detection is often utilized in chemical cytometry methods that employ a chemical sensor due to its inherently high sensitivity [3]. A wide range of single-cell assays using fluorescent chemical sensors have been described [18, 22–25]. To date, our group has developed fluorescent peptide-based sensors to measure the single-cell activity of kinases, phosphatases, proteases/peptidases, esterases, and other enzymes [26–35]. Detection of fluorescent species is especially powerful when paired with a separation method, enabling the analysis of closely related metabolic modifications to the parent fluorescent sensor [3, 18]. In particular, capillary electrophoresis employing fluorescence detection (CE-F) has been used extensively to quantify endogenous analytes and enzyme activity within individual cells [3, 18, 36–39]. CE is particularly well suited to handle the small sample volume of mammalian cells without significant sample dilution and loss of sensitivity [18]. Additionally, CE offers superb resolving power which enables separation of nearly identical molecular species. Furthermore, CE has a high peak capacity which facilitates simultaneous separation of multiple species and therefore enables multiplexed measurements of several biochemical pathways at once [18, 36, 37]. It is important to note that separations have been employed for chemical cytometry experiments utilizing high-performance and ultra performance liquid chromatography (HPLC and UPLC, respectively) in addition to CE. The benefits and drawbacks of utilizing different separation and detection schemes have been discussed in detail in previous reviews and depend on numerous factors such as analyte physicochemical properties, analyte concentration, method limits of detection (LOD), cell-sampling method, sample throughput, and inter-run variability, among others [6, 18, 40].

Our group has developed an automated CE-F chemical cytometry platform for performing measurements of enzyme activity within single cells [41–43]. The platform also facilitates the separation and/or collection of nucleic acids from single cells either simultaneously or immediately following the CE-F separation (Fig. 1a). This method takes advantage of the mobility differences between nucleic acids and a fluorescent kinase substrate due to the differences in size and net charge between the nucleic acids and fluorescent chemical sensors (Fig. 1b, c). The following section covers the fundamental principles that drive CE separations and will help elucidate the forces that facilitate the simultaneous electrophoretic separation of fluorescent enzyme sensors and ejection of nucleic acids from within the capillary.

1.3 CE Theory and Fundamentals for Chemical Cytometry

CE was first described in the early 1980s by Jorgenson and Lukacs [44]. Akin to other electrophoretic separation methods, separations in CE result from the differential migration rates of solutes under the influence of an applied electric field [44, 45]. A distinguishing feature of CE is that the separations occur within a small inner-diameter fused silica capillary (typically ≤50 μm) that is filled with a conductive solution known as the background electrolyte (BGE). To perform a separation, a sample solution is injected into the capillary inlet electrokinetically or via hydrodynamic forces [44, 45]. Once the sample injection is complete, both the capillary inlet and outlet are submerged in BGE, and a voltage (typically 10–30 kV) is applied across the capillary to drive analyte migration and separation. Analyte species are detected as they migrate past the detection region of the capillary, typically located near the capillary outlet (Fig. 2a).

Analyte mobility within the capillary is based on many factors. The experimentally observed mobility, known as the apparent mobility (μ_{app}), is a function of the analyte electrophoretic mobility (μ_{ep}) and the electroosmotic flow of the BGE (μ_{eo}); the sign of μ_{app} describes whether the analyte migration is toward the cathode (+) or the anode (−) when a positive voltage is applied to the capillary. μ_{app} may also be described in terms of net velocity (v_{net}) and the strength of the applied electric field (E). The parameter μ_{ep} describes the velocity and direction of a solute's electrophoretic migration and is a function of the size, shape, and charge of said solute and its hydration sphere [45]. For a rigid sphere, the μ_{ep} and μ_{app} can be approximated as shown in Eqs. 1, 2, and 3 [45].

$$\mu_{ep} = \frac{q}{f} = \frac{q}{6\pi\eta r} \tag{1}$$

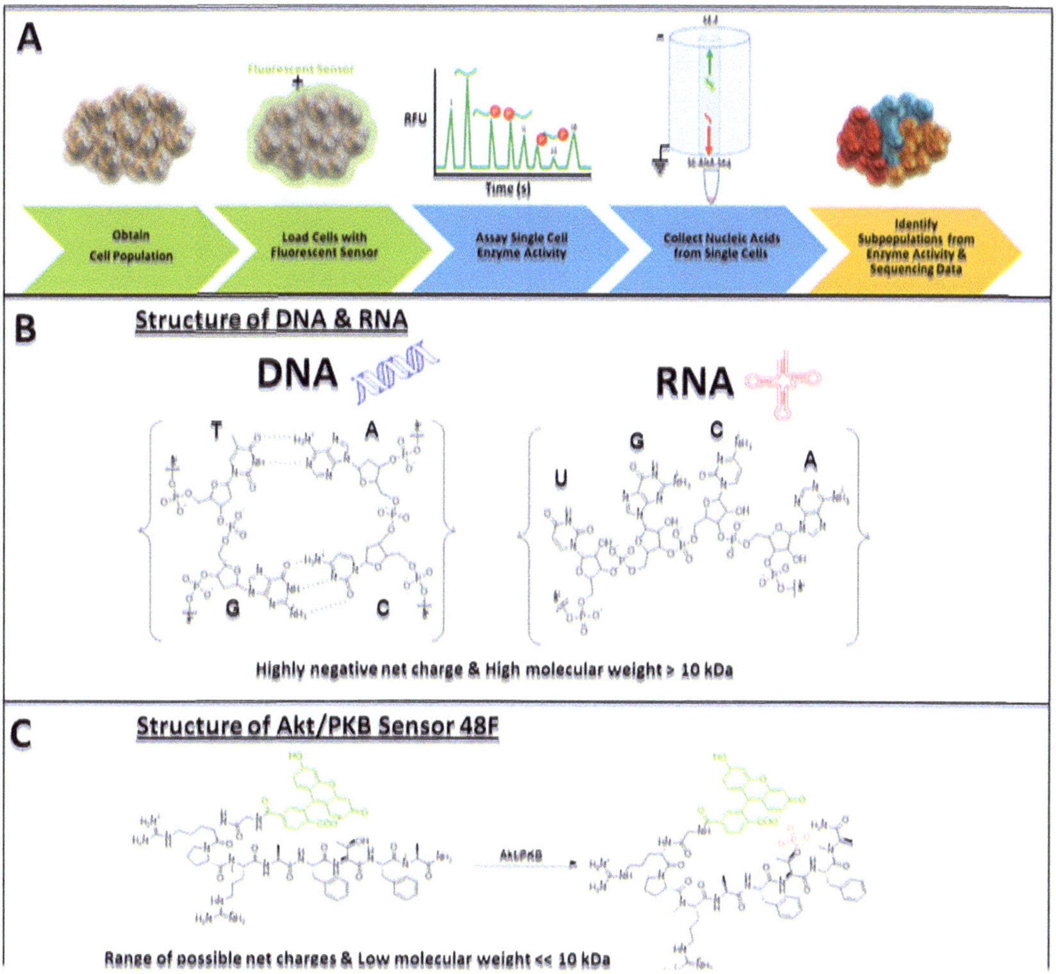

Fig. 1 (**a**) Single-cell CE-seq workflow. Cells are first loaded with a fluorescent enzyme sensor before being lysed and analyzed by CE-F to ratiometrically quantify the activity of Akt kinase. Additionally, the platform enables nucleic acids to be collected simultaneously or post CE-F analysis for sequencing experiments. Together, the enzyme activity and nucleic acid data facilitate the identification of subpopulations within the sample through correlations of gene expression and enzyme activity at the level of single cells. (**b**) Chemical structures of DNA and RNA. Their high negative net charge facilitates electrokinetic separation from fluorescent enzyme sensors and electrokinetic ejection from the capillary inlet under low electroosmotic flow conditions. Nucleic acids collected from single cells can then be analyzed using sequencing technologies to identify gene expression and transcript abundances for target enzymes within each cell. (**c**) Chemical structure of the 48F sensor for Akt [27]. The 48F sensor is selectively phosphorylated by Akt (orange phosphate group), which is detectable using CE-F since the addition of the phosphate moiety on the threonine residue results in the sensor having a greater net negative charge than the parent substrate, decreasing its electrophoretic mobility

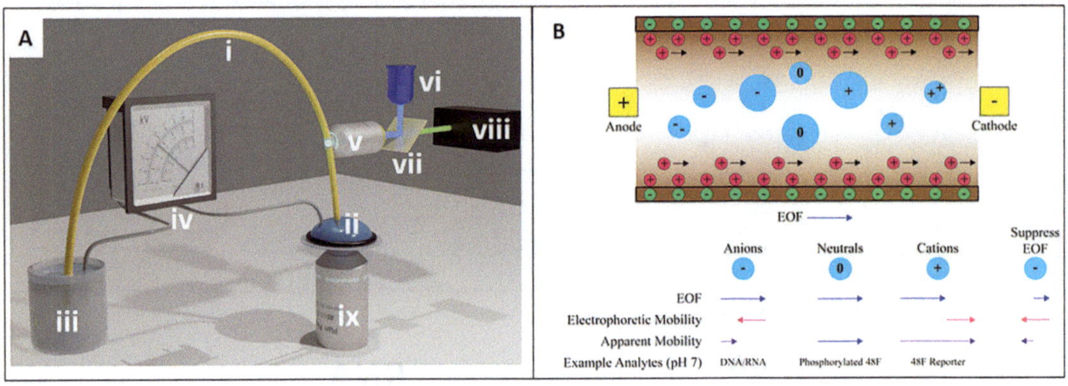

Fig. 2 (**a**) Simplified schematic of a single-cell CE-F instrument. This setup can be easily adapted for performing nucleic acid collections by applying pressure to the capillary outlet. The inlet (ii) of a fused silica capillary (i) is submerged in an incubation chamber containing single cells while the capillary outlet (iii) is submerged in a buffer reservoir. A high-voltage power supply (iv) is used to create an electric field across the capillary to facilitate analyte migration from the inlet toward the detection window. For instruments utilizing fluorescence detection, a microscope objective (v) focuses a 488 nm CW laser (vi) on the capillary lumen within the detection window, which excites fluorescent analytes as they migrate past the detection window. Emitted light from the fluorophore is collected by the same objective and passes through a dichroic mirror (vii) before reaching the detector. The photomultiplier detector (viii) transduces the signal, which is digitized into an electropherogram. An additional microscope objective (ix) is located below the cell chamber to facilitate cell imaging, capillary alignment, and cell lysis using a co-aligned 532 nm Nd:YAG pulsed laser (not shown). (**b**) Schematic (not to scale) outlining the concepts of electroosmotic flow, electrophoretic mobility, and apparent mobility. *Electroosmotic flow*: Negatively charged silanol groups (green) on the surface of the capillary cause a zeta potential to form at the surface when an electric field is applied to the channel using two electrodes (yellow). Cations (red) in the electrophoretic buffer interact with the channel surface, forming an inner layer of static cations called the Stern layer. Cations in the Stern layer do not completely neutralize the channel surface, and a second diffuse layer forms outside the Stern layer. Cations in the diffuse layer are drawn toward the cathode, initiating bulk fluid flow termed electroosmotic flow. *Electrophoretic mobility*: Electrophoretic mobility of analytes is based on charge and size. The relative position of analytes with different charges and sizes is demonstrated by the blue circles, where analytes with a greater net negative charge migrate more quickly toward the cathode. *Apparent mobility*: Apparent mobility of each analyte is the sum of the EOF of the system and electrophoretic mobility of the analyte. Note that if EOF is sufficiently suppressed or reversed, it is possible for some analytes to migrate toward the inlet rather than the outlet [45]

μ_{ep} = electrophoretic mobility

q = particle charge

f = translational frictional coefficient

η = BGE viscosity

r = Stokes radius

$$r = \sqrt[3]{\frac{3M}{4\pi V}} \tag{2}$$

M = mass

V = partial specific volume

$$\frac{V_{net}}{E} = \mu_{app} = \left(\mu_{ep} + \mu_{eo}\right) \tag{3}$$

V_{net} = net velocity

E = applied electric field

The μ_{eo} arises in CE due to the presence of the charged inner surface of the fused silica capillary, and the contribution of μ_{eo} to μ_{app} increases proportionally to the charge of the inner surface; the charge of the inner surface depends on the properties of the silica surface itself as well as the BGE (i.e., in a neutral or basic buffer, the net charge on the capillary wall is negative due to the deprotonated silanol moieties). The ionic attraction of the BGE cations to the negatively charged surface results in the formation of an electrical double layer [45]. Upon the application of a voltage, the solution cations will migrate toward the cathode. Due to hydrogen bonding and electrostatic interactions of these cations with other solution components, the bulk migration of all BGE contents occurs in the direction of the cathode. Therefore, the resulting bulk electro-osmotic flow displays a flat fluid flow profile (in contrast to the parabolic flow profile observed with pressure-driven flow). In this manner, all analytes are moved toward the detector at the same rate by μ_{eo}, while differences in μ_{ep} facilitate the separation of solutes into discrete bands or zones (Fig. 2b) [45].

Under the right conditions, this interplay of μ_{ep} and μ_{eo} can be used to facilitate separations of analytes that display opposite μ_{app} values. Under normal voltage polarity and low buffer conductivity (<5 mS), such bidirectional separations occur when a highly anionic analyte that displays a negative value for μ_{app} migrates backward out of the capillary inlet toward the anode, while the neutral and/or cationic fluorescent sensors display a positive value for μ_{app} and migrate forward toward the cathode. It is important to note that while our studies have been limited to nucleic acids and fluorescent sensors and dyes, this phenomenon can be applied to any classes of solutes which display opposite μ_{app} values [46].

1.4 Evaluating Enzyme Activity in Single Cells with CE-F Chemical Cytometry

Several important considerations must be considered when designing a sensor to facilitate reliable and robust measurements of enzyme activity within single cells. Genetically encoded sensors have been employed in imaging cytometry but are generally not of utility in CE-F due to their large size, difficult design, and requirement for genetically engineered cells. Chemical sensors for CE-F are often designed as modified fluorescent substrates for measurement of intracellular enzyme activity. Desirable traits of these sensors include the ability to (1) penetrate the plasma membrane to be retained within the targeted cell compartment, (2) distinguish between closely related biochemical events (i.e., enzyme selectivity), (3) retain structural integrity in the metabolically active intracellular environment during the time course of the experiment (min to h), (4) possess programmable start/stop times for the sensor measurement, and (5) provide an exquisitely sensitive readout as the quantities of intracellular sensor can routinely approach 10^{-18}–10^{-21} moles [18, 27, 47]. Furthermore, an ideal sensor should be usable in a diverse array of cell types to maximize the translatable utility of the sensor. In practice, it is extremely difficult to develop a sensor that displays all of the traits listed above, but significant progress continues to be made to improve the robustness of novel sensors. Most fluorescent enzyme sensors developed for CE-F-based chemical cytometry have been utilized for quantifying kinase, phosphatase, lipase, esterase, or protease/peptidase activity within cells by monitoring shifts in sensor migration time (t_m) following the desired enzymatic reaction. A difference in t_m can occur from a change in the net charge, size, or shape of the sensor following the addition/removal of phosphate moieties and/or amino acid residues [18, 27, 30, 34, 36]. Thus, the separation of these substrates and products in time enables ratiometric quantitation of intracellular enzyme activity by temporally resolving unique product species present within the sample. Enzyme activity is quantified by measuring the corrected area (CA), which is the peak area (A) divided by migration time, t_m, of the parent sensor peak and peaks resulting from unique enzymatic modifications to the sensor. We quantify the relative activity (RA) of enzymes in single cells by taking the ratio of the parent CA to that of an enzymatic modification peak CA, normalized to the total CA of the parent, phosphorylated, and proteolytic degradation peaks within the electropherogram (Eq. 4).

$$\mathrm{RA} = \frac{\mathrm{CA}_{\mathrm{Phosphorylated}}}{\mathrm{CA}_{\mathrm{Parent}} + \mathrm{CA}_{\mathrm{Phosphorylated}} + \mathrm{CA}_{\mathrm{Degraded}}} \quad (4)$$

$\mathrm{CA}_{\mathrm{Parent}}$ = Parent sensor corrected area

$\mathrm{CA}_{\mathrm{Phosphorylated}}$ = Phosphorylated sensor corrected area

$\mathrm{CA}_{\mathrm{Degraded}}$ = Degraded sensor corrected area

1.5 Utilizing CE to Collect Nucleic Acids from Single Cells

Chemical cytometry assays can provide high chemical specificity with each assay being able to differentiate between subtle structural modifications following a variety of metabolic reactions. However, as mentioned previously, chemical cytometry assays generally provide a limited scope into the biology of a cell due to the focus on a very small subset of analytes and/or enzymatic reactions. Additionally, sensor development for new chemical cytometry assays is often a research-intensive process that requires optimization of the sensor for cell loading, resistance to degradation, and target specificity, as discussed in Subheading 1.4 [27, 28, 35]. In comparison, nucleic acid-based single-cell assays often have a genome-, exosome-, or transcriptome-wide scope, and sensor development is either not necessary for next generation sequencing assays or is a relatively straightforward and a quick process for PCR-based assays [48]. While nucleic acid-based assays are convenient, genotype or gene expression is not tightly correlated with protein concentration or activity [49]. Correlation coefficients between mRNA and protein abundance are less than 0.7–0.8 for proteins that do not show significant temporal concentration variation in cells and are far worse for proteins with dynamically regulated concentrations in time [50]. Importantly, much cellular physiology cannot be understood by measuring cellular genotype or gene expression alone. As such, it is common when studying bulk cell lysates to perform chemical assays, such as enzyme activity assays, in addition to nucleic acid-based assays to connect the biochemical cellular phenotype cell to its genetic profile [51].

It is important to note that combining nucleic acid-based assays with chemical cytometry poses a significant challenge as many chemical cytometry assays are cell destructive. Additionally, splitting the contents of a single cell is not ideal as there can be significant spatial heterogeneity of analytes across the cell [52]. Another challenge is that many analytes of interest for single-cell analysis are present in extremely low abundance within a complex sample matrix and often approach the limits of detection for instrumentation. Thus, it is ideal to analyze all of the cellular contents without dilution and/or fractionation to ensure the reproducibility and robustness of the assay.

CE-F-based chemical cytometry utilizes a relatively straightforward analysis method that can readily be multiplexed with nucleic acid-based single-cell analyses. While CE-F-based chemical cytometry is destructive since each cell is lysed during the experiment, the contents of the cell are confined within the micron dimensions of the capillary, thereby avoiding significant lysate dilution. Therefore, we reasoned it would be quite feasible to collect the entire contents of the capillary into a vial or well plate following CE-F analysis. For example, the entire volume of a 45 cm long capillary with an inner diameter of 30 um is only 1.2 μL which can easily be incorporated into the sample preparation of many protocols for nucleic acid-based single-cell analysis [53]. Collecting nucleic acids in this manner requires minor (if any) changes to the traditional CE-F-based chemical cytometry instrumentation. Thus, post-separation collection of nucleic acids is compatible with most CE assays, only requiring prior knowledge regarding the mobility of the free nucleic acids in the CE separation buffer.

2 Materials

2.1 Capillary Electrophoresis

Boric acid, Trizma base, Tricine, 4-(2-hydroxyethyl)-1-piperazineethanesulfonic acid (HEPES), and Tween 20® were obtained from Millipore Sigma. All buffer solutions were prepared using deionized water from an ultrapure filtration system (18 MΩ). Glass slides, calcium chloride, sodium chloride, and sodium phosphate were purchased from Thermo Fisher Scientific. Polyimide-coated fused silica capillaries (30 and 50 μm i.d.) were purchased from Polymicro Technologies.

2.2 qPCR

RNA was reverse transcribed to cDNA using the SuperScript VILO master mix (Thermo Fisher Scientific, Waltham, MA). Three TaqMan gene expression probes (Thermo Fisher Scientific, Waltham, MA) were used to estimate the amount of RNA loaded into a cell: RPL8, COX5B, and RAD21. Gene expression qPCR was performed on a real-time PCR system (ViiA 7, Applied Biosystems, Waltham, MA) using these probes and a TaqMan fast advanced master mix. Digital droplet PCR was run on the QuantStudio 3D platform (Applied Biosystems) following manufacture recommendations. Endpoint PCR was performed on collected DNA and analyzed by polyacrylamide gel electrophoresis (PAGE). DreamTaq master mix was used for PCR amplification (Thermo Fisher Scientific, Waltham, MA). Seven percent polyacrylamide gels were prepared using 19:1 acrylamide/bis-acrylamide, N,N,N′,N-′-tetramethylethylenediamine (TEMED), ammonium persulfate, and 50 mM Tris-borate (all purchased from Millipore Sigma). Gels were imaged using an LED and blue light filter from miniPCR (blueBox, miniPCR, Cambridge, MA) with a digital camera (Pixel XL, Google, Mountain View, CA).

2.3 Cell Culture

PANC-1 cells were purchased from ATCC (CRL-1469). Bovine serum albumin was obtained from Calbiochem. Dulbecco's Modified Eagle Medium (DMEM) were purchased from Cellgro. Penicillin/streptomycin was obtained from Gibco, and fetal bovine serum (FBS) was obtained from Atlanta Biologicals. All other reagents and solvents were purchased from Sigma-Aldrich (now Millipore Sigma) or Thermo Fisher Scientific. Slide chambers were prepared with a 15 mm O-ring bonded with polydimethylsiloxane (PDMS) to a 25 mm diameter #1 glass coverslip. Approximately 10,000 cells were added to 600 μL of ECB on each analysis slide. Slides were incubated for 10 min at 37 °C in 5% CO_2 before placing on the custom CE instrument.

2.4 Akt Enzyme Sensors

Akt enzyme sensors C48F and 48F were synthesized using an Fmoc solid-phase peptide synthesis protocol [26–28]. O-(6-Chlorobenzotriazol-1-yl)-N,N,N′,N′-tetramethyluronium hexafluorophosphate (HCTU), 6-carboxyfluorescein (6FAM), Fmoc-protected amino acids, and NovaSyn TGR resin were purchased from ChemPep and NovaBiochem. Coupling validation was performed using a ninhydrin test (AnaSpec) for primary amines or a chloranil/acetaldehyde test for secondary amines. Sensors C48F and 48F were loaded intracellularly either by passive diffusion in culture medium or using an Influx pinocytic-loading kit from Thermo Fisher Scientific. All other reagents and solvents were purchased from Millipore Sigma or Thermo Fisher Scientific.

3 Methods

3.1 Constructing a Single-Cell CE-Seq Platform

Our group has developed a customized CE-F platform for performing nucleic acid collections and enzyme activity measurements from single cells [43]. A repository of the motorized hardware used in our systems as well as CAD and other design files are available at https://github.com/brae-pete/AutomatedCE, while a list of major components and their specifications have been included in Table 1. The platform consists of the following core components: an inverted fluorescence microscope with automated XY-stage, automated Z-stage, CCD camera, epifluorescence detection assembly, pulsed Nd:YAG laser, and a high-voltage power supply [41–43]. While the inverted microscope provides a large open area above cells for capillary access, buffer exchange, and reagent addition, the motorized stage is used to position single cells under the capillary with the assistance of a charge-coupled device (CCD) camera. Light-emitting diodes (LEDs) attached near the capillary inlet supply the bright-field light source to detect and locate cells. The inverted microscope is utilized to focus a pulsed light beam from a neodymium-doped yttrium aluminum garnet (Nd:YAG) laser to create a cavitation bubble near the surface of the glass that lyses individual cells. The *Z*-stage is used to control the height of the capillary over the microscope's *XY*-stage.

All of the CE-F optical components are mounted atop an optical post that is secured to the *Z*-stage and located adjacent to the inverted microscope such that the CE optical path is located directly above the *XY*-stage (Fig. 3). The on-capillary fluorescence detection pathway uses a constant wave (CW) 488 nm laser collimated from an optical fiber and focused to the capillary lumen using a 40× 0.6 numerical aperture (NA) objective. The emission light is collected through the same objective and subsequently passes through a dichroic, a 488 nm notch filter, a 520 nm band-pass filter, and a 5 mm iris window. Filtered light is detected and the signal is filtered with an analog elliptical low-pass filter (1 kHz cutoff) and is digitally oversampled (75 kHz sampling rate averaged to 8 Hz). The digital signal is then filtered with a digital Butterworth low-pass (1.5 kHz cutoff) filter and displayed as an

Table 1
Component list for the hardware used to build the single-cell CE system

Component	Company	Part number	Notes
40× objective	Nikon		
Four-way mounting cage	Thorlabs	C4W	
Cage dichroic filter mount	Thorlabs	FFM1	
Rotatable cover plate	Thorlabs	B3C	
Blank cover plate	Thorlabs	B1C	
10 mm lens tube	Thorlabs	SM1L10	
External SM1 threads to internal M25 × 0.75 thread (objective)	Thorlabs	SM1A12	
External C-mount to SM1 thread	Thorlabs	SM1A39	
Calibrated iris diaphragm	Thorlabs	SM1D12C	
SM1-threaded manual shutter	Thorlabs	SM1SH1	
End cap with SM1 threads	Thorlabs	SM1CP2	
Nonrotating adjustable SM1 lens tube	Thorlabs	SM1ZM	
Dichroic (488 nm)	Chroma	ZT488rdc-UF2 (IN036384)	25.5 × 36 × 2 mm
Band-pass filter (525 ± 25 nm)	Chroma	ET525/50m (IN005750)	25 mm dia mounted
Detent	McMaster-Carr	3408A65	#4–48 threads
Gasket tape	McMaster-Carr	95705K132	0.02″ thick
Lab jack	Thorlabs, optics focus	MLJ-150, M0Z-200-25	
XY-stage	Prior, Nikon		
Solenoid valve	Yosoo (Amazon)	2V025-1/4-12VDC	Must operate at 0–100 PSI
High-voltage power supply	Spellman, Bertan	CZE 100R, Custom	
Data acquisition card	National Instruments	PCI-6229, USB-6224	
3D printer	Prusa	MK3	
PLA filament	Matchbox	Black 1.75 mm	
Microcontroller	Arduino	Uno/Mega	
powerSTEP01 stepper driver	ST Microelectronics	X-NUCLEO-IHM03A1	

The component name, company manufacture, part number, and any notes relevant to the part are recorded in the table columns. Common components (i.e., stepper motors, lead screws) are not listed individually. A more comprehensive list of components and instrument schematics can be found in our chemical cytometry repository: https://github.com/braepete/CE_Hardware

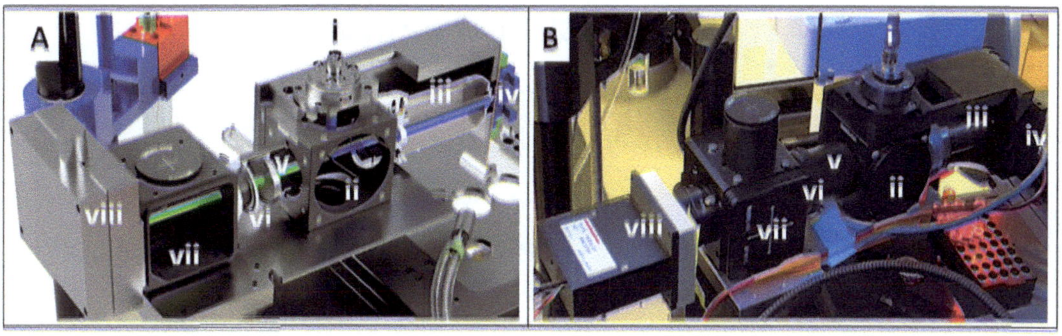

Fig. 3 The CE fluorescence detection assembly and excitation pathway. (**a**) CAD diagram and (**b**) photo. Excitation light from a 488 nm CW laser is collimated by a fiber-optic coupler (i). A dichroic beam splitter (ii) reflects the light to a 0.75 NA 40× objective (iii) which then focuses the light onto the capillary (iv). Emission light (from fluorophores within the capillary lumen) travels through the beam splitter (ii), and this emitted light is filtered using a 520 nm low-pass filter (v). Stray light is spatially filtered using an iris (vi). A four-way optic mount (vii) is used to mount the PMT detector (viii) to the optical path

electropherogram at an 8 Hz sampling rate. A high-voltage source is used to apply a potential to the capillary outlet for the CE separation, while the capillary inlet is held at ground. The platform also has a continuous-flow buffer system to ensure that the medium within the cell chambers is continuously refreshed.

3.2 Akt Sensor Cell-Loading Methods

The ability to deliver a chemical sensor through the plasma membrane is critical to performing CE-F chemical cytometry in the least cell-perturbative manner possible. Many delivery methods exist, including microinjection, electroporation, passive diffusion, pinocytosis, mechanical disruption, cell-penetrating peptides, nanostraws, and liposomes/exosomes, among others [54, 55]. Our lab utilizes a photoactivatable peptide-based sensor, C48F, for measuring Akt activity within single cells (Fig. 4) [27]. This sensor can be loaded intracellularly either by passive diffusion or pinocytosis (Fig. 4a). The C48F sensor sequence is as follows: 6FAM-GRP-(*R*)-AF-(T-DMNB)-F-(*A*)-Amide (where *R* and *A* represent N-methylated versions of amino acids and DMNB indicates the photocaging group on the sensor phosphorylation site). This sensor has previously been shown to display favorable phosphorylation kinetics, Akt selectivity, and resistance to proteolytic degradation in single cells [26, 27, 33]. C48F employs a 2-4,5-dimethoxy 2-nitrobenzyl (DMNB) moiety to mask the threonine phosphorylation site and block phosphorylation until the sensor is exposed to UV light at 360 nm [27, 56]. It is important to note that the non-photoactivatable variant, 48F, can still be used for measuring Akt activity in single cells; however, 48F necessitates the use of pinocytosis (or another of the cell-loading methods described above) in order to be efficiently delivered intracellularly and lacks the temporal control inherent to the photoactivatable variant [27].

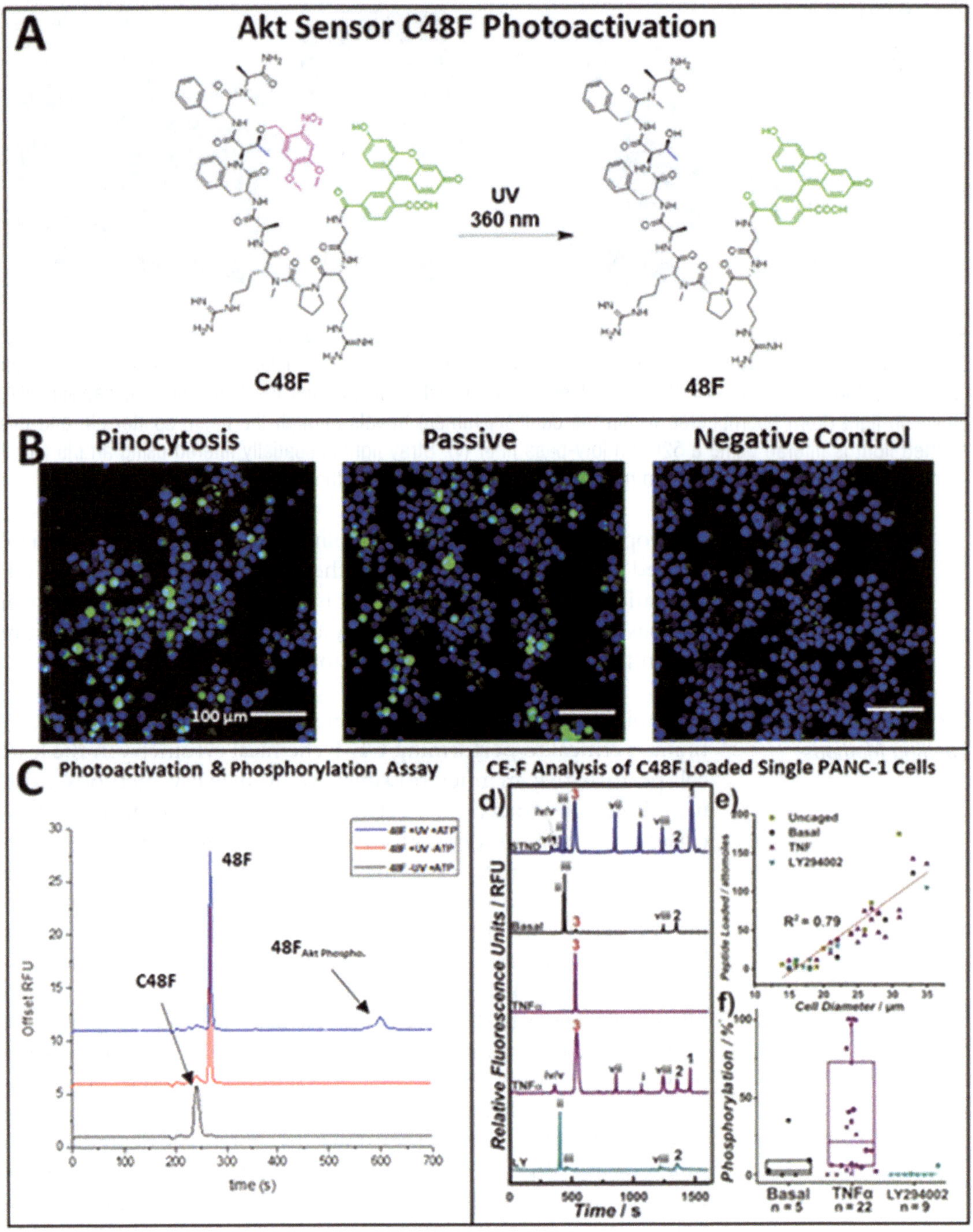

Fig. 4 Akt sensor photoactivation, intracellular loading, and single-cell CE-F analysis. (**a**) The photoactivatable sensor, C48F, is photoactivated (uncaged) following exposure to UV light at 360 nm to cleave the DMNB moiety (pink) to yield 48F, which can be phosphorylated by intracellular Akt [27]. (**b**) Fluorescent microscopy of PANC-1 cells that have been pinocytically or passively loaded with 100 μM C48F, compared to a negative C48F control (0.6 NA 20× objective). C48F fluorescence is shown in green, whereas cell nuclei have been stained with Hoechst 33342 (blue). (**c**) The photoactivation of C48F, and subsequent phosphorylation by purified Akt, is confirmed by CE-F ratiometrically, where the conversion of C48F to 48F is quantified using the corrected area (CA) of each peak. ((**d–f**) Reproduced from [27] with permission from John Wiley and Sons

3.2.1 C48F Loading by Passive Diffusion

Adherent PANC-1 cells were trypsinized and then resuspended in the desired culture medium at a density of 10^6 cells/mL. Next, cells were seeded on the desired substrate (glass coverslip, custom single-cell analysis culture chambers, or standard clear bottom 96-well plates) at a density of 10^4 cells/mL and were incubated overnight at 37 °C in a humidified atmosphere of 5% CO_2. Immediately prior to loading C48F, cells were washed twice with phosphate-buffered saline (PBS) and incubated for 10 min with the sensor at a concentration of 100 μM (in culture medium) at either 37 °C or 4 °C. In the case of peptide-based sensors, reduced loading temperatures have been demonstrated to mitigate the formation of punctate aggregations of the sensor within endosomes while simultaneously promoting a diffuse distribution throughout the cytosol (Fig. 4b) [27]. Next, the sensor solution was aspirated gently via pipet, and the culture vessel washed with three volumes of PBS or extracellular buffer (ECB). Thorough washing of sensor-loaded cells helps reduce the background fluorescence signal when imaging cells and/or analyzing enzyme activity of sensor-loaded cells by CE-F. However, care should be taken to avoid washing the cells too aggressively, which can result in mechanically displacing them from the glass surface.

3.2.2 C48F Loading by Pinocytosis

Cells to be loaded by pinocytosis underwent the same pretreatment steps described above for passive loading. A solution of the desired sensor (between 10 and 100 μM) was then prepared in hypertonic medium at a dilution of 10% (Influx, Pinocytosis Kit), and cells were incubated in hypertonic medium at either 37 °C or 4 °C for 20 min. Careful attention was given to ensure that enough hypertonic medium was added to completely cover the apical surface of the cells. The hypertonic loading solution was aspirated and replaced with 500 μL of hypotonic loading medium (60:40 (v/v) serum-free culture medium: deionized H_2O). The cells were incubated at the desired temperature for 2 min, then the hypotonic medium was gently aspirated, and cells were washed twice with 500 μL PBS, aspirated, and allowed to re-equilibrate in culture medium at 37 °C prior to analysis. Following re-equilibration, the medium was aspirated and cells were washed with PBS and allowed to incubate in 500 μL PBS at 37 °C until analysis. For pinocytic loading, cells should not be left in the hypotonic medium for longer

Fig. 4 (continued) through RightsLink 2020). (**d**) Single-cell Akt activity measured by chemical cytometry with the electrophoretic peaks labeled as 1, 2, and 3 corresponding to C48F, 48F, and phosphorylated 48F, respectively. 6FAM-labeled proteolytic products of 1 are marked as i–viii. The standard (STND) trace shows the migration time of a standard solution of the peptides. Those below are from individual PANC-1 cells which were deprived of serum (basal), stimulated with 100 ng/mL TNFα, or treated with 10 μM LY294002 + 100 ng/mL TNFα. (**e**) The attomoles of reporter loaded as a function of cell diameter (R2 = 0.79). (**f**) Akt activity for a population of single cells analyzed 20 min post-photolysis

than 2 min as this significantly increases the likelihood of cell damage and death [28]. Successfully loaded cells should display a diffuse fluorescence, indicating that the sensor is distributed relatively evenly throughout the cytosol. A punctate distribution of fluorescence may indicate sensor uptake by endocytosis and/or sensor surface adsorption.

3.3 Preparing CE Methods for Nucleic Acid Collection

While our method uses hydrodynamic forces to collect nucleic acids following a CE separation, nucleic acids can also be collected or ejected from the capillary inlet electrokinetically during the CE separation (*see* Subheading 1.2 for a review of analyte mobilities). Nucleic acids have a strong negative charge due to their repeating phosphodiester linkages (Fig. 1b). Due to the high density of negative charge, the anodic electrophoretic mobility of nucleic acids can be greater than the cathodic electroosmotic mobility when using buffers with moderate electroosmotic flow. Importantly, for a given assay's electrophoretic buffer, it is advisable to record the net mobility of a standard and/or purified nucleic acid. Unintended electrokinetic collection of nucleic acids during the collection could result in the loss of free nucleic acids from the cytoplasm or nucleus following cell lysis. It is possible to quantify nucleic acid in CE by employing UV-Vis detection at a wavelength of 260 nm and using the migration time to calculate the mobility by Eq. 5.

$$\mu_{\mathrm{ep}} = \frac{L_{\mathrm{path}}}{E t_{\mathrm{m}}} \tag{5}$$

μ_{ep} = electrophoretic mobility

L_{path} = length to the detector

E = electric field strength

t_{m} = migration time

Additionally, an approximate measure of nucleic acid mobility can be measured by adding SYTOX Green nucleic acid stain to the electrophoretic separation buffer. The added SYTOX Green binds with sample DNA to facilitate fluorescence detection, thereby improving the sensitivity and limits of quantitation for the assay [57]. However, note that the use of a fluorescent dye which binds to the nucleic acids will likely shift the mobility to be more cathodic due to changes in the net charge and increased size of the nucleic acid-dye complex [57].

To begin the assay, a 15 μL sample of 200 pg/nL purified DNA was prepared in 10 mM Tris-EDTA (pH 8.1). DNA was detected using a commercial CE instrument with a UV absorbance detector (MDQ+, AB-SCIEX). Separations were performed in normal and reverse polarity to identify the major DNA peak. Mobility was calculated using the center of the peak and the voltage applied (267 Vcm^{-1}), the total length of the capillary (30 cm), and the path length to the detector (10 cm).

Collections of purified nucleic acid are performed prior to analyzing single cells. The purified DNA sample can be analyzed alone or with the fluorescent sensor, but all conditions should be identical as would be performed for single-cell analysis, with the exception that there will be no laser lysis step. Additionally, if the nucleic acid mobility was shown to be net anodic, a hydrodynamic injection of nucleic acids should be employed instead of or in addition to electrokinetic injection [58].

Next, 50 μL samples of 100 nM 48F sensor and purified Hela-GFP DNA were prepared with ECB at a concentration of 10 pg/nL. The sample was added to a PCR tube and loaded into the capillary using gravity-driven flow for 3 s by applying a 3 cm height difference between the inlet sample fluid level and the outlet vial fluid level. Following sample injection, the capillary inlet was moved to a PCR tube with 5 μL of electrophoretic buffer. To collect the purified DNA, electrophoresis was conducted at 250 Vcm^{-1} for 120 s, followed by the outlet being pressurized for 30 s at 25 psi. Collected DNA was then frozen at −60 °C until ready to be assayed by PCR.

3.4 Measuring Enzyme Activity and Transcript Levels at the Single-Cell Level

All single-cell experiments were performed using our custom platform described in Subheading 3.1, and the experimental workflow is shown in Fig. 5. Cells were seeded and pretreated in a custom single-cell analysis chamber as described in Subheading 3.2. Next, the cell chamber was placed on the XY-stage of the customized single-cell CE-F analysis system. Individual cells were identified and brought into focus using bright-field microscopy via a live camera feed [43]. Once in focus, a cell was lysed with a single pulse from the Nd:YAG laser (532 nm) and immediately electrokinetically injected into the capillary (7 s, 250 Vcm^{-1}) [42, 44, 59, 60]. The capillary inlet was then repositioned into a reservoir of electrophoretic buffer (100 mM Tris/Tricine, 10 mM Tween 20, pH 8.2, 3.4 mS). For buffers where the free nucleic acids have a net-anodic mobility, the capillary inlet was placed into a collection vial or well that contains a small amount of electrophoretic buffer (5–20 μL). The injected cellular lysate was then separated under an applied field of 250 Vcm^{-1}. Following the separation of the fluorescent sensors, the cellular contents were collected from the capillary by pressurizing the capillary outlet vial (25 psi). The duration of the pressure collection should be sufficiently long to collect all of the residual nucleic acids in the capillary (the amount of time required to collect one capillary volume into the collection well). The volume of buffer in the collection vial prior to collection for each experiment was selected based on the sample requirements of the downstream assays. For many assays, lower volumes are ideal for reducing reagent use and sample dilution. It is important to account for the volume that will be collected from the capillary after the CE separation in addition to the volume added to the well

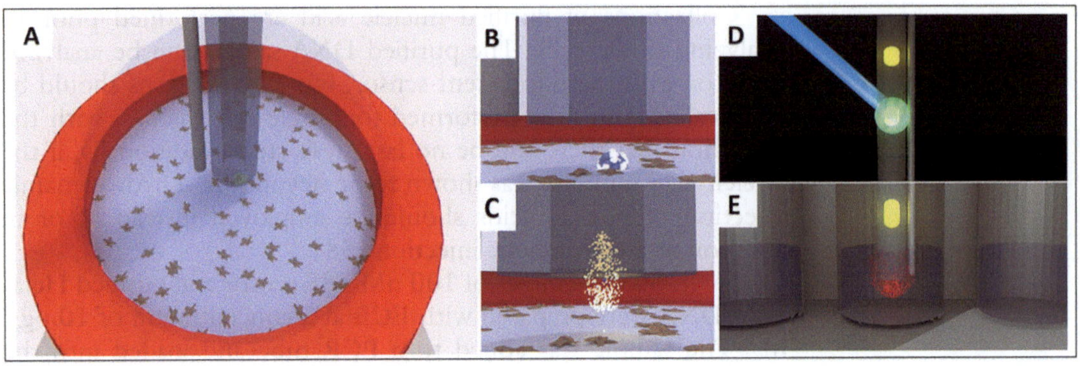

Fig. 5 Overview of the multiplexed chemical cytometry assay with nucleic acid collection. (**a**) Cells (brown stars) are loaded with the enzyme activity reporter that will be modified by the enzyme of interest. For example, an Akt reporter is phosphorylated by Akt. (**b**) A cell is lysed using a pulsed Nd:YAG laser that creates a cavitation bubble (gray and white sphere below capillary lumen). This lysis process stops enzymatic processes within the cell as intracellular contents are diluted into the sample buffer on microsecond timescales. (**c**) As the cell is being lysed, the contents of the lysate (white speckles) are loaded into the capillary by hydrodynamic flow through the capillary or using the EO flow to pull the contents into the capillary by applying a brief electric field across the capillary. (**d**) Once the cell is injected, the capillary moves to the separation buffer to start the electrokinetic sensor separation. In this example, phosphorylated Akt sensor in the detection zone and represented by its fluorescence (i.e., the green sphere) migrates more slowly than the unmodified sensor (yellow band), and both are detected by fluorescence as they pass the 488 nm detection laser. (**e**) Nucleic acids (red) are collected into a PCR vial or a well plate immediately following the separation by applying a pressure to the capillary outlet. Collected nucleic acids are then incorporated into the desired single-cell nucleic acid workflow (i.e., qPCR, NGS, etc.)

at the start of the experiment when planning for the downstream reactions. *See* Subheading 4.1 for information on troubleshooting CE-F experiments.

The capillary was rinsed between samples to eliminate cross-contamination with adsorbed nucleic acids that may have bound to the inner surface of the capillary from during a prior sample assay. A 0.5% solution of hypochlorite was used to rinse the capillary for 30 s, and the capillary was refilled with the electrophoretic buffer by rinsing for 60 s. Automated capillary rinsing can be performed by applying a vacuum to the capillary outlet vial for sufficient time to fill 50–70% of the capillary volume and letting the solution sit for the rinse duration. Care should be taken to avoid overfilling the capillary with hypochlorite rinse solution as it can alter the pH and conductivity of the outlet buffer. Collected nucleic acids can be analyzed using the same workflows optimized and tested in Subheading 3.2.

In addition to collecting the nucleic acid from single cells, the nucleic acid from multiple cells can be pooled together from populations with similar chemical profiles. The CE automation software developed in our laboratory incorporates a peak lookup table where the peak centers and expected deviations can be entered. As

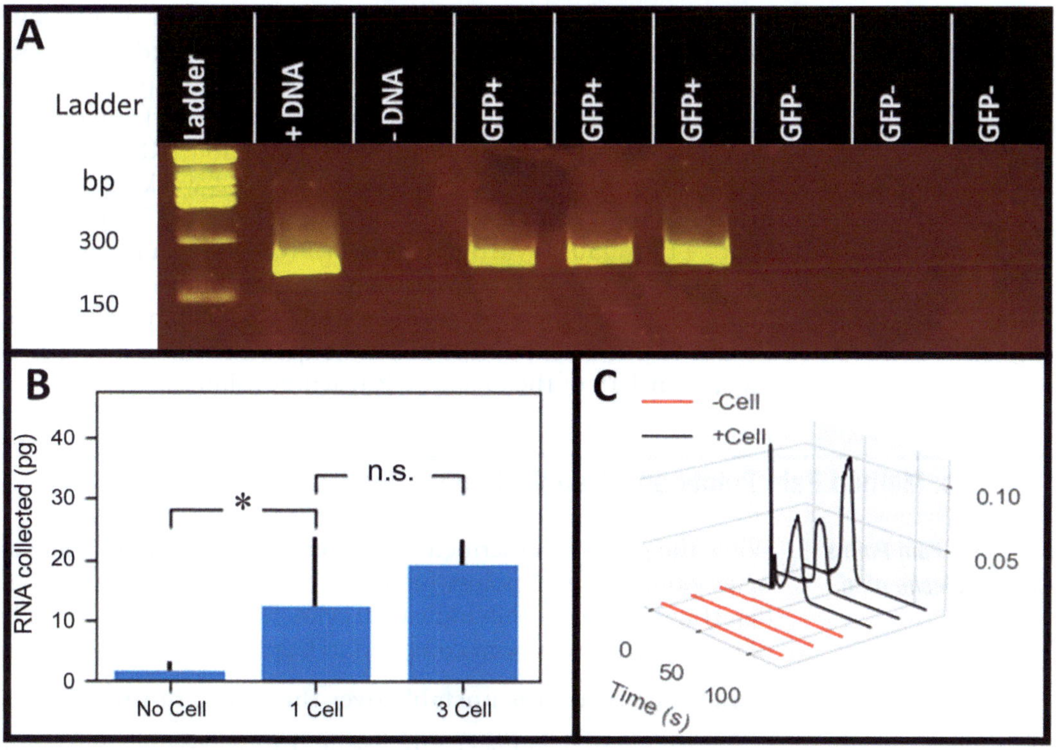

Fig. 6 PCR assays performed on collected single-cell DNA or RNA collected following a CE enzyme assay. (**a**) PAGE separation and fluorescence detection of a PCR amplicon targeting GFP gene between the DNA collected from the capillary following analysis of GFP and non-GFP transfected single cells. The PCR product forms at 210 bp for cells that had the GFP fusion proteins but not in cells that were not transfected. (**b**) RNA abundance measured by TaqMan qPCR on CE samples collected following analysis for buffer only (no cell), a single cell, and the contents of three individual cells pooled together. Abundance is calculated from the standard curve of two stably expressed genes, RPL8 and COX5B. (**c**) Representative electropherograms from B showing the fluorescence of calcein-AM detected for buffer only (-cell) and single-cell samples

separations progress and peaks are detected, peak centers are compared to the lookup table for identification, and the relative abundances of each peak are calculated from their corrected areas. The nucleotide contents from groups of cells can be pooled during collection or placed into individual vials. For example, cell contents with high and low PKB activity can be pooled into separate collection vials.

As a control for DNA analysis, a green fluorescent protein (GFP) gene was measured in a mixture of Hela cells with and without a GFP fusion to the histone gene H2B [61] (Fig. 6). The two different cell types were sequentially analyzed by CE-F and collected into individual PCR tubes. The total volume for each collection was 8 μL with the final PCR reaction volume being 20 μL.

For gene expression assays, the amount of RNA collected was estimated using the abundance of two stably expressed genes, and a calibration curve for each gene was generated using purified RNA. RNase inhibitor was added to the electrophoretic buffer in the PCR collection vials. Within 60 min after collection, RNA was snap frozen using liquid N_2 before storing for 1–2 weeks at −60 °C. Next, RNA was converted to cDNA using the VILO IV master mix (Cat# 11756050, Invitrogen) at a final reaction volume of 18 μL. The cDNA from each cell was split into four individual 4 μL aliquots with a final reaction volume of 40 μL so components from the RT reaction would not inhibit the qPCR reaction mix (i.e., no more than 10% of the final qPCR reaction volume).

4 Notes: Method Pain Points and Troubleshooting Guides

4.1 CE-F Pain Points and Troubleshooting

While the platform described is very sensitive and robust, there are many potential pain points involved in sampling, separating, and detecting analytes within individual cells. To help remedy this, a list of tips and tricks has been compiled to help resolve common issues.

Issue: The current is unstable over the course of the CE run.

Unstable current may be the result of an electric arc. To prevent arcing, ensure that all equipment is properly grounded, especially near the electrodes. Alternatively, it could be the result of an air bubble in the capillary. Apply pressure (>10 psi) to the capillary inlet for at least 2 min while it is submerged (>1 cm) in the BGE to flush the air from the capillary and ensure it is filled completely with BGE prior to analysis. Another problem may be that the capillary is cracked and/or damaged. Observe the entire length of the capillary for any damage or abnormalities. One should also observe the capillary inlet with a microscope to identify any noticeable cracks, chips, and/or biofouling; if none are observed, try replacing the capillary (always check the inlet for damage after cutting a new capillary). Also note that capillaries originating from either end of a spool may behave differently than the rest of the bulk spool.

Issue: Analyte migration times vary significantly from run to run and day to day, even when using the same buffer system and capillary.

An internal standard, such as an analyte with a significantly faster migration time than analytes of interest, can be used to track migration time variability. Our lab uses a HiLyte Fluor™ 488 dye (AnaSpec, AS-81160) that displays a high cathodic mobility. This internal standard can be mixed with a sensor standard or hydrodynamically loaded following the collection of a single cell. Variable migration time from day to day could be due to biofouling, differences in the capillary surface chemistry, and/or buffer aliquot composition, all of which can affect electroosmotic fluid flow (EOF). Importantly, buffer components that are volatile (including organic solvents such as 1-propanol) may evaporate over the time course of an experiment; therefore, nonvolatile alternatives that display similar intermolecular forces (such as Tween 20) should be utilized where available. If volatile buffer components are essential, replace the buffer frequently. Another possible cause for shifting migration times is buffer electrolysis, which can significantly alter the buffer pH over time [62]. A change in separation buffer pH can affect not only the surface chemistry of the capillary (and therefore the EOF) but also the charge state (and therefore the electrophoretic mobility) of analytes. To avoid these phenomena, it is recommended that each 4 mL buffer reservoir be refreshed after 2 h of total separation time. To produce a consistent capillary chemical surface, condition the capillary prior to experiments as well as after the experiments have concluded. A suggested rinse cycle for the beginning of the day consists of rinsing the capillary with 0.1 M HCl to remove any metal ions from the capillary surface, removing the HCl by rinsing with DI H_2O, rinsing extensively (>10 min) with 1.0 M NaOH to generate hydroxyls on the inner surface of the capillary lumen, rinsing with DI H_2O to remove any NaOH, and then filling the capillary with separation buffer. At the end of experiments, the capillary should be rinsed with DI H_2O and dried with air for overnight storage. If significant biofouling is observed, a 0.1% hypochlorite rinse can be performed between runs.

Issue: Limits of detection have significantly deteriorated.

A sample from a stock of a control solution (i.e., 10 nM fluorescein) was used at the beginning of each day to monitor the limits of detection over multiple days and/or weeks of the lifetime of a capillary. Degradation of the capillary surface due to biofouling is a major concern in single-cell CE methods. Therefore, we employ a 0.1% (v/v) aqueous hypochlorite rinse between samples to remove cellular debris from the capillary surface, followed by a 1.0 M NaOH rinse to restore the silanol surface.

Additionally, it is important to note that the concentrations of certain buffer additives can increase the background autofluorescence and/or result in light scattering (for micelles). Therefore, it is important to benchmark the BGE before and after the addition of any additives. Poorer limits of detection may be attributable to poor laser and capillary alignment. As all optical systems experience misalignment over time due to thermal drift, it is important to routinely calculate the mLOD of the system using a fluorescein standard and realign optical components as needed. It is also possible that optical pathways (excitation and/or emission) have become compromised. There are several possible failure points that need to be investigated. First, check that the excitation laser power at the capillary is sufficient (>2 mW). Next, check if the objectives, bandpass filters, and dichroic filter need to be cleaned. If fiber-optic cables are employed, verify that they are properly aligned and secured.

4.2 Tips and Tricks to Facilitate Robust RNA Collections and Single-Cell RNA-Seq Analyses

Single-cell analysis of DNA or RNA is challenging due to the very small amount of nucleic acid within a cell (10–50 pg) [59]. It is important to include appropriate technical controls during the experiment to verify any source of systematic noise. Negative controls can be used to identify sources of DNA or RNA contamination during the collection process. We have identified that the two most common sources of contamination are residual nucleic acids adsorbed to the capillary wall and nucleic acids present in the cell buffer. Rinsing the capillary as outlined in Subheading 3.4 minimizes intracapillary contamination between runs. Additionally, replenishing the cellular buffer, either manually before cell injection or in an automated fashion using a syringe pump and vacuum line, reduces cross-contamination within the cell chamber. Because of the low abundance of some genes or transcripts, it is important to be aware of any major source of nucleic acid contamination. Adding a known amount of purified RNA or DNA at various steps of the collection and nucleic analysis process (cell lysis and loading, nucleic acid collection, reverse transcription, or PCR) is a useful way to identify which steps may need to be optimized.

4.3 Hardware and Software Troubleshooting Guide for the Single-Cell Analysis Platform

Hardware recommendations can be found at https://github.com/brae-pete/CE_Hardware. Hardware components should be capable of serial communication, and appropriate drivers and libraries are available for full software control. Serial communication protocols are generally the most straightforward way to communicate where the user provides ASCII commands to the device's microcontroller. When using hardware with drivers and libraries available from the manufacturer, ensure that there is documentation for controlling the hardware in the programming language that will be used (i.e., Python for our automated CE platform).

The software for automated control of the CE system can be found at https://github.com/brae-pete/AutomatedCE. The code is open source and written primarily in Python. There is support for many common software and hardware packages including Micro-manager (via Pycromanager) [63], Thorlabs Kinesis, National Instruments DAQmx, and Digilent WaveForms. Configuration of a new instrument is relatively straightforward if hardware is selected from our recommendations. However, new hardware support can be added by creating new hardware control and utility classes. Test notebooks are written using Jupyter notebooks (Project Jupyter) to help introduce the API and confirm the basic functionality of the hardware [59]. Often, common errors arise from communication mishaps between the software and the hardware. In many cases, it is best to restart the hardware, software, or computer. If problems persist, verify functionality via the available Jupyter notebook tests.

4.4 Conclusion

Chemical cytometry and sequencing technologies have greatly improved our understanding of biochemical processes at the level of single cells. Our group has demonstrated the ability to combine these powerful techniques into a multiplexed platform that further deepens the amount of data that can be obtained from individual cells, thereby opening the door to novel opportunities in disease diagnostics and fundamental research. This technology is expected to be amenable to many different cell types, chemical sensors, and sequencing methods. Combined RNA and enzyme activity measurements appear to be a promising avenue to deepen the knowledge of cell subpopulation dynamics, disease progression, and drug resistance mechanisms at the single-cell level.

In conclusion, continued developments in multiplexed chemical cytometry are expected in the upcoming decade. These developments will likely be driven by improvements in chemical sensor multiplexing along with improved temporal control of intracellular reactions using photoactivatable sensors and selective cell-preservation reagents. Additionally, further integration of sequencing and cell imaging technologies and enhanced throughput from use of automation will make such platforms attractive to a large audience of researchers and clinicians [60]. The ideal single-cell analysis platforms of the future will provide information on nucleic acid expression and abundance, protein spatial distribution and abundance, protein activity, and metabolism. Such platforms are expected to provide an unprecedented level of insight into cellular processes.

Acknowledgments

Sensors C48F and 48F were synthesized and characterized in collaboration with Brianna Vickerman, Professor David Lawrence, and Dr. Qunzhao Wang at the University of North Carolina at Chapel Hill. Additional thanks to David Lawrence for reviewing the chemical structures in this manuscript. This work is supported by the National Institutes of Health grants R01 CA224763 (N.L.A.), R01 CA233811 (N.L.A.), F31 HL147500-01 (M.M.A.), and 5F31CA243312-02 (L.A.G.). The authors declare no conflicts of interest.

References

1. Milo R, Philips R. How big is a human cell? http://book.bionumbers.org/how-big-is-a-human-cell/. Accessed 28 Aug 2020
2. Characteristic cell volume – Human Homo sapiens – BNID 100434. https://bionumbers.hms.harvard.edu/bionumber.aspx?s=n&v=16&id=100434. Accessed 28 Aug 2020
3. Cohen D et al (2008) Chemical cytometry: fluorescence-based single-cell analysis. Annu Rev Anal Chem 1:165–190. https://doi.org/10.1146/annurev.anchem.1.031207.113104
4. Lähnemann D et al (2020) Eleven grand challenges in single-cell data science. Genome Biol 21(31):1–35. https://doi.org/10.1186/s13059-020-1926-6
5. Arvaniti E, Claassen M (2017) Sensitive detection of rare disease-associated cell subsets via representation learning. Nat Commun 8(14825):1–10. https://10.1038/ncomms14825
6. Hodzic E (2016) Single-cell analysis: advances and future perspectives. Bosnian J Basic Med Sci 16(4):313–314. https://doi.org/10.17305/bjbms.2016.1371
7. Abdallah BY et al (2013) Single cell heterogeneity. Cell Cycle 12:3640–3649. https://doi.org/10.4161/cc.26580
8. Goldman SL et al (2019) The impact of heterogeneity on single-cell sequencing. Front Genet 10:1–8. https://doi.org/10.3389/fgene.2019.00008
9. Qian Z, Bao L (2019) Cellular heterogeneity and single-cell omics. In: Single-cell omics. Elsevier, Amsterdam, Netherlands, pp 35–44
10. Sottoriva A et al (2013) Intratumor heterogeneity in human glioblastoma reflects cancer evolutionary dynamics. Proc Natl Acad Sci 110:4009–4014. https://doi.org/10.1073/pnas.1219747110
11. Qazi MA, Bakhshinyan D, Singh SK (2019) Deciphering brain tumor heterogeneity, one cell at a time. Nat Med 25:1474–1476. https://doi.org/10.1038/s41591-019-0605-1
12. Perus LJM, Walsh LA (2019) Microenvironmental heterogeneity in brain malignancies. Front Immunol 10:1–15. https://doi.org/10.3389/fimmu.2019.02294/full
13. Nowotarski HL, Attayek PJ, Allbritton NL (2020) Automated platform for cell selection and separation based on four-dimensional motility and matrix degradation. Analyst 145:2731–2742. https://doi.org/10.1039/c9an02224d
14. Dagogo-Jack I, Shaw AT (2018) Tumour heterogeneity and resistance to cancer therapies. Nat Rev Clin Oncol 15:81–94. https://doi.org/10.1038/nrclinonc.2017.166
15. Vembadi A, Menachery A, Qasaimeh MA (2019) Cell cytometry: review and perspective on biotechnological advances. Front Bioeng Biotechnol 7:1–20. https://doi.org/10.3389/fbioe.2019.00147
16. Doan M et al (2018) Diagnostic potential of imaging flow cytometry. Trends Biotechnol 36:649–652. https://doi.org/10.1016/j.tibtech.2017.12.008
17. Spitzer M, Nolan GP (2016) Mass cytometry: single cells, many features. Cell 165:780–791. https://doi.org/10.1016/j.cell.2016.04.019
18. Vickerman BM et al (2018) Design and application of sensors for chemical cytometry. ACS Chem Biol 13:1741–1751. https://doi.org/10.1021/acschembio.7b01009
19. Neumann EK et al (2019) Lipid heterogeneity between astrocytes and neurons revealed by single-cell MALDI-MS combined with Immunocytochemical classification. Angew Chem Int Ed 58:5910–5914. https://doi.org/10.1002/anie.201812892
20. Hsu C-C et al (2020) A single-cell Raman-based platform to identify developmental stages of human pluripotent stem cell-derived neurons. Proc Natl Acad Sci 117:18412–18423. https://doi.org/10.1073/pnas.2001906117
21. Baker LA (2018) Perspective and prospectus on single-entity electrochemistry. J Am Chem Soc 140:15549–15559. https://doi.org/10.1021/jacs.8b09747
22. Chan J, Dodani SC, Chang CJ (2012) Reaction-based small-molecule fluorescent probes for chemoselective bioimaging. Nat Chem 4:973–984. https://doi.org/10.1038/nchem.1500
23. Mullassery D, Horton C, Wood C, White M (2008) Single live cell imaging for systems biology. Essays Biochem 45:121–133. https://doi.org/10.1042/BSE0450121
24. Komatsu T, Urano Y (2015) Evaluation of enzymatic activities in living systems with small-molecular fluorescent substrate probes. Anal Sci 31:257–265. https://doi.org/10.2116/analsci.31.257
25. Ou Y, Wilson RE, Weber SG (2018) Methods of measuring enzyme activity ex vivo and in vivo. Annu Rev Anal Chem 11:509–533. https://doi.org/10.1146/annurev-anchem-061417-125619

26. Mainz ER et al (2016) Single cell chemical cytometry of Akt activity in rheumatoid arthritis and Normal fibroblast-like Synoviocytes in response to tumor necrosis factor α. Anal Chem 88:7786–7792. https://doi.org/10.1021/acs.analchem.6b01801
27. Mainz ER et al (2016) An integrated chemical cytometry method: shining a light on Akt activity in single cells. Angew Chem Int Ed 55: 13095–13098. https://doi.org/10.1002/anie.201606914
28. Turner AH et al (2016) Rational Design of a Dephosphorylation-Resistant Reporter Enables Single-Cell Measurement of tyrosine kinase activity. ACS Chem Biol 11:355–362. https://doi.org/10.1021/acschembio.5b00667
29. Phillips RM et al (2013) Measurement of protein tyrosine phosphatase activity in single cells by capillary electrophoresis. Anal Chem 85: 6136–6142. https://doi.org/10.1021/ac401106e
30. Phillips RM et al (2014) Ex vivo chemical cytometric analysis of protein tyrosine phosphatase activity in single human airway epithelial cells. Anal Chem 86:1291–1297. https://doi.org/10.1021/ac403705c
31. Dickinson AJ et al (2014) Single-cell sphingosine kinase activity measurements in primary leukemia. Anal Bioanal Chem 406: 7027–7036. https://doi.org/10.1007/s00216-014-7974-6
32. Dickinson AJ et al (2015) Analysis of sphingosine kinase activity in single natural killer cells from peripheral blood. Integr Biol (Camb) 7: 392–401. https://doi.org/10.1039/c5ib00007f
33. Proctor A et al (2014) Measurement of protein kinase B activity in single primary human pancreatic cancer cells. Anal Chem 86:4573–4580. https://doi.org/10.1021/ac500616q
34. Mainz ER, Dobes NC, Allbritton NL (2015) Pronase E-based generation of fluorescent peptide fragments: tracking intracellular peptide fate in single cells. Anal Chem 87:7987–7995. https://doi.org/10.1021/acs.analchem.5b01929
35. Proctor A et al (2016) Development of a protease-resistant reporter to quantify BCR–ABL activity in intact cells. Analyst 141: 6008–6017. https://doi.org/10.1039/C6AN01378C
36. Keithley RB et al (2013) Capillary electrophoresis with three-color fluorescence detection for the analysis of glycosphingolipid metabolism. Analyst 138:164–170. https://doi.org/10.1039/C2AN36286D
37. Essaka DC et al (2012) Metabolic cytometry: capillary electrophoresis with two-color fluorescence detection for the simultaneous study of two glycosphingolipid metabolic pathways in single primary neurons. Anal Chem 84: 2799–2804. https://doi.org/10.1021/ac2031892
38. Essaka DC et al (2012) Single cell ganglioside catabolism in primary cerebellar neurons and glia. Neurochem Res 37:1308–1314. https://doi.org/10.1007/s11064-012-0733-1
39. Keithley RB et al (2013) Manipulating ionic strength to improve single cell electrophoretic separations. Talanta 111:206–214. https://doi.org/10.1016/j.talanta.2013.03.012
40. Voeten RLC et al (2018) Capillary electrophoresis: trends and recent advances. Anal Chem 90:1464–1481. https://doi.org/10.1021/acs.analchem.8b00015
41. Sims CE et al (1998) Laser–micropipet combination for single-cell analysis. Anal Chem 70: 4570–4577. https://doi.org/10.1021/ac9802269
42. Dickinson AJ, Armistead PM, Allbritton NL (2013) Automated capillary electrophoresis system for fast single-cell analysis. Anal Chem 85:4797–4804. https://doi.org/10.1021/ac4005887
43. Abraham DH et al (2019) Design of an automated capillary electrophoresis platform for single-cell analysis. In: Enzyme activity in single cells. Elsevier, Cambridge, pp 191–122
44. Jorgenson JW, Lukacs KD (1983) Capillary zone electrophoresis. Science 222:266–272. http://www.jstor.org/stable/1691606
45. Grossman PD, Colburn JC (2012) Capillary electrophoresis: theory and practice. Academic Press, San Diego
46. Blackney DM, Foley JP (2017) Dual-opposite injection capillary electrophoresis: principles and misconceptions. Electrophoresis 38: 607–616. https://doi.org/10.1002/elps.201600337
47. Whitmore CD et al (2007) Yoctomole analysis of ganglioside metabolism in PC12 cellular homogenates. Electrophoresis 28:3100–3104. https://doi.org/10.1002/elps.200700202
48. Ding J et al (2020) Systematic comparison of single-cell and single-nucleus RNA-sequencing methods. Nat Biotechnol 38:737–746. https://doi.org/10.1038/s41587-020-0465-8
49. Peterson VM et al (2017) Multiplexed quantification of proteins and transcripts in single cells. Nat Biotechnol 35:936–939. https://doi.org/10.1038/nbt.3973

50. Becker K et al (2018) Quantifying post-transcriptional regulation in the development of Drosophila melanogaster. Nat Commun 4970:1–14. https://doi.org/10.1038/s41467-018-07455-9
51. Hamidi H et al (2014) KRAS mutational subtype and copy number predict in vitro response of human pancreatic cancer cell lines to MEK inhibition. Br J Cancer 111:1788–1801. https://doi.org/10.1038/bjc.2014.475
52. Martin KC, Ephrussi A (2009) mRNA localization: gene expression in the spatial dimension. Cell 136:719–730. https://doi.org/10.1016/j.cell.2009.01.044
53. Haque A et al (2017) A practical guide to single-cell RNA-sequencing for biomedical research and clinical applications. Genome Med 75:1–12. https://doi.org/10.1186/s13073-017-0467-4
54. Stewart MP, Langer R, Jensen KF (2018) Intracellular delivery by membrane disruption: mechanisms, strategies, and concepts. Chem Rev 118:7409–7531. https://doi.org/10.1021/acs.chemrev.7b00678
55. Meacham JM et al (2014) Physical methods for intracellular delivery: practical aspects from laboratory use to industrial-scale processing. J Lab Autom 19:1–18. https://doi.org/10.1177/2211068213494388
56. Klán P et al (2013) Photoremovable protecting groups in chemistry and biology: reaction mechanisms and efficacy. Chem Rev 113:119–191. https://doi.org/10.1021/cr300177k
57. Thakur S, Cattoni DI, Nöllmann M (2015) The fluorescence properties and binding mechanism of SYTOX green, a bright, low photo-damage DNA intercalating agent. Eur Biophys J 44:337–348. https://doi.org/10.1007/s00249-015-1027-8
58. Breadmore MC (2009) Electrokinetic and hydrodynamic injection: making the right choice for capillary electrophoresis. Bioanalysis 1:889–894. https://doi.org/10.4155/bio.09.73
59. Hellman AN et al (2008) Biophysical response to pulsed laser microbeam-induced cell lysis and molecular delivery. J Biophotonics 1:24–35. https://doi.org/10.1002/jbio.200710010
60. Rau KR et al (2006) Pulsed laser microbeam-induced cell lysis: time-resolved imaging and analysis of hydrodynamic effects. Biophys J 91:317–329. https://doi.org/10.1529/biophysj.105.079921
61. Kanda T, Sullivan KF, Wahl GM (1998) Histone–GFP fusion protein enables sensitive analysis of chromosome dynamics in living mammalian cells. Curr Biol 8:377–385. https://doi.org/10.1016/S0960-9822(98)70156-3
62. Hows MEP, Perrett D (1998) Effects of buffer depletion in capillary electrophoresis: development of a continuous flow cathode. Chromatographia 48:355–359. https://doi.org/10.1007/bf02467703
63. Edelstein A et al (2010) Computer control of microscopes using μManager. Curr Protoc Mol Biol 14(20):1–22. https://doi.org/10.1002/0471142727.mb1420s92

Chapter 9

Quantitative Chemical Imaging at the Cellular Level: SIMS, Fluorescence, and Correlative Techniques

Tho D. K. Nguyen, Alicia A. Lork, Andrew G. Ewing, and Nhu T. N. Phan

Abstract

The cell is a heterogeneous chemical structure designed to accommodate its complex cellular functions in a living organism. Quantitative chemical imaging at the cellular level enables the investigation of the structural and functional molecular relation underlying cellular processes. We describe here the detailed methodology of the state-of-the-art secondary ion mass spectrometry (SIMS, NanoSIMS) and fluorescence microscopy (confocal, STED), along with selected examples for quantitative imaging at the cellular level. Correlative imaging that combines different imaging techniques is also demonstrated for selected applications in cell imaging. This chapter serves as a guideline assisting readers from unfamiliar fields of research to obtain reliable imaging at the cellular level while highlighting the strengths, limitations, and potentials of these technologies for cell imaging.

Key words Single cell, Subcellular level, SIMS, NanoSIMS, Fluorescence, Correlative imaging

1 Introduction

The cell is the smallest unit of living organisms. Understanding the molecular mechanisms at the cellular level is fundamental to the understanding of the biological complexity of the entire organisms. Cells are heterogeneous structures comprising biomolecules and complex organelles which distribute unevenly across the cell with various abundances. The distribution of biomolecules in different cellular regions is functionally oriented. Therefore, it is essential to obtain the information of localization and local concentrations of the biomolecules at the cellular level in order to be able to assess the effects of any biochemical factors and to modulate cellular processes.

Imaging at the cellular level exhibits several challenges. First, the imaging techniques must have high spatial resolution to resolve

Authors "Tho D.K. Nguyen" and "Alicia A. Lork" have equally contributed to this chapter.

Jonathan V. Sweedler et al. (eds.), *Single Cell 'Omics of Neuronal Cells*,
Neuromethods, vol. 184, https://doi.org/10.1007/978-1-0716-2525-5_9,

very small structures and substructures of cells. Second, imaging techniques must have sufficient sensitivity to detect low abundance of chemicals inside cells. Unfortunately, there is a trade-off between sensitivity and spatial resolution as the amount of a detectable chemical is proportional to the volume of the imaged area. Third, sample preparation is critically important to preserve the localization and concentration of the biochemicals in specific cellular locations for measurements.

There has been a significant effort in the development of technology and sample preparation for quantitative imaging at the cellular level. Among those suitable for cellular imaging, secondary ion mass spectrometry (SIMS) and fluorescence microscopy have been increasingly employed for cellular imaging in biological research, neuroscience, medicine, and pharmaceutical science. These techniques provide high chemical specificity, high sensitivity, and high spatial and temporal resolution for imaging at the cellular and subcellular level. SIMS is well known as a surface-sensitive technique allowing the visualization of the distributions of various species, from metabolites to molecular lipids. Fluorescence microscopy has been commonly used to image cells tagged with fluorescent dyes. Especially in these fields, the emerging state-of-the-art nanoscale SIMS (NanoSIMS) and super-resolution microscopy both featured with spatial resolution that can be less than 50 nm, offering the possibility of imaging at the organelle and even single molecule levels. In addition, a combination of different imaging and quantitative techniques has been increasingly applied in an effort to obtain complementary chemical and structural information of cells.

In this chapter, we present detailed methods of quantitative imaging at the cellular level using secondary ion mass spectrometry (SIMS and NanoSIMS), fluorescence microscopy, and correlation of these techniques. For fluorescence, we chose to focus on two approaches, stimulated emission depletion (STED) and confocal microscopy, as these are the most applied in this area at this time. Information about materials, sample preparation, imaging procedures, and data analysis as well as selected examples of cell imaging using these technologies are discussed. Our aim is to provide readers from unfamiliar fields of research the information that is needed to carry out accurate imaging at the cellular level using these technologies in order to obtain reliable quantitative data. We focus on cells derived from the nervous system, keeping in mind the general procedures can be used for a wide variety of applications.

2 Materials

2.1 Cell Culture

Quantitative chemical imaging at the cellular level requires a suitable cellular model. For all cell culture experiments, a sterile working environment is needed. Therefore, culturing procedures are performed in a laminar flow hood, and equipment is sterilized beforehand with 70% ethanol, UV light, or in an autoclave with high temperature and pressure. Cellular models can be broadly classified into three groups: cell lines, primary cells, and stem cell-derived cells. All these groups offer suitable models for investigation of synaptic biology and exocytosis, although advantages and disadvantages are present for each group.

Cell lines are easy and cheap to culture, are well characterized, and do not require ethical evaluations. In addition, reproducible results can be obtained due to the purity of the cultures and the large number of cells, which can easily be acquired. However, since cell lines are often derived from cancerous tissue, they are less representative for the originating tissue than primary cells. For example, PC12 cells containing catecholaminergic vesicles [1] were first derived from a tumor in a rat adrenal medulla called pheochromocytoma in 1972 [2]. These cells are a suitable model in our lab to investigate the exocytosis mechanism of dopaminergic vesicles [3–5]. In brief, cells are stored at −150 °C, then rapidly thawed to 37 °C in a water bath, and added to RPMI-1640 culture medium supplemented with 10% donor equine serum and 5% fetal bovine serum. Cells are centrifuged and resuspended in culture medium and plated onto collagen type IV dishes (BD Biosciences, Bedford, MA). PC12 cells are kept in an incubator at 37 °C in the presence of CO_2 7% and propagated weekly by trypsinization.

Primary cells have the advantage of being highly representative for the originating tissue. They can be derived from animals with different genetic modifications, possibly disease models. However, large cell numbers are difficult to achieve. Primary cell culture is rather costly, mainly due to the need for space and handling of animals, and this also raises ethical questions for each experiment. Hippocampal neurons, for example, can be isolated from newborn rats (as described in [6]) by dissecting hippocampi in HBSS. To generate a single-cell suspension, hippocampi are incubated for 1 h with 2 mg cysteine, 100 mM $CaCl_2$, 50 mM EDTA, and 25 U papain/mL cysteine protease in Dulbecco's Modified Eagle Medium (DMEM) while bubbling with CO_2. Hippocampi are then transferred into 5% fetal calf serum, 25 mg bovine serum albumin, and 25 mg trypsin inhibitor in DMEM for 15 min to inhibit the enzymatic reaction. The digested hippocampi are triturated using a small opening pipette tip (e.g., 200 μL) in Neurobasal A with 20% B27, 10% Glutamax, 20 μg/mL penicillin, and 20 μg/mL streptomycin to mechanically dissociate the tissue into a single-cell suspension. Afterward, the cell suspension is plated on poly-L-lysine-coated dishes with a mixture of 3.3 mM glucose, 10% equine serum, and 2 mM glutamine in MEM. After 1–2 h when the neurons have settled down, the medium is exchanged to Neurobasal A and cells are kept in an incubator at 37 °C in the presence of 5% CO_2.

Stem cell-derived cells provide an elegant way to differentiate cells into theoretically any cell in an organism. This offers a broad variety of cellular models, wild-type and disease, and especially induced pluripotent stem cells can be generated from human tissue (fibroblasts). Stem cell culture is, however, costly and time consuming. Different types of neural progenitor cells, which are partially differentiated from stem cells, and their differentiation kits are commercially available helping to reduce the differentiation time as well as improve the differentiation reproducibility. Dopaminergic neurons can be differentiated from midbrain neuronal progenitor cells by culturing them on poly-L-ornithine/laminin-coated dishes in differentiation medium containing 2% B27 supplement, 1% N_2 supplement, 20 ng/mL GDNF and BDNF, and 1 μg/mL laminin for at least 20 days similar to the protocol by Bell et al. [7]. The neuronal progenitor cells reach a postmitotic stage after around 1 week. To purify the culture, cytosine arabinoside (Ara-C) is added to the medium where it will be incorporated into the DNA of dividing cells, and this will lead to apoptosis of these dividing non-neuronal cells. Immunostaining against marker proteins such as tyrosine hydroxylase for differentiated dopaminergic neurons is eventually performed in order to verify the specific cell type with fluorescence microscopy.

2.2 Materials for SIMS Imaging

2.2.1 ToF-SIMS

Instrument and Data Analysis

In our laboratories, ToF-SIMS imaging is performed mainly on two commercial instruments, the IONTOF V (IONTOF GmbH, Münster, Germany) equipped with a 25 keV Bi_3 liquid metal ion gun (LMIG) and the J105 ToF-SIMS (Ionoptika Ltd., Eastleigh, UK) equipped with a 40 keV C_{60} gun and a 40 keV CO_2 gas cluster ion beam (GCIB). The primary ion source, however, can be changed depending on the target molecules.

Data processing for ToF-SIMS images can be carried out with software provided by the manufacturers of the instruments, for example, Surface Lab version 6.3 (IONTOF GmbH) from IONTOF V and Ionoptika Image Analyzer 2D for the J105 (Ionoptika Ltd.).

Reagents for Sample Preparation

ToF-SIMS is known as a label-free imaging technique; however, labeling is sometimes useful, depending on the research questions. In practice, isotopically labeled molecules, such as deuterated fatty acids, can be used for relative quantification.

To prepare cells for freeze-drying, a common washing buffer such as 50 mM ammonium formate (pH 7.4) is required. Liquid nitrogen-cooled isopentane or liquid propane is used for flash-freezing cells to avoid vitreous ice formation on the sample surface. For safety, these chemicals should be handled in a hood.

To prepare cells for freeze-fracturing, HEPES buffer (37 °C, 10 mM, pH 7.4) is used to suspend the cells after washing. Liquid nitrogen-cooled isopentane or liquid propane is again used for flash-freezing to avoid vitreous ice formation on the sample surface. A freeze-fracture device is required to fracture the cells at cryogenic temperature.

2.2.2 NanoSIMS

Instrument and Data Analysis

In our laboratories, NanoSIMS imaging is carried out on a commercial instrument, the NanoSIMS 50L ion microprobe (CAMECA, France) equipped with two primary ion sources, Cs^+ and O_n^-. The primary ion source is chosen depending on the target molecules.

There are many choices of software for NanoSIMS image processing, including WinImage (CAMECA), ImageJ with the plugins OpenMIMS, L'Image, and Look@NanoSIMS [8].

Reagents for Sample Preparation

Stable isotope-labeled species such as ^{13}C-labeled precursors of lipids or metabolites or ^{15}N amino acids are often used to reveal the metabolic processes in cells. Additionally, probes containing stable isotopes detectable in SIMS for labeling specific molecules in cells (e.g., boron-containing probes, fluorinated nanobodies [9, 10]) can also be used.

For sample fixation and embedding, common fixatives are glutaraldehyde (GA) 25% stock solution in PBS, formaldehyde 4% in PBS, and osmium tetroxide (OsO_4) 4% stock solution in water. The working solution 1% OsO_4 is often prepared in a 1:1 dilution in 0.2 M cacodylate buffer. A saturated dilution of uranyl acetate in ethanol 70% is often used for staining samples. This solution should be kept in the dark and filtered before use. For dehydration, ethanol 99.9% is used to prepare a series of solutions of ethanol 70%, 85%, and 95–99.5% in ddH_2O. Acetone 100% is also needed to exchange the dehydrated sample to embedding solution. Agar 100 (Agar Scientific), or LR White (Sigma-Aldrich), can be used to embed the sample in the embedding capsules (Beem Inc., West Chester, PA, USA). The resins and their accelerators are stored at 4°C and need to be warmed up to room temperature before use to avoid condensation. This fixation and embedding step should be carried out in a chemical hood for safety. An oven with temperature ~60 °C for a 16 h period is required for resin polymerization.

To cut a sample into ultrathin sections, the cells are embedded in the resin block and are cut with an EM UC6 ultramicrotome (Leica Microsystems, Wetzlar, Germany) equipped with a diamond knife. The sections are then transferred onto a silicon wafer, an indium tin oxide-coated glass, or a grid for electron microscopy (TEM grid) using an eyelash tool.

2.3 Materials for Fluorescence Imaging

2.3.1 Microscopy

Confocal and STED imaging can be performed on commercial microscopes; the most common are from Leica, Abberior, PicoQuant, or on an in-house built setup. Fixed samples on #1.5 cover glasses are mounted onto a 76 × 26 mm and 1 mm thick object glass with cut edges (VWR). Live samples are imaged in glass bottom dishes (#1.5 thickness) or in a commercial imaging chamber. Fluorescent molecules are excited and detected using their specified wavelengths.

2.3.2 Fixation Reagents

To preserve the original state for fluorescence imaging, cells can be chemically fixed by different fixation reagents. Aldehydes (e.g., paraformaldehyde, glutaraldehyde) or alcohols (e.g., methanol) are the most commonly used to crosslink or dehydrate/precipitate proteins in the cells. These are all toxic chemicals; thus, they should be handled with great care using protective equipment in a fume hood, and they need to be discarded separately after use.

2.3.3 Fluorescent Probes

To visualize a particular cellular target, labeling with a fluorescent compound which is specific for that target is needed. Immunochemical approaches, expression of fluorescent proteins [11], and live-cell labeling are commonly used methods to investigate cellular structures. For STED imaging, the excitation and emission wavelengths of the fluorescent probes should be in the same regions of the excitation and emission of the laser beams available on the instrument. Common fluorescent reagents used for cellular labeling for fluorescence microscopy imaging are listed in Table 1.

3 Data Analysis

3.1 ToF-SIMS

ToF-SIMS data are often very complex with 2D or 3D arrays of spectra containing hundreds of peaks per spectrum. To extract meaningful chemical information from SIMS ion images, multivariate analysis (MVA) approaches such as principal component analysis (PCA) and maximum autocorrelation factor (MAF) are often applied. MVA helps to deconvolute complex data sets, identify the chemistry present, and improve the image contrast, resulting in better identification of regions of interest (ROIs) [18, 19]. However, each method is suitable for a particular goal of the analysis. PCA analyzes the variance of each mass channel through the data set, looking for the major directions of variance. The output as the score images and peak loading can reveal the areas with distinct chemistry and the peaks responsible for the differences. In MAF analysis, besides finding the variance in mass channels corresponding to spectral contributions as in PCA, the relationship between pixels in the image is also taken into account.

The pros and cons of common practices with data pretreatment and multivariate analysis approaches in ToF-SIMS image

Table 1
List of common fluorescent reagents used for cellular labeling

Probes	Properties	Examples	References
Antibodies	Large size (~150 kDa), fixed samples, immunostaining, high specificity and sensitivity, antibodies for various targets commercially available	Anti-Synaptotagmin, anti-SNAP-25, anti-Syntaxin	[6, 12, 13]
Nanobodies	Very small size (~15 kDa), high affinity labeling, immunostaining, fixed or live samples	Anti-SNAP-25-Atto647, anti-Syntaxin 1A-Atto647	[14]
Fluorescent proteins	Specific labeling, reversible or irreversible photoactivation, live cells, overexpression of fluorescent proteins through transfection or transduction	pHluorine, GFP, RFP, mCherry	[12]
Live-cell labeling dyes	Live-cell imaging, small molecules, high specificity, direct incubation into live cells	Fluorescent false neurotransmitter (FFN), mCLING, silicon rhodamines (SiR)	[15–17]

Table 2
Summary of the efficacy of pretreatments and multivariate analysis approaches in ToF-SIMS image interpretation

		Outcome images	
Input	Pretreatment	PCA	MAF
Low signal intensity spectra	Poisson scaling	+	×
	Mass binning	±	+
	Pixel summation	+	–

Notes: (+) improve, (±) may impair, (–) severe blurring, (×) redundant

interpretation have been shown [19], which are summarized in Table 2. In brief, MAF shows greater benefit over PCA, especially in the case of low spectral intensity. Poisson scaling prior to PCA can improve the outcome, while any data scaling before MAF is redundant as it will be canceled out during the analysis due to the nature of the method. Pretreatments such as mass channel summation (mass binning) and pixel summation are also often used to increase pixel signal intensity; however, these can lead to a reduction in mass resolution and spatial resolution, respectively, affecting the outcomes of PCA and MAF. Mass binning may impair the PCA result but insignificantly change the MAF outcome. On the other hand, pixel summation prior to MVA causes severe blurring of the MAF images, which draws a misleading result, while PCA performs well.

3.2 NanoSIMS

Similar to ToF-SIMS, NanoSIMS provides ion images and depth profiles but often with better lateral resolution. The data sets, however, contain fewer mass channels (up to 7 mass channels in the NanoSIMS 50L, CAMECA), which diminishes the need for MVA for image interpretation. With the high sensitivity of dynamic SIMS, low relative abundance isotopes can be measured, allowing isotopic ratios (e.g., $^{15}N/^{14}N$, $^{13}C/^{12}C$) to be determined.

To analyze NanoSIMS images, the first step is to screen for possible errors in all images such as lines, charging spots, and missing/unrecorded pixels within the images. Next, image drift correction is performed to correct for the primary beam drift during the measurement. Dead-time correction is also required; otherwise, an underestimation of count rate can occur. This is because some secondary ions might not be detected during the dead time (44 ns), the period when the electron multiplier (EM) detector is occupied processing the signal from the former secondary ion [20]. Afterward, image accumulation from different layers of the selected ions is carried out to improve the image contrast. Subsequently, image normalization of the species of interest to a major species, which is present abundantly and evenly in the sample, is performed to produce ratio images (e.g., $^{12}C^{13}C^-/^{12}C_2^-$). This allows further evaluation of the isotopic enrichment in samples.

For quantitative imaging, count rates and isotopic ratios of the selected ion species can be extracted and interpreted from designated ROIs. Importantly, the count rate and isotopic ratio data should be extracted from drift and dead-time corrected layer images to obtain the average values, but not from an accumulated image. Since an accumulated image is considered to be a single layer image, this will produce the sum of the count rate from all the layers. Additionally, ROI selection can be assisted by correlating images with electron microscopy or fluorescence images to confirm the relation between the ROI on NanoSIMS images with a specific organelle or cellular region.

Considering that precise isotope abundance is very difficult to quantify, relative isotope abundance is preferably reported in comparison to internationally agreed-upon reference values and is described in terms of parts per thousand or per mille [21]. The result of relative abundance measurement (δ) can be expressed in per mille as follows:

$$\delta^x = \left(\frac{{}^xR_{\text{sample}}}{{}^xR_{\text{reference}}} - 1\right) \times 1000‰$$

where $x = {}^{13}C$, ^{15}N, ^2H, etc., R_{sample} is the isotopic ratio of the sample of interest, and $R_{\text{reference}}$ is the isotopic ratio of the reference value: ${}^{13}C_{R_{\text{VPDB}}} = \frac{^{13}C}{^{12}C} = 0.0112372$, ${}^{15}N_{R_{\text{Air}}} = \frac{^{15}N}{^{14}N} = 0.0003676$,

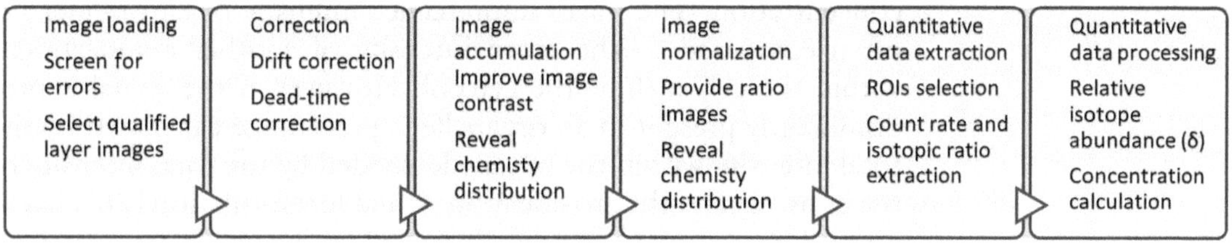

Fig. 1 General NanoSIMS data analysis workflow

${}^{2}\mathrm{H}_{\mathrm{R_{VSMOW}}} = \frac{{}^{2}\mathrm{H}}{{}^{1}\mathrm{H}} = 0.00015576$, ${}^{18}\mathrm{O}_{\mathrm{R_{VSMOW}}} = \frac{{}^{18}\mathrm{O}}{{}^{16}\mathrm{O}} = 0.0020052$, ${}^{34}\mathrm{S}_{\mathrm{R_{VCDT}}} = \frac{{}^{34}\mathrm{S}}{{}^{32}\mathrm{S}} = 0.045005$ [21, 22].

Notes: VSMOW, Vienna Standard Mean Ocean Water; VPDB, Vienna Pee Dee Belemnite; VCDT, Vienna Canyon Diablo Troilite.

Regarding the quantification of NanoSIMS data, it is possible to convert secondary ion signal intensities and the isotopic ratio data into concentrations. To do so, matrix effects need to be corrected properly, and the concentration calibration should be done using a well-constructed standard, which is a sample with a known concentration of the element of interest, and containing the same matrix chemical composition as the sample being studied [23]. This topic has been well described in the recent work by Thomen et al. [24], providing a platform for absolute quantification in single cells using NanoSIMS. The general workflow for NanoSIMS data analysis is summarized in Fig. 1.

3.3 Fluorescence Microscopy

For the analysis of fluorescence images, which consist of a large number of pixels and the fluorescence intensity of each pixel, great care is required. Spatial resolution typically depends on the localization uncertainty, labeling density, and spatial structure of a sample. Nieuwenhuizen et al. [25] presented a way to compare resolution between different nanoscopic methods based on experimental data using a Fourier ring correlation (FRC) method. A plugin for the commonly used image analysis software Fiji and a MATLAB code is available [25]. For STED, a FRC-based method was developed to estimate an absolute resolution directly from an image [26]. In addition, a plugin SQUIRREL for Fiji was developed by Culley [27] for quantitatively mapping image artifacts. The analysis was performed using an image at the diffraction-limited resolution and an image at the super-resolution which functioned as the resolution scaling. Common measurements of cells and organelles are length, area, circularity, fluorescence intensity, etc., which can be measured using ImageJ/Fiji. Fluorescence intensity is used for quantitative measurements of chemical concentration within the cell and its organelles. A useful measurement includes mean fluorescence intensity, which is the total intensity divided by the number of pixels and therefore is proportional to concentration. The mean fluorescence intensity of an organelle is often compared to the mean intensity of another compartment within the cell such as the cytosol. However, if the signal is predominantly present in an organelle, a percentage calculated by the total intensity within the organelle divided by the total intensity of the entire cell might provide more valuable information [28]. Background correction to remove the autofluorescence can be obtained by the 'subtract background' command or (pseudo-) flat-field correction in ImageJ/Fiji. Appropriate thresholding and other computing approaches, such as image segmentation to identify the structures of interest, should be considered to avoid possible human bias with manual selection.

4 Methods

4.1 ToF-SIMS Imaging

ToF-SIMS allows visualization of chemical distribution across the sample often at sub-micrometer spatial resolution. Having the practical mass range limited to ~1000 Da, ToF-SIMS has been used to detect a diverse population of molecules (e.g., lipids [29–33], DNA bases [33], pharmaceuticals [34, 35]) in various cell types. Different types of target molecules are detected depending on the choice of the primary ion source, which also defines the sensitivity and spatial resolution. In short, the higher the energy of primary ions, the higher the secondary ion yield. This helps to improve the sensitivity but also causes more fragmentation, often abolishing the molecular information needed to identify the chemistry present. In general, the smaller the size of the primary ions, the better the spatial resolution. Critical parameters of primary ion sources that may benefit the analysts in choosing a suitable source for their applications are presented in Table 3.

The experimental conditions, including sample preparation and instrumental parameters, are crucial in determining the success of the measurement. ToF-SIMS requires samples to be compatible with high vacuum, and the nature of the sample should be well preserved. Several sample preparation techniques have been applied to ToF-SIMS imaging at the cellular level, namely, chemical fixation [33], frozen hydration [33], freeze-fracturing [30], and the most common freeze-drying [33–35]. Chemical fixation using reagents such as glutaraldehyde (GA) and paraformaldehyde (PFA) helps preserve cellular structures by condensing amines. GA and PFA are known as crosslinking reagents for proteins. Additionally, GA can also react with phospholipids containing free amino groups (e.g., phosphatidylethanolamine, phosphatidylserine) [37]. Thus, a combination of fixatives can also be applied in order to keep the cell morphology and localization of organelles and molecules as close to the living state as possible.

Table 3
Important features of common primary ion sources in ToF-SIMS imaging [36]

Primary ion source		Energy	Spatial resolution	Mass range	Detected molecules
LMIG	Bi_3^{2+}	20 keV	≤500 nm	≤600 Da	Lipid fragments, small molecules
Fullerene	C_{60}^+	40 keV	~1 μm	≤600 Da	Lipid fragments, small molecules
GCIB	Ar_{4000}^+, $\|(CO_2)_{2000}^+$, $(H_2O)_{1000}^+\|$	40 keV	~3 μm	>600 Da	Intact lipids, small peptides

Notes: *LMIG* liquid metal ion guns, *GCIB* gas cluster ion beams

Both frozen hydration and freeze-fracturing methods require the ToF-SIMS instrument to be equipped with a cold sample stage, allowing the analysis at cryogenic temperatures. For frozen hydration, the cells are cryogenically fixed by flash-freezing with liquid propane or liquid nitrogen-cooled isopentane and then stored in liquid nitrogen. For freeze-fracturing, the cells are placed between two Si wafers which are then flash-frozen immediately with liquid propane or liquid nitrogen-cooled isopentane. The Si wafers containing frozen cells are then inserted into the SIMS instrument where they are physically separated at cryogenic temperature in order to expose the fractured cells within the ice matrix [38].

Freeze-drying also requires flash-freezing the cells followed by slow warming under high vacuum to let the water sublimate while retaining the cell morphology. This method allows analysis at room temperature. The general workflows of typical sample preparation techniques for ToF-SIMS imaging are presented in Fig. 2. The protocols of chemical fixation and frozen hydration for ToF-SIMS imaging have been described elsewhere [38]. In this book chapter, we focus on the freeze-fracturing and freeze-drying methods. Examples are presented of research on lipid accumulation [30] and lipid turnover [39] in the plasma membrane of PC12 cell, a well-characterized biological model for studying exocytosis as well as neuronal functions [40, 41].

4.2 Relative Quantification of Phospholipid Accumulation in the Cellular Plasma Membrane

Exocytosis is a neuronal communication process in which synaptic vesicles dock, fuse to the cellular plasma membrane, and release their contents (e.g., neurotransmitters, small peptides/proteins, etc.) into the extracellular matrix. The shape of the membrane lipids, which is affected by the proportionate size of their head and tail groups, is known to influence membrane curvature. It is believed that during exocytosis, the lipid composition of the vesicular membrane and the plasma membrane at the active zone rapidly adapt to support the formation of high curvature regions for

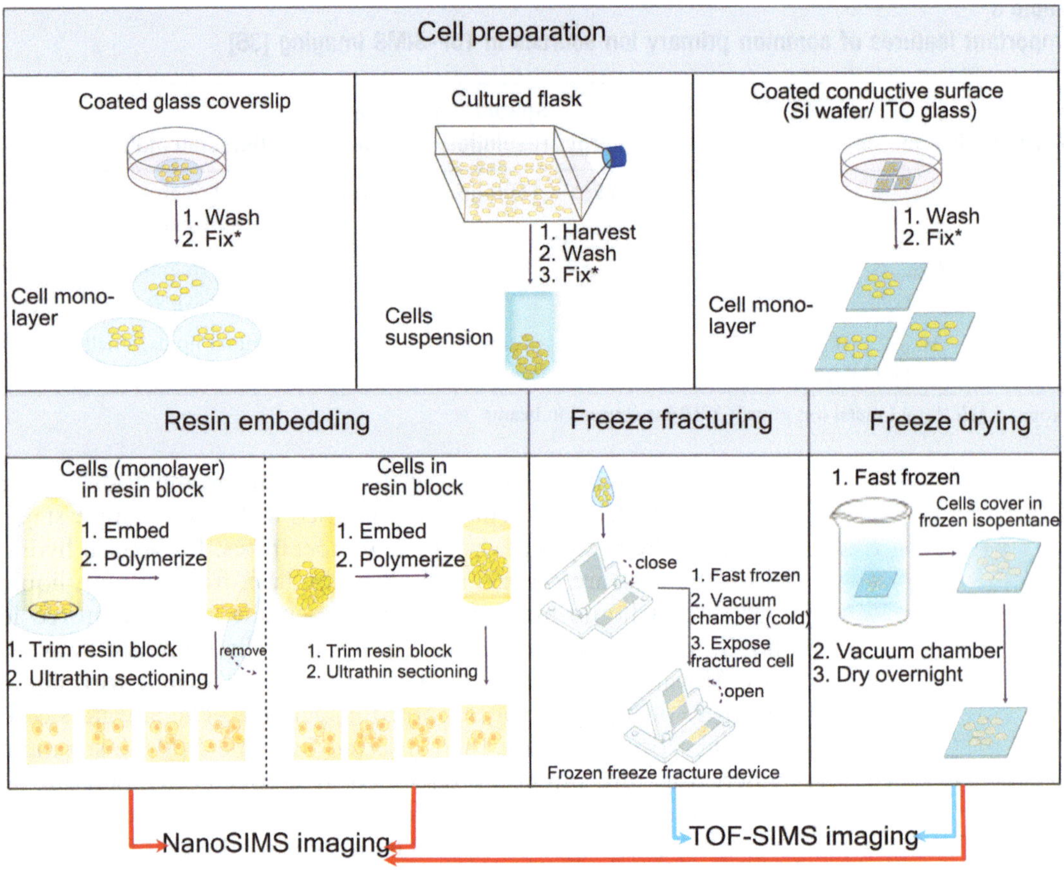

Fig. 2 Workflow of sample preparation for SIMS imaging of single cell. (Schematic of frozen freeze-fracture device is adapted from Lanekoff [30] with permission from the American Chemical Society)

membrane fusion. Moreover, we have seen that the rate of exocytosis in PC12 cells changes significantly after the incubation with lipids [42]. Thus, finding "how much" of "which" exogenous phospholipids have been incorporated in the plasma membrane to affect exocytosis would provide a better understanding of the nature of this process. In this research, ToF-SIMS imaging was used to relatively quantify the amount of deuterated phospholipids labeled in freeze-fracture PC12 cells. The sample preparation and imaging procedure of this work are described as follows.

4.2.1 Sample Preparation

PC12 cells were cultured until confluent and were then incubated with fresh medium, which was added with liposomes containing deuterated PC or PE for 19 h. Afterward, 2 μL of the cell suspension was sandwiched between two silicon shards mounted on a freeze-fracture device, which was then immediately flash-frozen in liquid propane at −185 °C (Fig. 2). The device was mounted onto the cold stage sample holder and transferred immediately into the

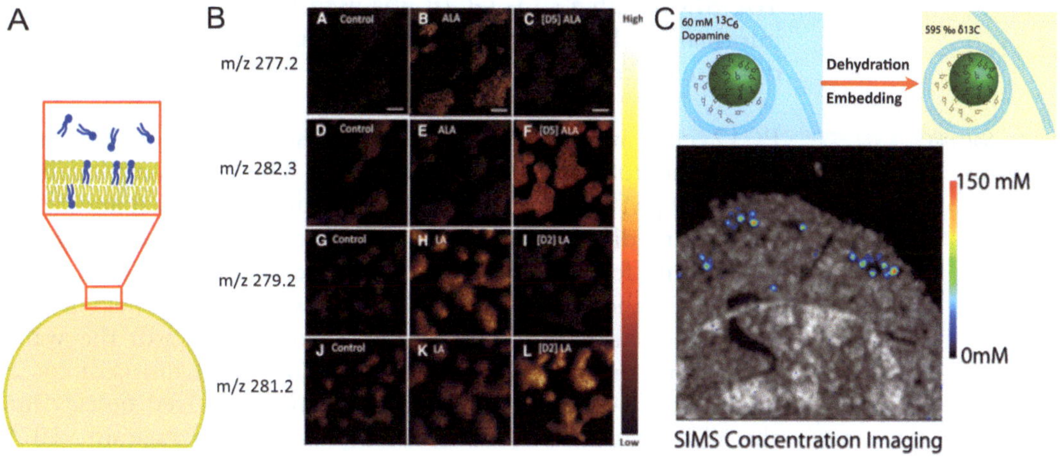

Fig. 3 (**a**) Incorporation of exogenous phospholipids into the cellular plasma membrane. (**b**) ToF-SIMS ion images showing the enrichment of exogenous FAs into the cell membrane. Scale bar, 40 μm. (**c**) Absolute quantification ^{13}C-labeled dopamine with NanoSIMS imaging. (Panel A is adapted from Lanekoff et al. [30], with permission from the American Chemical Society. Panel B is reproduced from Philipsen et al. [39], with permission from the American Society for Biochemistry and Molecular Biology, Inc. Panel C is adapted from Thomen et al. [24]. Copyright 2020 American Chemical Society, https:/pubs.acs.org/doi/10.1021/acsnano.9b09804, and further permissions related to the material excerpted should be directed to the ACS)

analysis chamber with a cryo-cold finger. Finally, the freeze-fracture device was opened at −110 °C to fracture the cells within the ice matrix. The analysis was performed at cryo-temperature.

4.2.2 Instrumental Setup for Imaging Process

An IONTOF IV instrument (IONTOF GmbH, Germany) equipped with a 25 keV Bi_3^+ LMIG was used to acquire spectra and ion images in the bunched mode. The pulsed current was set at 0.1 pA. Low-energy electron flooding was used for the charging neutralization. The primary ion dose density was set below 4×10^{12} ions/cm^2 maintaining the static SIMS mode. A lateral resolution of ~4 μm was achieved over analysis areas between 70×70 and 80×80 μm^2 with 256×256 pixels.

This work showed that after the 100 uM incubation, exogenous phosphatidylcholine (PC) and phosphatidylethanolamine (PE) were integrated into the plasma membrane at levels of only around 0.5% and 1.3%, respectively [30]. The concept of how exogenous phospholipids incorporate into the plasma membrane is graphically presented in Fig. 3a.

4.3 Relative Quantification of Lipid Turnover in PC12 Cell Plasma Membrane

Another study focusing on lipid turnover in the plasma membrane of PC12 cells using deuterated α-linolenic acid (ALA) and linoleic acid (LA) was carried out to follow phospholipid synthesis and transport into the plasma membrane [39]. In this experiment, the IONTOF V with a Bi_3^+ LMIG was used to detect deuterated low mass ionic species at high spatial resolution. In contrast, the J105

ToF-SIMS with a CO_2 GCIB was used to achieve molecular lipid information. Details of the sample preparation and imaging procedure are described as follows.

4.3.1 Sample Preparation

PC12 cells were grown on poly-L-lysine-coated silicon wafers under the conditions described above. The cells were incubated with 100 μM deuterated or non-deuterated FAs (ALA and LA) in the cell media for 1 day. Cells were then rinsed with 50 mM ammonium formate (pH 7.4) three times. All excess liquid on the surface was then removed as much as possible with the help of tissue papers placed at the edge of the wafers. Afterward, the wafers containing the cells were flash-frozen in liquid nitrogen-cooled isopentane (−185 °C) and subsequently transferred quickly into the vacuum chamber for freeze-drying in high vacuum (10^{-2}–10^{-3} mbar) overnight (Fig. 2).

4.3.2 Instrumental Setup for Imaging Process

The IONTOF V with a 25 keV Bi^{3+} LMIG was used in bunched mode, with a pulsed current of 0.3 pA. The primary ion dose density was ~2×10^{12} ions/cm^2 in both positive and negative ion modes. Imaging areas were 255 × 255 μm^2 with 256 × 256 pixels. For the J105 ToF-SIMS with the 40 keV CO_2 GCIB, the primary ion dose density was ~1×10^{13} ions/cm^2. Imaging areas were 800 × 800 μm^2 with 128 × 128 pixels. A low-energy electron gun was used for charge neutralization.

As expected, the fatty acid treatments increased the amounts of ALA and LA as well as their long-chain conversion products. Importantly, an enhanced level of turnover of these FAs in PCs, PEs, and phosphatidylinositols (PIs) in the cellular plasma membrane was observed (Fig. 3b). The data from this work might provide important information about specific phospholipids that could be involved in neurodegenerative diseases and brain aging.

4.4 NanoSIMS Imaging

NanoSIMS is a dynamic SIMS method generating highly fragmented secondary ions with limited molecular information. It can be used to analyze both elemental distribution and isotopic identification. NanoSIMS instruments are often equipped with two primary ion sources, Cs^+ and O_n^-, for analyzing negative ions and positive ions, respectively [21].

The technique requires labeling with stable isotopes such as ^2D, ^{15}N, ^{13}C, ^{18}O, or rare elements, although in some cases, the samples can be analyzed without labeling (e.g., studying subcellular localization of trace elements [43]). The isotopes are often introduced to the cells by incubating the cells with isotopic amino acids [44], lipids [45], and DNA bases [46], enabling tracking of different biological processes, for example, cellular metabolism, and protein and lipid turnover. Additionally, NanoSIMS imaging can also be utilized to reveal the distribution, exchange, and dynamics of different pharmaceuticals, such as studying the cisplatin levels within ovarian cancer cells [47] and assessing the phospholipidosis effect of amiodarone in macrophages [48]. Another approach is to use click chemistry or immunostaining to integrate chemical labeling probes to specific proteins that can be subsequently detected by NanoSIMS. Typical examples of this approach are the boron probe [9] and fluorine probe [10] targeting directly green fluorescent proteins (GFPs) or indirectly via a light chain of mouse antibody. It should be noted that multiple labels on the same molecule can also be used.

Depending on the cells (adherent or suspension cell culture) and the research questions, cells can be cultured on different substrates. Adherent cells can be cultured and labeled in the culture flasks or coated coverslip/ITO glass/Si wafer to achieve a monolayer, which can help to omit the harvesting step. Suspended cells can be analyzed directly after washing or centrifuging down to the coverslip for further fixation and embedding. These different alternatives are graphically presented in the workflow in Fig. 2.

NanoSIMS requires samples to be compatible with ultrahigh vacuum, limited topography, and to be conductive to reduce sample charging during the analysis. The most common sample preparation method for biological samples is adopted from that for transmission electron microscopy (TEM), including chemical fixation or flash-freezing, dehydration, resin embedding, and sectioning. Regarding chemical fixation, besides GA and PFA, osmium tetroxide (OsO_4) is often used to crosslink lipids by reacting with C=C in unsaturated fatty acid chains of phospholipids and also to react with some lipoprotein complexes. Between chemical fixation steps, samples need to be washed carefully with PIPES in addition to glycine to remove excess aldehydes as much as possible to avoid precipitates in the presence of OsO_4. Uranyl acetate (UA) can be then added to improve the contrast if TEM imaging is also performed. As UA can precipitate at pH ~7.4 and in the presence of many salts, the solution should be filtered, and the staining step should be carried out in the dark. Extra washing steps with water also help prevent precipitation by removing salts or OsO_4 residue from the previous fixation steps. Afterward, samples are dehydrated in cold ethanol solution at low concentration and sequentially with increasing concentrations to reduce the shrinking effects. The cell samples are then embedded in resin and sectioned. NanoSIMS imaging requires a thicker section ($\geq$100 nm) compared to TEM (~70 nm) to obtain the steady state of the sputtering rate and avoid the breaking of the samples during the measurement. For absolute quantitative analysis, samples of 300 nm thickness are desired [24]. On the other hand, freeze-drying can also be applied for NanoSIMS [49]. Once prepared, the sections are often coated with a conductive material such as Au, Pt, or Pd. The general workflows of typical sample preparation used for NanoSIMS imaging are summarized in Fig. 2.

4.5 Absolute Quantification of ^{13}C-Labeled Dopamine Within Secretory Vesicles of PC12 Cells

In this section, we elaborate on the fixation and resin embedding sample preparation from the recent work to develop a methodology for absolute quantification of an *organelle-associated*, *isotopically labeled prodrug* in single cells using NanoSIMS imaging [24].

To convert secondary ion intensities and isotopic ratio data obtained by NanoSIMS into concentrations, there are strict requirements for matrix effect correction and well-constructed control samples. Especially in the case of single-cell analysis, it is extremely challenging to reproduce such a complex matrix material in the cell environment. Thomen et al. [24] tackled this challenge by deriving a formula for the ratio of $^{13}C/^{12}C$ measured at each ROI considering all possible contributions from the biomass, the epoxy resin infiltrated in the cell, and the ^{13}C dopamine detected at the *ROI* (Eq. a). A short summary of all the rearrangement of equations for quantification of dopamine concentration in a single vesicle (ROI) is presented in Eqs. a, b, and c [24, 50].

PC12 cells are incubated with ^{13}C L-DOPA, which will be converted to ^{13}C dopamine by DOPA decarboxylase and actively loaded into vesicles through vesicular monoamine transporter 2 (VMAT2) located on the vesicle membrane, resulting in highly enriched ^{13}C inside the vesicles but not in the biomass and epoxy. These enrichment values are known, and the ^{13}C enrichment values of biomass and epoxy can be referred to as the terrestrial abundance of ^{13}C. Therefore, the fraction of the signal contributed by ^{13}C dopamine can be calculated from the $^{13}C/^{12}C$ measured at the ROI, taken into account also the value in control samples (Eq. b).

$$\left(\frac{^{13}\mathrm{C}}{^{12}\mathrm{C}}\right)_{\mathrm{ROI}} = \frac{^{13}\mathrm{C}_{\mathrm{biomas}} + {}^{13}\mathrm{C}_{\mathrm{epoxy}} + {}^{13}\mathrm{C}_{\mathrm{Dopa}}}{^{12}\mathrm{C}_{\mathrm{biomas}} + {}^{12}\mathrm{C}_{\mathrm{epoxy}} + {}^{12}\mathrm{C}_{\mathrm{Dopa}}} \tag{a}$$

$$f_{\mathrm{Dopa}} = \frac{\left(\frac{^{13}\mathrm{C}}{^{12}\mathrm{C}}\right)_{\mathrm{ROI}} - \left(\frac{^{13}\mathrm{C}}{^{12}\mathrm{C}}\right)_{\mathrm{control}}}{\left(\frac{^{13}\mathrm{C}_{\mathrm{atoms}}}{^{12}\mathrm{C}_{\mathrm{atoms}}}\right)_{\mathrm{Dopa}} - \left(\frac{^{13}\mathrm{C}_{\mathrm{atoms}}}{^{12}\mathrm{C}_{\mathrm{atoms}}}\right)_{\mathrm{control}}} \tag{b}$$

$$[\mathrm{Dopa}] = \frac{f_{\mathrm{Dopa}} \times \left[^{12}\mathrm{C}\right]_{\mathrm{sample}}}{^{12}\mathrm{C}_{\mathrm{atoms\,Dopa}}} \tag{c}$$

To convert the fraction (f_{Dopa}) to absolute concentration, ^{12}C concentration at the ROI needs to be quantified. The elemental analysis and NanoSIMS ion yield measurements showed that the epoxy resin (Agar 100) and the biomass contained comparable ^{12}C content and matrix effects. This allowed calculation of the ^{13}C dopamine concentration using Eq. c.

The protocols for sample preparation and imaging procedure are described as follows.

4.5.1 Sample Preparation

PC12 cells were grown on collagen-coated flasks, as previously described. Dopamine precursor $^{13}C_6$ L-3,4-dihydroxyphenylalanine (L-DOPA) (Cambridge Isotope Laboratories Inc., MA, USA) was used to label the neurotransmitter content of vesicles. $^{13}C_6$ L-DOPA stock solution was prepared in PBS purged with argon and kept in the dark to prevent oxidation. The cells were treated with $^{13}C_6$ L-DOPA 150 μM prepared in culture media for 12 h, followed by harvesting and washing two times with DPBS without calcium or magnesium.

The cells were fixed with a fixative mixture containing 0.01% sodium azide, 1% formaldehyde, and 1.25% GA at 4 °C overnight. After washing with 150 mM sodium cacodylate buffer, the cells were postfixed in 1% osmium tetroxide at 4 °C for 2 h. Contrasting and tertiary fixation was performed using 0.5% uranyl acetate (RT, in the dark) for 1 h. Dehydration was carried out using a series of EtOH/water rinses at increasing concentration (70%, 85%, 95%, and 99.5%) and finally 100% acetone. Cells were then embedded in Agar 100 resin and were subsequently cut into sections of 300 nm thickness by an ultramicrotome. The sections were transferred onto coated copper grids and were ready for NanoSIMS measurement.

4.5.2 Instrumental Setup for Imaging Process

The NanoSIMS 50L with a 16 keV Cs^+ primary ion beam, which has a primary ion current of ~2 pA (D1_2) or ~0.5 pA (D1_5), provided the spatial resolution of 150 nm and <100 nm, respectively. Prior to each measurement, implantation with 10^{17} Cs^+. cm^{-2} was performed at the targeted area until reaching the steady state of sputtering. The transmission was set at 35%, an entrance slit was of 15 μm width, an aperture slit was of 150 μm width, and the energy slit was fully open.

In this work, the calculated concentration of ^{13}C dopamine in individual vesicles of these cells was in the range of 60 mM, which was in agreement with the value calculated from independent electrochemical experiments. With a precision of 10‰ for isotopic measurement, this translated to a concentration limit of approximately 1 mM in the case of $^{13}C_6$-dopamine. The methodology of this work is graphically summarized in Fig. 3c.

4.6 Challenges and Possible Solutions in SIMS Imaging

SIMS imaging is an excellent analytical tool for single-cell analysis with all of the advantages regarding high spatial resolution, high mass resolution, and chemical specificity. However, sensitivity is still a challenge for imaging at the cellular level, which could be due to the low abundance of the molecules of interest in the sample itself or the ionization probability of the molecules [51]. Therefore, several issues need to be considered in the overall experiment, including sample preparation, imaging, and data interpretation to ensure the proper use of SIMS imaging. First, the signals of the molecules of interest might be diluted during chemical fixation and embedding [21]. This can be limited by using a suitable fixation method, suitable fixative, and well-constructed control or standard. Second, the delocalization of the molecules of interest is also an issue that often occurs in the freeze-drying procedure if the temperature and vacuum are not well controlled during the flash-freezing and drying steps. Ice crystals can form if the sample freezing is too slow, or water removal from the samples is too rapid during vacuum drying, which results in sample cracking and molecular delocalization. In addition, sample dehydration with ethanol can cause the loss of some lipids if the previous fixation steps are not efficient. Analysis of easily relocated molecules, such as cholesterol [52], at room temperature can also lead to misleading results. In this case, freeze-fracture or frozen hydration techniques can be useful.

4.7 Fluorescence Imaging

Fluorescence microscopy is used to visualize specific biomolecules labeled with fluorophores for detection upon excitation by appropriate laser beams. Specific labeling of the target molecules can be achieved by immunolabeling with antibodies or nanobodies, expression of fluorescent proteins, or tagging with specific labeling dyes (Table 1). Fluorescence microscopy has many advantages including optical sectioning of samples, high specificity of labeling, multicolor imaging, high sensitivity, and possible live-cell imaging. In confocal laser scanning microscopy, the fluorophores in the sample are locally excited, pixel by pixel, by an excitation laser, and the emitted fluorescence is detected after passing through a pinhole. The pinhole helps filter out the out-of-focus fluorescence signal from surrounding areas and thus increases imaging resolution compared to conventional fluorescence microscopy to the diffraction limit (~200 nm). The development of several super-resolution techniques such as STED, RESOLFT, PALM/STORM, and MINFLUX has brought the resolution down to the size of molecules. Stimulated emission depletion microscopy, for example, breaks the diffraction limit by depleting the fluorescence emission around the focal spot. The setup is similar to a confocal microscope with an additional donut-shaped depletion laser, achieving a resolution of ~50 nm on commercial instruments. In this chapter, we focus on confocal and STED microscopy for chemical imaging at the cellular level.

4.7.1 Sample Preparation

Sample preparation depends on the fluorescent probe used and whether the cells are imaged in the fixed or live state.

To image fixed cells, cells are grown on coverslips and are fixed with PFA 4% in PBS at 37 °C for 15 min. The addition of GA 0.1% to the fixation solution improves crosslinking and thus fixation efficiency; however, this can lead to a loss of epitopes and therefore should not be used when all epitopes are supposed to be labeled [53]. After fixation, cells are washed with PBS three times and subsequently permeabilized and blocked with Triton X-100 0.1%

and bovine serum albumin (BSA) 1.5% for 20 min. Higher or lower concentrations of Triton X-100 lead to changes in morphology or improper permeabilization. BSA inhibits unspecific binding, but increasing concentrations will lead to a decrease in specific binding. The incubation with primary antibody is sufficient for labeling all epitopes with the antibody concentrations in the range of 1–10 μg/mL in blocking/permeabilization buffer for 1 h at room temperature. After the incubation, samples are washed with PBS three times, each for 5 min followed by incubation with secondary antibody in blocking/permeabilization buffer for 1 h at room temperature. The fixed cells subsequently mounted on coverslips are mounted onto the microscopic glasses with embedding reagent Mowiol (Sigma-Aldrich, Sweden) and are ready for imaging.

For live imaging, cells are usually labeled shortly before the experiments. Cells are grown in transparent petri dishes which are washed with warm PBS. The fluorescent dye is then added to the cell medium or imaging solution in which cells are incubated at 37°C for a period of time. Afterward, cells are washed again with PBS, and subsequently a fresh imaging solution is added to the cells for measurement.

4.7.2 Imaging

For imaging in the confocal mode, multichannel imaging can be performed in parallel depending on the number of excitation lasers available on the microscope. The common excitation lasers are at the wavelengths 405 nm, 488 nm, 585 nm, and 635 nm, which produce fluorescence emission at 430–470 nm (blue), 510–545 nm (green), 600–650 nm (red), and 665–700 nm (far red), respectively. In STED mode, the most common excitation laser used is the 635 nm, and the high-intensity STED laser is at 750 nm. Images of different channels can be recorded by sequentially scanning line by line or image by image. Scanning frequency depends on the number of pixels of the image and the number of the averaged time lines.

Confocal microscopy is a valuable tool to investigate the localization of different proteins within a neuron as well as its morphology. Dominguez et al. [12] used confocal microscopy to study the effects of a double knockout of chromogranins A and B in cultured hippocampal neurons. Chromogranins are part of the intravesicular matrix to promote cargo aggregation and had previously been shown to influence dense core vesicle biogenesis in endocrine chromaffin cells [54]. First, Dominguez et al. determined neuronal morphology based on the dendritic and axonal length, branching, and synaptic morphology. Fixed hippocampal neurons were immunostained for microtubule-associated protein 2, axonal marker SMI312, and vesicular protein synaptophysin 1 (Fig. 4). To look at possible changes in the cargo of dense core vesicles and its release, neuropeptide Y-pHluorine, a pH-sensitive reporter, was

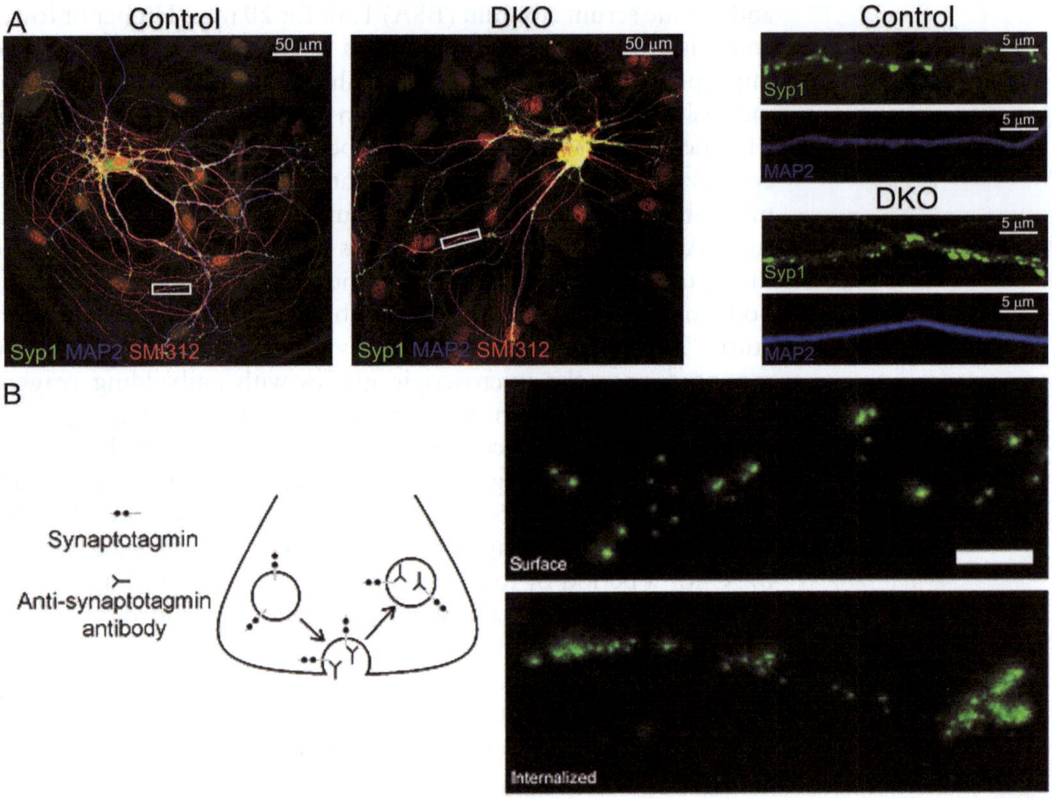

Fig. 4 (**a**) Confocal images of control and double chromogranin A and B knockout hippocampal neurons stained against synaptophysin 1, MAP2, and SMI312 (left). Zoom of the areas indicated on the left for synaptophysin and MAP2 (right). (**b**) Schematic of the internalization of the Synaptotagmin antibody into hippocampal neurons (left). STED images showing clusters of Synaptotagmin at the plasma membrane when hippocampal neurons were stained on the plasma membrane (surface) and inside the cell when antibodies were internalized (right). Scale bar, 1 µm. (Panel A is reproduced from Dominguez et al. [12], with permission under the Creative Commons Attribution License. Panel B is reproduced from Willig et al. [13], with permission from Springer Nature)

expressed in hippocampal neurons via lentiviral transduction. This showed a similar number of neuropeptide Y-pHluorine puncta and only small changes in the cargo loading and number of fusion events, but the duration and amplitude of fusion events remained similar in chromogranin knockout cells. Overall, these data led to the conclusion that the effects of chromogranin double knockout are cell type specific.

A study from Willig et al. [13] shows how antibodies can be used on primary neurons to investigate mechanisms of vesicle recycling with STED microscopy. Hippocampal neurons were incubated with a monoclonal antibody against the intravesicular domain of Synaptotagmin. Monoclonal antibodies were used to avoid antibody clustering. When a vesicle opens, the primary antibody binds to the intravesicular domain of Synaptotagmin. Once the vesicle is retrieved from the membrane during the endocytosis, the antibody will be internalized into the vesicle (Fig. 4). Cells were incubated with primary antibody at 37 °C for 5 min in Tyrode's solution, hence allowing endocytosis. Afterward, cells were fixed and permeabilized, and the Synaptotagmin antibodies were visualized using a secondary antibody, Atto532. This showed clustering of Synaptotagmin, and total fluorescence increased upon stimulation with 70 mM KCl, whereas the staining pattern itself remained similar. When primary antibody incubation was carried out at 4 °C in the absence of Ca^{2+}, endocytosis was inhibited, and thus only staining at the plasma membrane was achieved.

In a similar approach, Richter et al. [6] developed an immunolabeling essay (Fig. 5) to investigate the interactions of synaptic proteins on the plasma membrane during the exocytosis and endocytosis of synaptic vesicles. The question they sought to answer is whether the synaptic proteins co-localize on the plasma membrane after exocytosis and prior to the following retrieval process to reform the vesicles. This was explored by capturing the newly exocytosed vesicular protein Synaptotagmin 1 on the plasma membrane of hippocampal neurons using large antibody clusters. Its localization was then correlated with those of other synaptic proteins residing on the plasma membrane. First, Synaptotagmin 1 epitopes on the plasma membrane were blocked with anti-Synaptotagmin 1 antibody in the presence of 0.5 μM tetrodotoxin. The excess of the antibody was then washed off. Second, the newly exocytosed Synaptotagmin 1 was labeled with biotinylated anti-Synaptotagmin antibody which was then further labeled with aggregates of goat anti-biotin antibody and anti-goat antibody. Finally, cells were fixed and stained for several different synaptic proteins to assess their localizations in correlation with Synaptotagmin 1 on the plasma membrane. These included proteins of the SNARE complex on the membrane and the vesicles, vesicular proteins, an early endosome marker, and endocytosis cofactors. The results showed that Synaptotagmin 1 co-localized with all of the studied proteins except for the early endosome marker (Fig. 5).

4.8 Challenges and Possible Solutions in Fluorescence Imaging

There are many advantages of fluorescence microscopy; however, these are accompanied by challenges. Many of them account for live-cell imaging and the fixation of samples. Challenges and optimization strategies have recently been reviewed by Jahr et al. [55]. Aromatic side chains and nucleic acids, for example, absorb light, and therefore biological samples generally do. In addition, endogenous fluorophores present in the samples cause autofluorescence, which can reduce the signal-to-noise ratio. This needs to be considered and controlled with a negative control sample. Both autofluorescence and absorption effects are present in live as well as fixed cell samples, and these decrease from UV to redshifted spectra, making the red spectrum the preferred imaging region

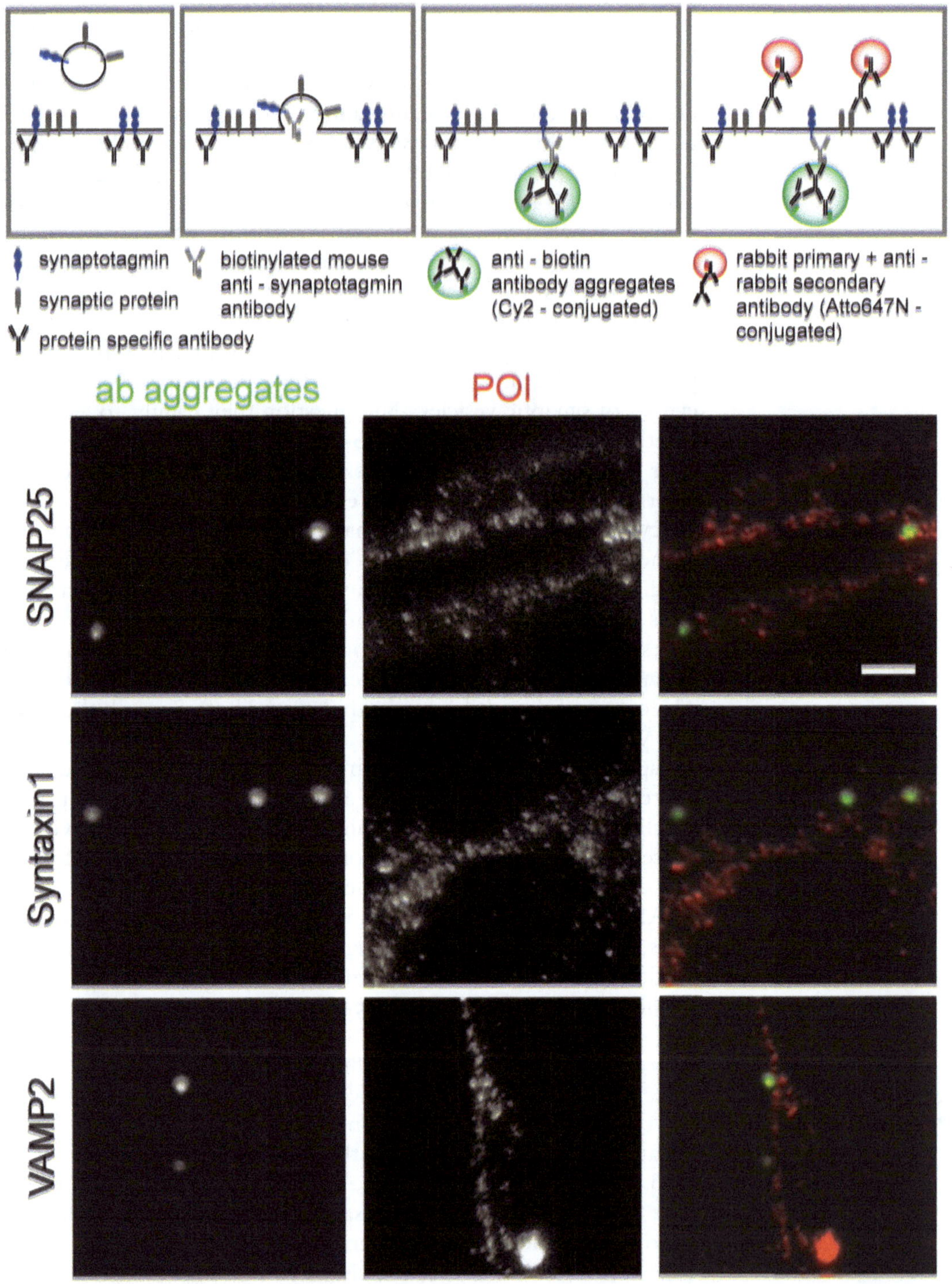

Fig. 5 Immunolabeling assay to investigate the interaction of synaptic proteins on the plasma membrane after exocytosis in hippocampal neurons. Schematic of the labeling scheme (top). Confocal and STED images of Synaptotagmin 1 antibody aggregates and proteins of interest (POI) SNAP-25, Syntaxin 1, and VAMP2 of the SNARE complex showing their co-localization (bottom). Scale bar, 1 μm. (Reproduced from Richter et al. [6], with permission under the Creative Commons Attribution License)

[55]. Photobleaching is another adverse effect, which increases with resolution, as higher imaging resolution requires higher laser powers. It has been shown that most fluorescent proteins undergo supralinear bleaching [56], and this reduces the number of fluorophores and ultimately leads to a decrease in the labeling density and signal-to-noise ratio. Photobleaching is a particular concern in STED imaging due to the use of powerful laser beams. Newly developed fluorescent probes are designed to be more photostable [57] and fluorophore exchangeable [58]. Additionally, the size of the probe becomes a factor influencing resolution especially when super-resolution is desired. IgG molecules are up to 11 times larger than a nanobody [59]. Nanobodies are within the resolution limit of super-resolution microscopy. Therefore, nanobodies and even smaller-sized probes are increasingly used. Proper preservation of samples is crucial for fixed samples. Glyoxal was introduced in 2017 as a less toxic, fast-acting alternative to PFA, improving sample preservation for many targets [60]. Moreover, high irradiation intensities in super-resolution microscopy have been shown to be toxic to cells by generating radical oxygen species and influencing actin and β-tubulin structures inside the cell. Optimization strategies for live-cell imaging include reducing the penetration depth of illumination, imaging in the red spectrum, and adding ascorbic acid to the imaging buffer, and these should be considered to prevent the phototoxicity [61].

4.9 Correlative Imaging

A recent trend in bio-imaging is to combine different imaging techniques in order to answer more complex biological questions that need multidimensional information which cannot be obtained by a single technique. This imaging approach helps gain the strength of all techniques while overcoming their limitations. Correlative imaging at the cellular level provides the means to connect different properties, structures, and functions of cells in order to better understand the underlying molecular and cellular mechanisms. The approach, however, is challenging, particularly the complicated instrumental setup and the compatibility of the sample for measurement by different techniques. Samples can be measured simultaneously or sequentially by individual techniques. In the latter case, it is important that the samples are kept intact for all measurements. Highly technical demands in correlative imaging therefore require well-trained and experienced scientists to ensure accurate analysis.

There have been various combinations of SIMS and fluorescence microscopy and with other techniques such as electron microscopy and electrochemistry [53, 62–64]. We elaborate here a few commonly used approaches along with their selected applications in imaging at the cellular level.

4.10 Correlative SIMS and Fluorescence Microscopy

Correlative imaging that combines NanoSIMS and STED microscopy has been successfully applied to study the molecular turnover of synapses in hippocampal neurons [53]. NanoSIMS allows the visualization and quantification of biomolecules labeled with stable isotopes in biological samples, thereby enabling in vivo tracking of molecular turnover in a particular cellular pathway. However, the technique exhibits a weakness that specific structures of the samples cannot be accurately identified. The combination of NanoSIMS and STED microscopy offers the advantages of visualization of specific protein complexes and organelles by STED and quantification of corresponding molecular turnover by NanoSIMS at high lateral resolution (~50 nm) and depth resolution (~5.7 nm).

To prepare samples for measurements with both techniques, the neurons were first pulsed with an isotopic precursor like ^{15}N-leucine for 3 days, which was incorporated into newly synthesized proteins as a marker for protein turnover in neurons. The cells were then fixed with a mixture of 4% PFA and 0.1–0.2% GA and then immunostained for different organelle markers such as the Golgi apparatus, endoplasmic reticulum, mitochondria, active zone, or synaptic vesicles. The cells were subsequently embedded in LR White resin (London Resin Company Ltd.) and cut into thin (200 nm) slices with an ultramicrotome (Leica Microsystems) and placed on silicon wafers. There were several adjustments in the sample preparation, compared to the conventional preparation procedures for individual techniques, so that the samples were suitable for analysis by both modalities. First, GA for fixation has the tendency to block the epitopes which are the binding sites of primary antibodies; therefore, it is important to use this reagent at very low concentration (0.1–0.2% in PBS) to not affecting the fluorescence signal. The samples can be alternatively fixed with PFA and postfixed with GA in PBS after the immunostaining [9]. Here, the postfixation with GA helps to better preserve the localization of the cell structures. Second, LR White resin was replaced with Epon, the common resin material for TEM imaging, because of its low level of autofluorescence which was suitable for fluorescence microscopy. Third, silicon wafers were used to replace microscopic glass slides due to the conditions for NanoSIMS measurements which must be acquired on a conducting substrate. However, as silicon wafers are not transparent to light in fluorescence microscopy, the samples are imaged upside down facing to the objective lens and covered with immersion oil.

To carry out these experiments, the imaging procedure was performed sequentially, first by STED followed by NanoSIMS. The positions of synapses indicated by the localizations of the active zone marker Bassoon were identified on the STED channel. Other organelles were imaged concurrently on the confocal channels. To locate the areas previously imaged by STED on the NanoSIMS instrument, it is very useful to create several marks on the regions nearby the imaged areas using high-intensity multiphoton laser beam on the microscope. After STED imaging, the immersion oil on the samples was removed by letting it drift down to the discard plate while the silicon wafers were placed facing sideward. The samples were subsequently inserted into the NanoSIMS instrument. Signals of different isotopic ions, particularly $^{12}C^{14}N$ and $^{12}C^{15}N$, were recorded across the scanned area producing ion images. The enrichment of ^{15}N was calculated based on the ratio $^{12}C^{15}N/^{12}C^{14}N$ compared to their natural abundance ratio. The NanoSIMS ion images were correlated with the corresponding STED images of the same area in order to determine the local enrichment of ^{15}N at the synaptic regions. It was shown that the protein turnover in the synapse was significantly higher than that in the axonal regions (Fig. 6a). Correlation of SIMS and fluorescence microscopy has also been applied to study the effects of erythropoietin in neurogenesis in the hippocampal neurons [65] and to investigate the lipid compositional changes in epithelial mouse lung cells under the exposure of nanoparticles [66].

4.11 Correlative SIMS, Electron Microscopy (EM) Imaging, and Electrochemical Analysis

Another combination approach employing NanoSIMS, TEM, and quantitative single-cell amperometry and intracellular vesicle impact electrochemical cytometry (IVIEC) is also very attractive to cell and biological research. The method was used to investigate the distribution of dopamine and its transfer kinetics across the substructures of single vesicles [67]. This approach allows the correlation of the localization of newly synthesized dopamine visualized by NanoSIMS with the morphological structure of the vesicles and cells obtained by TEM and the concentration of electroactive dopamine in single vesicles quantified by the electrochemistry.

For the experiments carried out here, the PC12 cells were incubated with dopamine precursor ^{13}C L-DOPA for the desired periods. L-DOPA is easily oxidized in the presence of oxygen; thus, the stock solution was prepared by dissolving the compound in PBS solution (Sigma-Aldrich, Sweden) while purging continuously with argon in the dark. The stock solution was diluted with warm media to obtain a working solution which was used immediately afterward. In the case of additional treatment with drug which was prepared in DMSO, for example, reserpine, the final concentration of DMSO for cell incubation was not higher than 0.05% in order to avoid any toxicity to the cells.

For TEM and NanoSIMS measurements, the cells were fixed following the conventional fixation approach for TEM. Cells were harvested and suspended after the trypsinization with TrypLE Express (GIBCO, Fisher Scientific, Sweden). Cell suspensions were preliminarily fixed overnight at 4°C with Karnovsky solution containing 0.01% sodium azide (BDH, UK), 1% PFA (Sigma-Aldrich, Sweden), and 1.25% GA (Agar Scientific Ltd., UK). The cells were then washed with cacodylate buffer 150 mM followed by

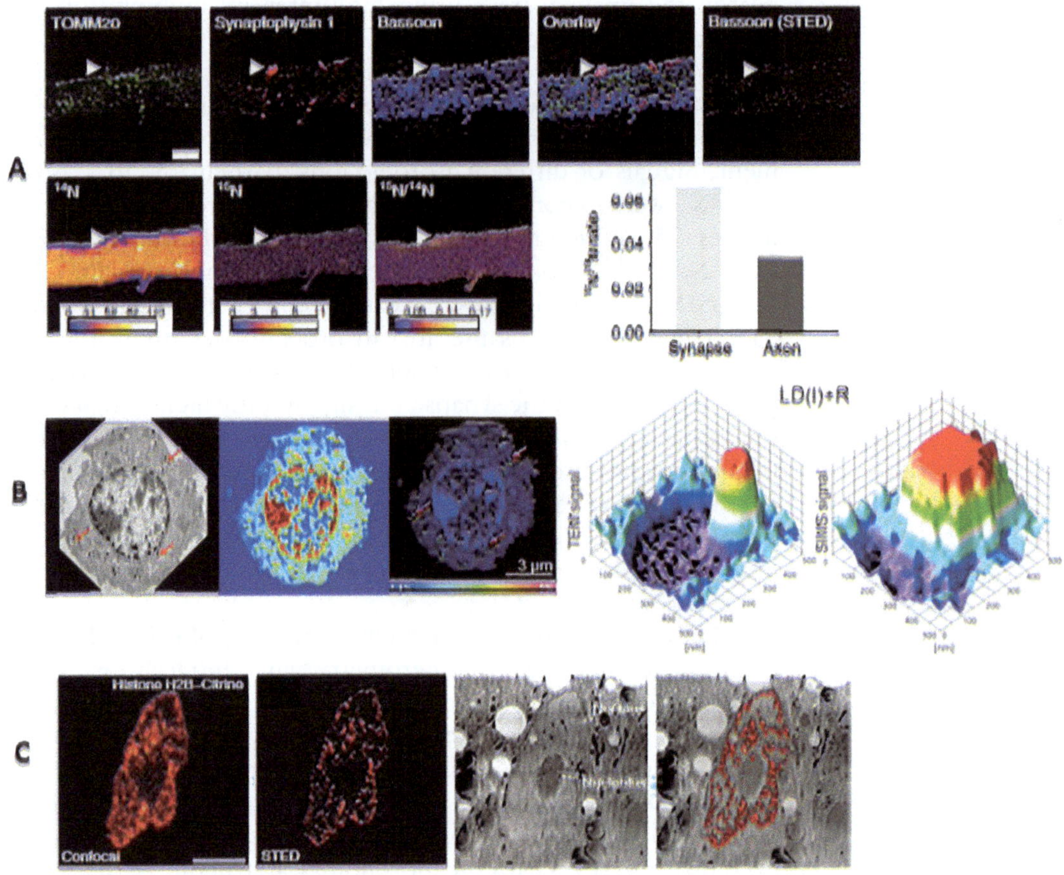

Fig. 6 Correlative imaging at the cellular level with SIMS and fluorescence microscopy. (**a**) Correlative STED and NanoSIMS imaging of synaptic proteins to study protein turnover in hippocampal neurons. Top row from left to right, fluorescence images of the mitochondrial marker TOMM20, vesicular protein synaptophysin 1, active zone marker Bassoon, an overlay of these proteins, and STED image of Bassoon. Bottom row from left to right, NanoSIMS images of ^{14}N, ^{15}N, ratio $^{15}N/^{14}N$, and a chart showing the turnover rate at the synapse and axon areas. Scale bar, 2 μm. (**b**) Correlative TEM and NanoSIMS imaging to study dopamine distribution inside single vesicles. From left to right, TEM image, NanoSIMS image of $^{12}C^{14}N$, ratio $^{13}C^{14}N/^{12}C^{14}N$, and the 3D plots of TEM and NanoSIMS $^{13}C^{14}N$ ion images showing the distribution of ^{13}C dopamine between the halo and dense core of single vesicle. Conditions: 12 h incubation with ^{13}C L-DOPA and 1 h exposure to reserpine. Scale bar, 3 μm. (**c**) Correlative fluorescence and SEM imaging using nucleus protein histone H2B fused with Citrine in *C. elegans*. From left to right, confocal image, STED image of histone H2B-Citrine, SEM image of the same area, and an overlay of STED and SEM images showing histone H2B localizes around the nucleus. Scale bar, 3 μm. (Panels A, B, and C are reproduced with permissions from Saka et al. [53], with permission from Springer Nature; Lovric et al. [67], with permission from the American Chemical Society; Watanabe et al. [62], with permission from Springer Nature, respectively)

postfixation with 1% OsO_4 (Agar Scientific Ltd., UK) and with 0.5% uranyl acetate (Merck, Sigma-Aldrich, Sweden). The fixed cells were subsequently dehydrated and embedded in Agar 100 resin (Agar Scientific Ltd., UK) and cut into sections of 70 nm thickness by ultramicrotome. The cell sections were placed on Formvar-coated copper grids (FCF200F1-Cu, EMS, USA) and poststained with uranyl acetate and Reynolds lead acetate. Here, due to the requirement of the thickness of sample to obtain optimum TEM imaging, the sections should not be thicker than 70 nm, unless being compromised by the quality of TEM images. In addition, conductive copper grids with coordination values were used as a substrate facilitating location of the imaged areas for the correlation between TEM and NanoSIMS.

The cell sections were first measured with TEM to obtain morphological images of cells containing vesicles. The imaged areas on the same samples were then measured on NanoSIMS. Signals for ions $^{13}C^{14}N$ and $^{12}C^{14}N$ were recorded from which the localization of ^{13}C enrichment was calculated based on the ratio $^{13}C^{14}N/^{12}C^{14}N$. The image of the isotopic enrichment was then correlated with TEM images of the same sample area in order to determine the distribution of newly synthesized dopamine in single vesicles.

Another set of cells which were prepared in parallel with the ones for TEM-NanoSIMS experiments were rinsed with warm isotonic saline buffer for electrochemical analysis. Here, measurements were carried out on an inverted microscope with a heated stage (37°C). Microelectrodes were prepared in laboratory as described [67]. The working electrode was held at an applied potential of +700 mV vs. a Ag/AgCl reference electrode, which selectively oxidizes dopamine neurotransmitter. Dopamine oxidation for individual vesicle content, represented as spikes along the time axis, was subsequently analyzed with Igor Pro 6 to obtain the concentration value of dopamine inside single vesicles. In order to obtain accurate and reproducible values of dopamine concentration in vesicles, it is essential to select the microelectrodes with high signal-to-noise and stable electrochemical response. This was ensured by testing the electrodes with cyclic voltammetry before and after the measurement.

The results demonstrated that dopamine indeed is enriched in the vesicles, and its distribution in the vesicular substructures is affected by the length of time for L-DOPA incubation or treatment with reserpine – a drug used to deplete dopamine from vesicles (Fig. 6b). Furthermore, dopamine was shown to localize more in the halo with short incubation time (1.5 h), whereas the transmitter accumulated more in the dense core with long incubation time (12 h). In addition, a higher concentration of dopamine is retained in vesicles when incubated with L-DOPA for longer period followed by reserpine treatment. The correlation between NanoSIMS and TEM imaging enabled the observation of dopamine distribution at the nano-compartments of single vesicles.

In another correlative example, EM was combined with ToF-SIMS imaging for analysis [68]. This correlation approach adds valuable information of the morphology of the fly brain related to the organization of lipids in the brain obtained by ToF-SIMS.

4.12 Correlative Fluorescence Microscopy and EM Imaging

The combination of super-resolution microscopy and scanning electron microscopy (SEM) has been optimized and applied to localize several proteins in *C. elegans* tissues, particularly the nucleus protein histone H2B, mitochondrial outer membrane protein TOM20, and presynaptic dense projection α-liprin [62]. The approach enables the correlation between the precise localization of proteins and the subcellular contexts. Immunocytochemical electron microscopy (immuno-EM) was previously used for this purpose; however, this exhibits several limitations mainly accounting for the lack of suitable antibodies, inaccessibility of antigens, and nonspecific binding of antibodies. The correlative microscopy approach tackles these weaknesses since protein labeling and sample embedding can be carried out separately to ensure the optimum labeling result for any proteins.

In this example of correlative imaging of *C. elegans*, the fluorescent protein Citrine was chosen for STED imaging due to its high and stable quantum yield. Transgenic *C. elegans* was expressed with Citrine fused to histone H2B, TOM20, or liprin. The worms were then quickly frozen using a high-pressure freezer where the worms were dehydrated and fixed with a fixative mixture of 0.001% OsO_4, 0.1% potassium permanganate, and acetone. Afterward, the fixed tissues were infiltrated with the embedding resin glycol methacrylate (GMA) (SPI, US) and gradually polymerized. The resin tissue samples were subsequently cut into thin sections (~100 nm) and mounted on cover glass slides. The possibility to use cover glass in SEM presents an advantage over TEM as it allows a large number of consecutive sections to be mounted on the same cover glass for imaging, which produces a three-dimensional image assembled from consecutive two-dimensional images.

Crosslinked fixation protects the sample morphology from the dehydration and embedding. GA is an effective fixative for proteins but can induce autofluorescence in the samples. In contrast, OsO_4 crosslinks unsaturated membrane lipids but can break peptide bonds that quench the fluorescence in the samples. The optimum fixative is a mixture of 0.001% OsO_4 and 0.1% potassium permanganate which preserves both the morphology and the fluorescence of the tissue samples. For embedding, several embedding materials, such as Lowicryl K4M (EMS), LR Gold (EMS), LR White (EMS), and GMA, were tested based on the tissue morphology and the preserved fluorescence signal. LR White and GMA provided the best results. GMA preserved fluorescence slightly better compared to LR White; however, the section thickness must be above 70 nm to avoid breaking during sample cutting.

The samples were first measured with STED to obtain fluorescence images of protein localizations before SEM imaging. To facilitate an accurate alignment of the STED images and SEM images of the same sample areas, fluorescent silica beads were used as fiducial markers. These beads produce fluorescence in UV light having excitation and emission wavelength at 354 nm and 450 nm, respectively, and dark circles in the SEM as a result of their electron reflection.

It was shown that STED imaging nicely resolved the positions of the labeled proteins with a resolution of 60–80 nm. Correlation between STED and SEM showed significant features, particularly the histone H2B was aligned with chromatin around the nucleus (Fig. 6c), whereas TOM20 appeared as circular rings overlayed with the outer membranes of mitochondria. α-Liprin, on the other hand, was distributed in the presynaptic dense projection.

5 Summary and Future Perspective

In this chapter, we have presented protocols for quantitative imaging at the cellular level using secondary ion mass spectrometry and fluorescence microscopy. Additionally, we have summarized with limited examples of correlative imaging with these techniques and in one case correlation to quantitative analysis of a metabolite in nanoscopic cellular vesicles with electrochemistry. We focused on cells derived from the nervous system; however, the general procedures can be used for a wide variety of cells and applications.

The challenges in single-cell and subcellular imaging with these approaches include the need for high spatial resolution, exceptional sensitivity in extremely small volumes, and sample preparation for extremely small structures while maintaining spatial localization. A trade-off between sensitivity and spatial resolution occurs as detectable analyte amount is proportional to the imaged volume.

These imaging approaches have only scratched the surface of what we can accomplish at the subcellular level. Future goals will be to expand these measurements to single molecules in cells, to a large range of chemical species that can be imaged at once, and to accomplish this with high temporal resolution. Most STED and all mass spectrometric imaging is now carried out on frozen or fixed samples. Live-cell STED and other optical methods are improving and will be a large part of the future of this area. Rapid freezing and fixing will, in the future, allow time frames of dynamic events on a series of cells to be better handled with SIMS or NanoSIMS imaging. Other techniques such as matrix-assisted laser desorption/ionization (MALDI) MS imaging and related methods are improving and will become more important in cell and subcellular analysis.

Likewise, new methods in single-molecule fluorescence are now past the horizon. In the future, we will be able to track and map a wide range of molecules and molecular processes across cell surfaces and in the cellular compartments with exquisite spatial and temporal resolution. For some proteins and other molecules, this is already being carried out.

References

1. DeLellis RA et al (1973) Ultrastructure and in vitro growth characteristics of a transplantable rat pheochromocytoma. Cancer 32: 227–235
2. Warren S, Chute RN (1972) Pheochromocytoma. Cancer 29:327–331
3. Gu C, Larsson A, Ewing Andrew G (2019) Plasticity in exocytosis revealed through the effects of repetitive stimuli affect the content of nanometer vesicles and the fraction of transmitter released. Proc Natl Acad Sci 116: 21409–21415
4. Phan NTN, Li X, Ewing AG (2017) Measuring synaptic vesicles using cellular electrochemistry and nanoscale molecular imaging. Nat Rev Chem 1
5. Li X, Duneval J, Ewing Andrew G (2016) Quantitative chemical measurements of vesicular transmitters with electrochemical cytometry. Acc Chem Res 49:2347–2354
6. Richter KN et al (2019) Antibody-driven capture of synaptic vesicle proteins on the plasma membrane enables the analysis of their interactions with other synaptic proteins. Sci Rep:1–12
7. Bell S, et al (2019). www.bio-protocol.org/e3188. 9:1–18
8. Polerecky L et al (2012) Look@NanoSIMS – a tool for the analysis of nanoSIMS data in environmental microbiology. Environ Microbiol 14:1009–1023
9. Kabatas S et al (2019) Boron-containing probes for non-optical high-resolution imaging of biological samples. Angew Chem Int Ed 58:3438–3443
10. Kabatas S et al (2019) Fluorinated nanobodies for targeted molecular imaging of biological samples using nanoscale secondary ion mass spectrometry. J Analyt Atomic Spec 34: 1083–1087
11. Thorn K (2017) Genetically encoded fluorescent tags. Mol Biol Cell 28:848–857
12. Dominguez N et al (2018) Dense-core vesicle biogenesis and exocytosis in neurons lacking chromogranins A and B. J Neurochem 144: 241–254
13. Willig KI et al (2006) STED microscopy reveals that synaptotagmin remains clustered after synaptic vesicle exocytosis. Nature 440: 935–939
14. Maidorn M et al (2019) Nanobodies reveal an extra-synaptic population of SNAP-25 and Syntaxin 1A in hippocampal neurons. MAbs 11:305–321
15. Lukinavičius G et al (2016) Fluorogenic probes for multicolor imaging in living cells. J Am Chem Soc 138:9365–9368
16. Revelo NH et al (2014) A new probe for super-resolution imaging of membranes elucidates trafficking pathways. J Cell Biol 205:591–606
17. Gubernator NG et al (2009) Fluorescent false neurotransmitters visualize dopamine release from individual presynaptic terminals. Science 324:1441–1444
18. Biesinger MC et al (2002) Principal component analysis of TOF-SIMS images of organic monolayers. Anal Chem 74:5711–5716
19. Henderson A, Fletcher JS, Vickerman JC (2009) A comparison of PCA and MAF for ToF-SIMS image interpretation. Surf Interface Anal 41:666–674
20. Kilburn MR, Wacey D (2015) Nanoscale secondary ion mass spectrometry (NanoSIMS) as an analytical tool in the geosciences. In: Grice K (ed) Principles and practice of analytical techniques in geosciences, vol 4. The Royal Society of Chemistry, Cambridge, pp 1–34
21. Nuñez J et al (2018) NanoSIMS for biological applications: current practices and analyses. Biointerphases 13:03B301
22. de Laeter JR et al (2003) Atomic weights of the elements. Review 2000 (IUPAC technical report). Pure Appl Chem 75:683–800
23. Davission M et al (2008) Development of standards for NanoSIMS analyses of biological materials; LLNL-TR-406039. Lawrence Livermore National Lab (LLNL), Livermore
24. Thomen A et al (2020) Subcellular mass spectrometry imaging and absolute quantitative analysis across organelles. ACS Nano 14: 4316–4325

25. Nieuwenhuizen RPJ et al (2013) Measuring image resolution in optical nanoscopy. Nat Methods 10:557–562
26. Tortarolo G et al (2018) Evaluating image resolution in stimulated emission depletion microscopy. Optica 5:32
27. Culley S et al (2018) Quantitative mapping and minimization of super-resolution optical imaging artifacts. Nat Methods 15:263–266
28. Kelley JB, Paschal BM (2019) Fluorescence-based quantification of nucleocytoplasmic transport. Methods 157:106–114
29. Leefmann T et al (2013) Biomarker imaging of single diatom cells in a microbial mat using time-of-flight secondary ion mass spectrometry (ToF-SIMS). Org Geochem 57:23–33
30. Lanekoff I, Sjövall P, Ewing AG (2011) Relative quantification of phospholipid accumulation in the PC12 cell plasma membrane following phospholipid incubation using TOF-SIMS imaging. Anal Chem 83: 5337–5343
31. Sjövall P et al (2003) Imaging of membrane lipids in single cells by imprint-imaging time-of-flight secondary ion mass spectrometry. Anal Chem 75:3429–3434
32. Ostrowski S et al (2004) Mass spectrometric imaging of highly curved membranes during Tetrahymena mating. Science 305:71–73
33. Fletcher JS et al (2011) Three-dimensional mass spectral imaging of HeLa-M cells – sample preparation, data interpretation and visualisation. Rapid Commun Mass Spectrom 25: 925–932
34. Newman CF et al (2017) Intracellular drug uptake—a comparison of single cell measurements using ToF-SIMS imaging and quantification from cell populations with LC/MS/MS. Anal Chem 89:11944–11953
35. Tian H et al (2017) Subcellular chemical imaging of antibiotics in single bacteria using C60-secondary ion mass spectrometry. Anal Chem 89:5050–5057
36. Phan NTN, Rizzoli SO (2020) High-resolution molecular imaging and its applications in brain and synapses. In: Wacker I, Hummel E, Burgold S, Schröder R (eds) Volume microscopy: multiscale imaging with photons, electrons, and ions. Springer US, New York, pp 37–58
37. Roozemond RC (1969) The effect of fixation with formaldehyde and glutaraldehyde on the composition of phospholipids extractable from rat hypothalamus. J Histochem Cytochem 17: 482–486
38. Winograd N, Bloom A (2015) Sample preparation for 3D SIMS chemical imaging of cells. Methods Mol Biol 1203:9–19
39. Philipsen MH et al (2018) Relative quantification of deuterated omega-3 and -6 fatty acids and their lipid turnover in PC12 cell membranes using TOF-SIMS. J Lipid Res 59: 2098–2107
40. Greene LA, Tischler AS (1976) Establishment of a noradrenergic clonal line of rat adrenal pheochromocytoma cells which respond to nerve growth factor. Proc Natl Acad Sci U S A 73:2424–2428
41. Westerink RHS, Ewing AG (2008) The PC12 cell as model for neurosecretion. Acta Physiol 192:273
42. Uchiyama Y et al (2007) Phospholipid mediated plasticity in exocytosis observed in PC12 cells. Brain Res 1151:46
43. Penen F et al (2016) Chemical bioimaging for the subcellular localization of trace elements by high contrast TEM, TEM/X-EDS, and NanoSIMS. J Trace Elem Med Biol 37:62–68
44. Zhang D-S et al (2012) Multi-isotope imaging mass spectrometry reveals slow protein turnover in hair-cell stereocilia. Nature 481: 520–524
45. Frisz JF et al (2013) Direct chemical evidence for sphingolipid domains in the plasma membranes of fibroblasts. Proc Natl Acad Sci 110: E613–E622
46. Steinhauser ML et al (2012) Multi-isotope imaging mass spectrometry quantifies stem cell division and metabolism. Nature 481: 516–519
47. Lee RFS et al (2017) Differences in cisplatin distribution in sensitive and resistant ovarian cancer cells: a TEM/NanoSIMS study. Metallomics 9:1413–1420
48. Jiang H et al (2017) High-resolution sub-cellular imaging by correlative NanoSIMS and electron microscopy of amiodarone internalisation by lung macrophages as evidence for drug-induced phospholipidosis. Chem Commun 53: 1506–1509
49. Grovenor CRM et al (2006) Specimen preparation for NanoSIMS analysis of biological materials. Appl Surf Sci 252:6917–6924
50. Gorman BL et al (2020) Measurement of absolute concentration at the subcellular scale. ACS Nano 14:6414–6419
51. Passarelli MK, Ewing AG (2013) Single-cell imaging mass spectrometry. Curr Opin Chem Biol 17:854–859
52. Sjövall P, Johansson B, Lausmaa J (2006) Localization of lipids in freeze-dried mouse

brain sections by imaging TOF-SIMS. Appl Surf Sci 252:6966–6974
53. Saka SK et al (2014) Correlated optical and isotopic nanoscopy. Nat Commun 5:3664
54. Kim T et al (2005) Chromogranin A deficiency in transgenic mice leads to aberrant chromaffin granule biogenesis. J Neurosci 25:6958–6961
55. Jahr W, Velicky P, Danzl JG (2020) Strategies to maximize performance in STimulated Emission Depletion (STED) nanoscopy of biological specimens. Methods 174:27–41
56. Cranfill PJ et al (2016) Quantitative assessment of fluorescent proteins. Nat Methods 13: 557–562
57. Zheng Q, Lavis LD (2017) Development of photostable fluorophores for molecular imaging. Curr Opin Chem Biol 39:32–38
58. Spahn C et al (2019) Whole-cell, 3D, and multicolor STED imaging with exchangeable fluorophores. Nano Lett 19:500–505
59. Sahl SJ, Hell SW, Jakobs S (2017) Fluorescence nanoscopy in cell biology. Nat Rev Mol Cell Biol 18:685–701
60. Richter KN et al (2018) Glyoxal as an alternative fixative to formaldehyde in immunostaining and super-resolution microscopy. EMBO J 37:139–159
61. Waldchen S et al (2015) Light-induced cell damage in live-cell super-resolution microscopy. Sci Rep 5:15348
62. Watanabe S et al (2011) Protein localization in electron micrographs using fluorescence nanoscopy. Nat Methods 8:80–84
63. Wilhelm BG et al (2014) Composition of isolated synaptic boutons reveals the amounts of vesicle trafficking proteins. Science 344: 1023–1028
64. Salamifar ES, Lai RY (2013) Use of combined scanning electrochemical and fluorescence microscopy for detection of reactive oxygen species in prostate cancer cells. Anal Chem 85: 9417–9421
65. Hassouna I et al (2016) Revisiting adult neurogenesis and the role of erythropoietin for neuronal and oligodendroglial differentiation in the hippocampus. Mol Psychiatry 21: 1752–1767
66. Hua X et al (2016) Chemical imaging of molecular changes in a hydrated single cell by dynamic secondary ion mass spectrometry and super-resolution microscopy. Integr Biol (Camb) 8:635–644
67. Lovric J et al (2016) Nano secondary ion mass spectrometry imaging of dopamine distribution across nanometer vesicles. ACS Nano 11: 3446–3455
68. Phan NTN, Fletcher JS, Ewing AG (2015) Lipid structural effects of oral administration of methylphenidate in *drosophila* brain by secondary ion mass spectrometry imaging. Anal Chem 87:4063–4071

Chapter 10

Electrophysiology and Single Cells

Gábor Juhász, Dániel Mittli, Vanda Tukacs, and Katalin Adrienna Kékesi

Abstract

Information processing, storage, and retrieval in the central nervous system are largely dependent on the electrical activity of single neurons and on the propagation of electrical signals in synaptic networks. The most important protein complexes, underlying cellular electrical activity, are the voltage-gated ion channels. These proteins form controllable ion-conducting pores across the cell membrane, enabling the rapid changes of the membrane potential. In addition, the expressed set of functional ion channels determines the stereotypical firing pattern (the electrophysiological phenotype) of single neurons. The amount, subcellular localization, and biophysical behavior of channel proteins can be regulated via different extracellular stimuli acting on short-term and long-term signaling processes. A functionally relevant aspect of the modulation of neuronal electrical activity is the regulation of ion channel gene expression. The expression pattern of ion channel genes can be effectively studied by single-cell transcriptomics, which enables the qualitative and quantitative analysis of the mRNA content of a neuron. Furthermore, the electrophysiology-combined single-cell mRNA sequencing largely elucidated the causal relationship between the transcriptomic profile of ion channel genes and the electrical behavior of certain neuronal cell types. Based on currently available data, it can be strongly assumed that the transcriptomic control of ion channels aims the maintenance of the appropriate level of cellular electrical activity, rather than, solely, the amounts of different channel protein complexes. This combinatorial regulation of ion channel gene expression significantly contributes to the functional plasticity of neuronal networks.

Key words Single-cell electrophysiology, Single-cell sequencing, Ion channels, Neuronal activity, Neuron classification, Ion channel regulation, Transcriptomics

1 Introduction

Fundamental brain functions as information processing, memory formation, and retrieval are based on dynamic cooperation of neurons in local and far-field neuronal networks. Information flow – through a network – is performed by synapses continuously tuned by the transmitted information via changes in molecular composition of the synapse. There is a complex feedback mechanism involving local synaptic protein synthesis and the incorporation of proteins supplied by the neuronal protein trafficking system. It is supported by the fact that memory trace formation requires de

Jonathan V. Sweedler et al. (eds.), *Single Cell 'Omics of Neuronal Cells*,
Neuromethods, vol. 184, https://doi.org/10.1007/978-1-0716-2525-5_10,

novo protein synthesis [1]. Rapid adjustment of synaptic transmission efficiency could be performed by translocation of synaptic protein assemblies of the pre- or postsynaptic membrane, resulting in functional changes in the molecular scaffolding and conformational changes of synaptic proteins [2]. In a certain sense, each neuron in a network should be molecularly individual because of its unique position in the network, resulting in specific composition of its molecular assembly. At the same time, there are clearly distinguishable neuronal cell types having similar morphological and functional characteristics such as pyramidal cells, interneurons, etc. Based on these, it can be assumed that a part of the molecular phenotype of a neuron determines the cell type; however, there is a variable part of the molecular phenotype making each cell individual.

The basic principles of neurophysiology are that the electrical properties of the neuronal cell membrane can be changed by synaptic communication and the large number of functional voltage-sensitive ion channels allows to generate a rapidly propagating membrane potential change, called action potential. There are several thousands of presynaptic terminals on a neuron [3]; their inputs are integrated by the spatial and temporal summation of potential changes at the postsynaptic membrane. These integrated potential changes can trigger action potentials at the low-threshold areas of the cell surface, i.e., the axon initial segment [4] or basal dendritic areas [5]. Consequently, the number and type of ion channels, their cell surface distribution, and the posttranslational modifications of certain channel proteins determine the functional characteristics of individual neurons. It means that the functional heterogeneity of neurons is the consequence of their molecular heterogeneity. In turn, the understanding of the molecular composition of individual cells could greatly facilitate the understanding of neuronal functions. At the current state of technical development of molecular neurobiology, the only possible large-scale approach to study the molecular composition of individual neurons is single-cell mRNA sequencing, the qualitative and quantitative analysis of the mRNA content of a neuron [6]. There is no predictable and trivial correlation between cellular mRNA and protein levels [7], moreover, the abundance of ion channel proteins is relatively low compared to scaffold proteins and metabolic enzymes, thus the complete set of ion channel transcripts can be obtained only by ultra-deep sequencing. In this chapter, we intend to give a review of the actual state of knowledge about ion channel composition of major types of neurons and its functional aspects emphasizing the importance of ultra-deep mRNA sequencing method. Because of the small number of single-cell sequencing studies on neuronal ion channel sub-transcriptome, we have to focus on the classical experiments. Thus, we provide a brief review of available knowledge about causal relationship between ion channel composition and neuronal excitability established by classical methods of neurochemistry and electrophysiology. Additionally, we give suggestions for single-cell transcriptomic studies in the future.

2 The Basic Structural and Functional Properties of Neuronal Ion Channels

Regulated changes in the electrical properties of the neuronal cell membrane play a fundamental role in the direct communication between nerve cells, which is thought to underlie information processing in the central nervous system (CNS) [8–10]. According to Ohm's law, the transmembrane electrical potential (commonly referred to as membrane potential) can be altered by transmembrane currents mediated by the controlled flow of charge carriers, such as ions. When ions enter or leave the cell, they have to pass through the cell membrane, an electrically insulating phospholipid bilayer. Certain (mainly small and/or hydrophobic) molecules can cross the cell membrane without hindrance; some molecules need primary or secondary active transporters to get across the cell membrane. However, ions require controllable molecular gates to pass through; these are the ion channels. They are passive conduits and composed of multidomain or multisubunit transmembrane proteins that form hydrophilic pores permeable for certain ions. The movement of ions through the channel proteins is controlled by the transmembrane electrochemical potential difference of each ion; its magnitude depends on the presence of specific ionic conductances and on the concentrations of particular ions on both sides of the membrane. Ion channels provide specific ionic permeability, and their opening results in a flow of ions down their particular electrochemical potential gradient. The transmembrane ion gradients are maintained by active ion transport mediated by ion pumps (e.g., Na^+/K^+-ATPase), which are primary active transporters consuming energy (i.e., ATP) to move ions against their electrochemical potential gradient [11, 12]. Due to their opposite tasks and different speeds, the channels and pumps are traditionally considered to be functionally different entities [13]. The most important ion channels, regulating neuronal activity, are probably the voltage-gated (or voltage-dependent) ion channels, since the neuronal firing rate and pattern, the spike morphology, and the dynamics of synaptic transmission are largely determined by the expressed set of voltage-gated ion channels [9, 14]. These channel proteins open in response to changes in membrane potential and are highly selective for specific ions. The ion currents and, thus, membrane potential can be effectively controlled by the fine-tuning of channel permeability, the number of functional ion channels, and mitochondrial metabolism (indirect action), which are targeted by several signaling systems operating on different time scales [14–18].

We cannot go without mentioning the existence of noncanonical functions of ion channels. Namely, many ion channels can themselves directly influence biochemical processes and are likely involved in numerous cellular events. Kaczmarek, in his excellent summary written in the early 2000s, shows that voltage-dependent ion channels participate in intracellular signaling pathways. They can directly activate enzymes involved in cellular signal transduction, serve as cell adhesion molecules, or be supportive elements of the cytoskeleton, and their activity may regulate the expression of specific genes [19]. Such regulation of cellular signaling has been reported for each of the major classes of ion channels, including the sodium [20], potassium [21], calcium [22], and nonselective cation channels [23]. Sodium channels are expressed in different types of excitable (neurons and myocytes) and non-excitable cells (glial cells, macrophages, dendritic cells, endothelial cells, and fibroblasts, among others). They are involved, for example, in the regulation of cell migration during development, cell adhesion, phagocytosis, and Na^+/K^+-ATPase activity [24, 25]. Potassium channels expressed in glial cells may be crucial in modulating glial proliferation [26] and the indirect regulation of neuronal excitability [27]. Bates et al. underlined the role of several ion channel proteins in the regulation of cell proliferation behind normal development and cancer metastasis as well as in the determination of cellular differentiation and morphology [28]. The mechanisms by which ion channels fulfill their noncanonical functions may be related to ion conduction, affecting osmotic conditions and cell volume, or to protein-protein interactions not directly coupled to ionic transport [24, 28].

Voltage-gated cation channels are evolutionarily related in structure and function. They possess three important properties that allow neurons to transmit signals via electrical pulses: (1) they open in response to certain changes in membrane potential (voltage gating), (2) followed by a subsequent channel closure and inactivation, and finally (3) these processes are selective for certain ions due to biophysical constraints. In this chapter, we focus primarily on voltage-gated cation channels due to their prominent role in the regulation of neuronal activity. The proteins that build up these cation channels form complex molecular structures consisting of a central pore-forming region (or α subunit) and additional auxiliary subunits [29, 30], each with a number of sequence motifs regulating subcellular targeting, protein-protein interactions, posttranslational modifications, and channel function [14, 31]. In the case of voltage-gated sodium (Nav) and calcium (Cav) channels, the pore-forming region consists of a single α polypeptide chain, having four homologous domains (I–IV), while the voltage-gated potassium channels (Kv) consist of four α subunits [32–34]. The pore-forming α subunits can be associated with different auxiliary subunits (e.g., β1, β2, γ, α2δ), which modulate biophysical properties, cell surface expression, trafficking, targeting, and anchoring of channel protein complexes [14, 30, 35]. However, the presence or absence of auxiliary subunits within the same subfamily of ion channels has also been demonstrated. For example, the high-voltage-activated (HVA) Cav channels form such hetero-multimeric complexes, while the low-voltage-activated (LVA) Cav channels lack the auxiliary subunits [33], highlighting the functional heterogeneity of voltage-gated cation channels. Kv channels form functional homo- and heterotetramers as well; moreover, they are combinatorially associated with different auxiliary subunits. Thus, together with the inwardly rectifying (Kir) and two-pore domain (K2P) potassium channels, they are the most diverse group of ion channels modulating neuronal activity [31, 36]. The hyperpolarization-activated cyclic nucleotide-gated (HCN) channels (permeable to sodium and potassium ions) also belong to the superfamily of Kv channels and play a complex role in the fine-tuning of electrical activity of cortical neurons [37]. Thus, the considerably high number of neuronal voltage-gated ion channels and the ionic currents mediated by them largely contribute to the heterogeneity of electrophysiologically identified neuronal phenotypes. Furthermore, the number of well-defined, reproducible electrophysiological cell types (see below) is much less than the number of possible ion channel combinations. Thus, it can be strongly presumed that (1) a certain firing pattern may be derived from several different sets of functional ion channels and/or (2) there are some ion channels that preferentially show common expression and regulation and are dedicated to certain electrophysiological phenotypes. The latter assumption is consistent with the findings that suggest that metabolic cost (e.g., energy minimization) may be a guiding principle determining the actual expressed set of ion channels [11].

The intrinsic excitability of a neuron and its ability to convert synaptic inputs of the somatodendritic region to axonal outputs strongly correlate with the ratio and distribution of expressed ion channels and the interactions between functional channel complexes and regulatory signaling systems. The emergence of electrophysiology-combined single-cell mRNA sequencing (the patch-seq technique) greatly facilitated the understanding of the causal relationship between the transcriptomic profile and the electrophysiological phenotype of mammalian central neurons [38–40]. Based on such transcriptomic studies and earlier electrophysiological classification of CNS neuronsii under in vitro and ex vivo conditions [41, 42], the molecular correlates of neuronal activity became broadly elucidated. In addition, the sequencing of whole transcriptome of single neurons will certainly further support the uncovering of ion channel expressional regulation at the single-cell level, in particular, on issues such as the co-expression of pore-forming and auxiliary subunits, the stoichiometric expression of different subunits, the contribution of certain splice variants to the electrophysiological properties of neurons, and the differences in the ion channel composition of functionally distinct neuronal cell types.

3 Ion Channels Determining Firing Pattern

Intercellular synaptic communication, underlying information processing in the CNS, can largely be correlated with the generation, propagation, and termination of neuronal electrical signals. The mammalian central neurons express several different voltage-gated ion channels in large numbers, enabling the generation of action potentials and the modulation of its pattern, frequency, and shape by combining voltage-dependent ionic conductances [9]. Propagation of action potentials at the cell membrane depends on the density of voltage-gated sodium and potassium channels, resulting in an avalanche-type depolarization followed by a slightly slower repolarization (Fig. 1). The resting membrane potential and the ionic distribution underlying de- and repolarization are dependent mainly on the Na^+/K^+ pump activity, using ATP for active transport. Based on resting membrane properties (i.e., the balance of active ion transport and leakage current via spontaneously opened ion channels) and firing pattern, the electrophysiological classification of mammalian neocortical neurons is well established [41–43]. Nevertheless, it should be emphasized that the early neuron classification studies were carried out mainly under artificial in vitro and ex vivo circumstances (in cell cultures and brain slice preparations, respectively), which may cause functional cellular differences in comparison with the intact nervous tissue [44, 45]. However, it is also known that the basic electrical features of CNS neurons are quite similar both in vitro and in vivo [46, 47], proposing that the intrinsic physiological properties of neurons are determined by the current molecular composition and state of cells, rather than by synaptic inputs and neuromodulatory signaling effects.

The main electrophysiological neuron types of the mammalian brain were described mostly in acute slice preparations based on whole-cell patch clamp recordings [48–50]. It is a widely used method to study bioelectricity at the single-cell level by injecting different current waveforms (e.g., square pulses, ramps, artificial noises) via a patch pipette and measuring the subsequent changes in membrane potential. However, it should be mentioned that during the application of this method, the dialysis of the intracellular molecular components by the patch pipette and its downstream effects should be considered as an artificial modulatory factor. The most widely studied cortical neurons of different mammal species can be sorted into some major groups based on subthreshold (e.g., resting membrane potential, membrane time constant, membrane

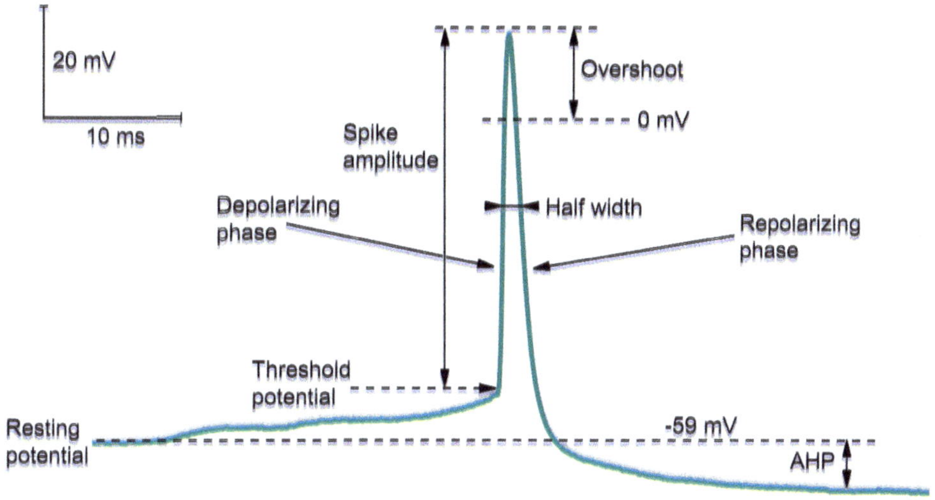

Fig. 1 A spontaneous action potential recorded from a cortical pyramidal cell in a mouse brain slice (our own measurement). Spontaneous action potentials (i.e., evoked not by current pulse stimulation) may be triggered by ion channel fluctuations and/or synaptic activity when the membrane potential reaches the threshold potential. The rapid depolarizing phase is mediated by the synchronized opening of voltage-gated sodium channels, while the slower repolarizing phase and the afterhyperpolarization (AHP) are largely determined by the activity of voltage-gated potassium channels. However, the fine morphology of action potentials (e.g., the amplitude, the extent of overshoot, the spike width) is shaped by the regulated interplay of several other voltage-dependent and voltage-independent ion channels

resistance, and capacitance) and suprathreshold (spike morphology and stereotypical firing pattern) electrical properties. Excitatory pyramidal cells form groups of regular-spiking (RS), intrinsically bursting (IB), fast repetitive bursting (FRB), and intermediate firing (IM) neurons [41, 51, 52]. On the other hand, inhibitory interneurons are classified into fast-spiking (FS), irregular-spiking (IS), and late-spiking (LS) neurons [41, 42, 53]. However, it should be mentioned that the electrophysiological classification of CNS neurons and the compatibility of these functional classes with molecular or anatomical cell types are not absolutely complete. For example, low-threshold-spiking (LTS) neurons can be either excitatory pyramidal cells in the monkey [52] or inhibitory interneurons in the mouse neocortex [54]. It is noteworthy as well that some cortical excitatory neurons can show both regular- and burst-spiking activity, depending on the temperature [55]. It is well established that the electrical activity of single neurons depends also on the developmental state of the nervous system. It was reported for FS interneurons of the mouse neocortex that their electrical behavior and the underlying transcriptomic profile of ion channel genes show significant alterations during the first 40 postnatal days [56]. It also has practical significance, since the widely used ex vivo slice preparations are usually obtained from laboratory rodents less than a month old. Thus, it can be assumed that probably not all artificially established electrophysiological cell types represent stable functional (in vivo) categories that can be reproduced in different experimental species and conditions. Apart from how far these functional cell types reflect the in vivo conditions, it could certainly be pointed out that the electrophysiological phenotype of a neuron is largely determined by the expressed set of functional ion channel proteins, which may be ultimately regulated by cell-type-specific transcriptional mechanisms [56].

The voltage-dependent sodium and potassium channels play the most important role in shaping the action potential, as they mediate inward sodium currents and outward potassium currents during the depolarizing and repolarizing phase of action potentials, respectively (Fig. 1). The voltage-dependent calcium channels, together with large and small conductance calcium-activated potassium channels (BK and SK channels), may contribute to the fine-tuning of spike morphology and firing pattern. However, they do not control the spike genesis directly, since their activation kinetics is remarkably slower than in the case of Nav channels [9]. Several earlier studies using electrophysiological, pharmacological, and imaging approaches have identified those ion channels or channel subunits that predominantly determine the characteristic spiking pattern of certain CNS neurons. The most important function of Nav channels in cellular electrophysiology is related to the initiation, propagation, and backpropagation of action potentials [35, 57]. The low-threshold Nav1.6 isoforms (encoded by *Scn8a*) at the distal region of the axon initial segment contribute to action potential initiation, while at the nodes of Ranvier, together with Nav1.1 α subunits (encoded by *Scn1a*) contribute to the saltatory propagation of action potentials. In turn, the high-threshold Nav1.2 isoforms (encoded by *Scn2a*) at the proximal region of the axon initial segment, together with Nav1.1 channels, promote the somatodendritic backpropagation of action potentials [35, 57, 58]. The dendritic backpropagation of somatic/axonal action potentials may influence the firing pattern of some central neurons. It was revealed in hippocampal slices of rats that the activation of dendritic Cav channels (especially T- and R-type channels) by back-propagating spikes generates a prolonged inward current inducing burst firing in CA1 pyramidal cells [59]. The burst firing mechanism of some CNS neurons was thoroughly elucidated. It was described for the inhibitory interneurons of the murine dorsal cochlear nucleus that the burst genesis is controlled by the interplay between the P/Q calcium channels of the somatic plasma membrane, the endoplasmic reticulum ryanodine receptors, and the large conductance calcium-activated potassium channels coupled by calcium nanodomain signaling [60].

It has become apparent relatively early that the two electrophysiologically distinguishable cortical neurons, the FS interneurons and RS pyramidal cells, show functionally relevant differences in the expressed set of Kv channels. Martina et al. found in hippocampal slices of rats that pharmacological differences between voltage-dependent potassium conductances of basket cells and pyramidal neurons derive from cell-type-specific expression of Kv3 and Kv4 potassium channel subunits. The single-cell reverse transcription PCR (RT-PCR) analysis revealed that Kv3.1 and Kv3.2 subunit transcripts (encoded by *Kcnc1* and *Kcnc2*, respectively) showed enriched expression in FS interneurons, while Kv4.2 and Kv4.3 transcripts (encoded by *Kcnd2* and *Kcnd3*, respectively) were expressed to a greater extent by RS pyramidal cells [61]. In our previous study, we also found an enriched expression of *Kcnc2* in FS cells of the mouse prefrontal cortex [62]. The opening characteristics of Kv3 channels allow the generation of currents that facilitate the fast repolarization of membrane potential without interfering with spike genesis or reducing its amplitude [63]. A subsequent study using a pharmacological and modeling approach revealed that Kv channels containing Kv3.1-Kv3.2 α subunits enable the firing of high-frequency trains of narrow action potentials by promoting the recovery of sodium channel inactivation and by reducing the duration of afterhyperpolarization (AHP) in neocortical interneurons in mice [64]. Since Kv3 channels are expressed not only in the somatodendritic region but also at the presynaptic terminals of FS cells, their role in the regulation of synaptic transmission can be assumed as revealed in murine neocortical slices [65]. Moreover, it should be noted that Kv3 channels are required not only in neocortical or hippocampal inhibitory interneurons for FS character. For example, the Kv3.1, Kv3.3, and Kv3.4 subunits are expressed also in Purkinje cells, the main GABAergic projection neurons of the cerebellar cortex [66], indicating the role of Kv3 channels in high-frequency repetitive firing in different types of neurons [63]. However, there may be species-specific differences in the expression pattern of Kv3 channels by cell type. Soares et al. demonstrated by immunohistochemistry that different pyramidal cell types in macaque motor cortex showed dense staining to Kv3.1b on their soma membranes, while in rat motor cortex the number of Kv3.1b immunopositive pyramidal cells was very low. This observation can be related to the electrophysiological differences between primate and rodent pyramidal cells [67]. Contrary to rats, a significant portion of macaque pyramidal cells can fire action potentials with short duration [68]. Evidently, it can be presumed that Kv3 channels are not solely responsible for the firing pattern of FS cells, since some sodium channels (Nav1.6 encoded by *Scn8a* and sodium channel subunit beta-4 encoded by *Scn4b*), two-pore domain potassium channels (TWIK-1 encoded by *Kcnk1* and TASK-1 encoded by *Kcnk3*), and other types of Kv channel subunits (Kv1.1 encoded by *Kcna1* and Kv1.6 encoded by *Kcna6*) may also contribute to the unique electrophysiological properties of these neurons [56].

In conclusion, it is clear from classical physiological and pharmacological studies that the two major functional classes of cortical neurons, the pyramidal and FS cells, can be distinguished on the basis of their electrophysiological features, which is largely determined by their ion channel transcriptome and proteome. A major question is how many functionally relevant neuronal subtypes can be separated using single-cell mRNA sequencing beyond the classical types of CNS neurons. The vast majority of single-cell transcriptomic analyses aiming the separation of different neuronal subtypes [38, 39, 69–71] established many novel cell classes based on the expression pattern of several genes in several investigated neurons. However, the main problem of transcriptome-based cell clustering is that the in vivo (i.e., in the intact nervous tissue of a living animal) investigation of neuronal clusters defined on the basis of a large number of transcripts is often challenging. Thus, their interpretation as functionally unique cell types is rather uncertain.

4 Receptors and Secondary Messenger Systems Modulating Neuronal Activity

The essence of network function is synaptic communication, which largely determines the electrical activity of each individual neuron in a network. A considerable portion of neurotransmitter receptors are ion channels (i.e., ionotropic receptors), while other types of receptors (e.g., metabotropic receptors) regulate the function of ion channels via secondary messenger systems. In addition, neurons are interconnected by gap junctions, which form direct connections between cells, enabling the transmission of ions, metabolites, and small signaling molecules. On the other hand, there are non-synaptic forms of intercellular communication, such as paracrine and endocrine interactions, contributing to the regulation of neuronal activity through different signaling pathways. In addition, mitochondrial metabolism can modulate the electrical activity as well, since it supplies the energy demand of cellular processes, maintaining and regulating membrane potential and synaptic transmission (e.g., the function of Na^+/K^+-ATPase and vesicular transporters).

Ionotropic glutamate (Glu) and γ-aminobutyric acid (GABA) receptors are expressed by several neuronal and non-neuronal mammalian cell types; however, their most important function is to mediate fast synaptic communication between CNS neurons [72, 73]. Ionotropic Glu receptors are ligand-gated cation channels localized both synaptically and extrasynaptically and are involved in excitatory neurotransmission between neurons [72]. The three members of the ionotropic Glu receptor family are the AMPA, kainate, and NMDA receptors; all of them are homo- or hetero-tetramers, forming a central ion-conducting pore permeable for sodium, potassium, and/or calcium ions [74, 75]. However, the sodium and calcium influx compensates the potassium efflux; thus, the net effect of Glu binding is the depolarization of the cell membrane [76]. On the other hand, the activation of ionotropic GABA receptors (the $GABA_A$ receptors) results in fast synaptic inhibition in adult mammalian neurons, as they are ligand-gated chloride channels that hyperpolarize neuronal membranes [73]. $GABA_A$ receptors form heteropentameric protein complexes comprised of α, β, γ, and δ subunits. Similar to ionotropic Glu receptors, they are located at both synaptic and extrasynaptic sites of the cell membrane, where they mediate phasic and tonic inhibition, respectively [77, 78]. There are many other types of ligand-gated ion channels (e.g., ionotropic glycine, acetylcholine, and serotonin receptors) contributing to the rapid modulation of neuronal activity. However, the backbone of fast synaptic communication in the neocortex is largely determined by the physiologically regulated balance of excitatory glutamatergic and inhibitory GABAergic transmission [79].

Neuronal electrical activity can be fine-tuned also indirectly via short-term and long-term signaling processes, acting on gene transcription, protein synthesis, posttranslational modification, and subcellular protein localization (for a comprehensive review, see [17]). The basic regulation of neuronal activity takes place at the level of ion channel gene expression. The transcription of ion channel genes is regulated homeostatically in order to maintain an appropriate ratio of ionic conductances, underlying the unique electrophysiological properties of individual neurons [80]. The exact mechanism of these regulatory processes remains to be elucidated; however, it can be strongly assumed that activity-dependent feedback control largely contributes to the proper ratio of mRNAs encoding key ion channels, as it was proposed by modeling methods [81]. In accordance with this hypothesis, it has been demonstrated by quantitative single-cell RT-PCR in crab stomatogastric ganglion neurons that the correlated expression of Cav and Kv channel mRNAs is actively maintained by a feedback mechanism and can be abolished by blocking neuronal activity with pharmacological treatment or denervation [82]. Therefore, it can be supposed that the homeostatic control of neuronal activity is ensured by the continuous adjustment of optimal ratio of different ion channels, rather than by the absolute quantity of individual ion channels separately [80, 83]. Accordingly, it was established that homeostatic signaling systems are involved in the restoration of ion channel gene transcription upon a shift in its physiological balance. Parrish et al. demonstrated in *Drosophila* motoneurons that the loss of Kv4.2 ortholog *Shal* gene induces compensatory overexpression of *Sh*, *slo*, and I_h transcripts, encoding two potassium channels and a hyperpolarization-activated cation channel, respectively [84]. The expression of ion channel genes may be regulated combinatorially (i.e., by the combined effect of more than one transcription factor), as it was revealed in developing *Drosophila* neurons for the gene *Sh*, which is transcribed under the common control of Isl and Lim3 [85]. Based on these observations, it can be assumed that ultimately the level and pattern of neuronal activity and the responsiveness of neurons to stimulation are homeostatically regulated by coordinated expression of different ion channels, rather than the independent adjustment of certain channel proteins. Subsequently, it should be emphasized that the same firing pattern can be generated by different compositions of ion channels. It has been shown in the lateral pyloric neurons of crabs that as long as the firing pattern of different neurons is the same, individual potassium conductances can show up to threefold differences between neurons from different animals [86]. There are also data from mammalian neurons that are consistent with these observations in invertebrate animals. For example, Dueck et al. showed by deep single-cell RNA-sequencing in different rodent cells (including CNS neurons) that ion channel genes exhibit the largest degree of expression variation, likely suggesting a homeostatic modulatory regulation [87]. Furthermore, the expression of Nav channels in mouse sensory neurons appears to be activity dependent, since it was demonstrated by quantitative PCR and immunocytochemistry that Nav1.8 and Nav1.9 undergo mRNA- and protein-level downregulation following electrical stimulation [88]. In accordance with the findings of Schulz et al. [86], it was revealed by analyzing single-cell RT-PCR data from 203 neocortical neurons of juvenile rats that the expression pattern of ion channel genes is different in the same morpho-electrical neuron types located in the same cortical layer [89]. In addition, Khazen et al. constructed a combinatorial rule extractor model, which uncovered preferred combinations of ion channel transcripts in different neuronal cell types, and found that the expression of genes encoding HCN4, Kv2.2, and Kv3.2 can reliably predict the expression of other genes [89]. Taking all these findings into account, the understanding of cellular electrical activity at the molecular level appears to be feasible using single-cell omics, as these can focus not only on a few genes, transcripts, or proteins but on thousands of molecules and their quantitative correlations, which shape neuronal activity.

Extracellular signals largely contribute to the mRNA- and protein-level regulation of ion channel expression and function and to the modulation of neuronal activity. Brain-derived neurotrophic factor (BDNF), a pleiotropic CNS signaling molecule acting on TrkB and p75NTR receptors, influences the expression of several ion channel and receptor genes via a noncanonical JAK/-STAT signaling pathway as revealed by deep RNA-sequencing of primary rat neurons [90]. Peripheral signaling molecules (e.g., hormones, cytokines) may also modulate neuronal excitability through the regulation of ion channel gene expression. Corticosterone can alter the mRNA levels of P/Q- and L-type calcium channels, AMPA, and NMDA receptor subunits in dissociated hippocampal neurons as revealed by single-cell RNA amplification and expression profiling [91]. The observed transcriptome changes were found to depend on the type of activated corticosteroid receptors (only mineralocorticoid receptors or both mineralocorticoid and glucocorticoid receptors). The functional effects of the observed gene expression changes were demonstrated by calcium current recordings. There are evidences regarding the potential modulatory role of cytokines in neuronal ion channel gene expression as well. It was established in cortical neuron culture using real-time PCR that interleukin-1β (IL-1β), a multifunctional proinflammatory cytokine, can significantly reduce the expression of *Scn1a* [92]. The transcriptomic and the associated electrophysiological effects of IL-1β may be related to PKC and p38 MAPK-dependent signaling processes as shown by the application of specific inhibitors. Furthermore, it should be strongly emphasized that not only the expressed set of ion channels and receptors determines neuronal activity, but vice versa, the pattern of cellular electrical activity also efficiently regulates gene expression processes via hundreds of activity-regulated enhancers [93, 94]. Coupling between neuronal activity pattern and gene transcription is partially mediated by MAPK/ERK signaling as shown by Tyssowski et al. in neuronal cultures [94]. In conclusion, we can establish that ion channel gene expression is activity dependent in nerve cells; therefore, it is clear that all neurons are molecularly individual, and single-cell mRNA sequencing is actually the most useful way to understand the molecular background of neuronal activity regulation and to get closer to the working principle of neuronal networks.

The posttranslational modification of ion channel proteins is also effectively controlled by receptors and secondary messenger systems responding to diverse extracellular signals. Heterotrimeric G proteins (Gαβγ) and the associated receptors (G protein-coupled receptors (GPCRs)), activated by several neurotransmitters and hormones, are potent regulators of ion channel function [15]. The modulation of ion channels by G proteins can be mediated indirectly by signaling systems or directly by protein-protein interactions between G protein and ion channel subunits [95]. Depending on the type of the activated heterotrimeric G protein, the mechanisms underlying indirect ion channel modulation can be related to reversible covalent modifications (e.g., phosphorylation, palmitoylation) of channel proteins or different secondary messengers (e.g., Ca^{2+}, cAMP, IP_3, PIP_2); thus, signaling pathways have various effects on neuronal activity [15, 96]. For example, the G_q protein-coupled muscarinic receptors of ACh (e.g., M_1R) can enhance neuronal excitability by reducing the opening probability of Kv7 channels, which is mediated by the phospholipase Cβ-dependent hydrolysis of plasma membrane PIP_2. On the other hand, the M_1 receptors of ACh can also hyperpolarize neurons via IP_3-dependent elevation of intracellular $[Ca^{2+}]$, which activates small conductance calcium-activated potassium channels [96]. The direct modulation of biophysical properties of ion channels by the βγ subunit (Gβγ) of G proteins is well established in the case of some calcium (e.g., Cav2.2) and potassium (e.g., Kir3.1, Kir3.2) channels, which are inhibited and activated by Gβγ, respectively [95, 96]. It is worth mentioning that in addition to heterotrimeric G proteins, small monomeric G proteins (e.g., Ras, Rho, and Rab GTPases) are also physiologically important regulators of ion channels that can exert their effects both directly and indirectly [97].

The reversible covalent modification of ion channel proteins by controlled enzymatic reactions is a thoroughly studied topic of neuronal activity modulation. Protein phosphorylation, initiated and regulated by diverse signaling processes, is a key mechanism underlying the modification of ion channel properties. All of the major voltage-gated ion channels contain amino acid sequences that enable the direct interaction with protein kinases and the subsequent phosphorylation on the suitable serine, threonine, or tyrosine residues [30]. The most important kinases modulating neuronal activity are protein kinase A (PKA), protein kinase C (PKC), calcium/calmodulin-dependent protein kinase II (CaMKII), extracellular signal-regulated kinase (ERK), and Src kinase [17, 30]. The phosphorylation of pore-forming α subunit can significantly alter the sub- and suprathreshold electrical properties of plasma membrane and, thus, neuronal firing pattern [17]. In addition to the changes of electrophysiological features of ion channels due to phosphorylation on α subunit, the auxiliary subunits can also be phosphorylated, which may influence channel protein trafficking and subcellular localization as in the case of Kv1 and Kv4 channels [31]. Dephosphorylation, catalyzed by various protein phosphatases (e.g., PP1, PP2A, PP2B), is closely related to the phosphorylation-based modulation of ion channels, and it further improves the plasticity of regulatory processes and fine-tunes neuronal activity [30, 98]. Ion channels can be regulated by other forms of reversible posttranslational modifications as well. Protein *S*-palmitoylation, the linkage of a palmitic acid to an intracellular cysteine residue by palmitoyl acyltransferases, seems to be a physiologically relevant mechanism underlying ion channel modulation. Ion channel maturation (e.g., in the case of Nav1.2, Kv1.5), cell surface expression (e.g., in the case of Kv4.3), and the biophysical properties of channel proteins (e.g., Kv1.1 and N-type calcium channels) can also be regulated by protein palmitoylation [99]. Finally, it should be noted that neuronal activity can be modulated by the controlled internalization and degradation of ion channel proteins as well. Among others, the ubiquitination of intracellular channel domains by ubiquitin ligases (e.g., Nedd4 and Nedd4-2) is a well-established posttranslational modification that largely contributes to the regulation of cell surface channel density [18].

5 The Correlation of Transcriptome with Neuronal Electrophysiology

The development of single-cell mRNA sequencing and the transcriptomic analysis of a large number of automatically separated neurons significantly contributed to the understanding of neuronal activity regulation at the transcriptomic level (for mouse and humane datasets, see the RNA-Seq Data of Allen Institute: https://portal.brain-map.org/atlases-and-data/rnaseq). In addition, the combination of single-cell transcriptomics with standard patch clamp recordings (the patch-seq technique) enabled the investigation of causality between the transcriptomic and electrophysiological properties of neurons. However, earlier studies based on RT-PCR analysis of single neurons after whole-cell patch clamp recordings successfully identified some ion channel and receptor subunit transcripts related to certain electrophysiological phenotypes [100]. For example, Toledo-Rodriguez et al. reported a coherent relationship between the co-expression of defined ion channel genes and the electrical properties of neocortical rat neurons based on patch clamp recordings and subsequent single-cell multiplex RT-PCR for 26 ion channel genes [101]. However, in parallel with the development of omic sciences, single-cell mRNA sequencing made it possible to study thousands of cytoplasmic transcripts harvested from single, electrophysiologically identified neurons.

The whole-genome expression profiling of 18 cortical neurons from mouse brain slices revealed that the transcriptome-based unsupervised cell clustering broadly reflects the electrophysiological classification of RS and FS neurons. However, the molecular and functional clustering of some neurons proved to be controversial, even based on marker genes [102]. Tasic et al. performed the cellular classification of more than 1600 FACS-isolated cells of the mouse primary visual cortex by single-cell mRNA sequencing [69]. The authors found that neuronal cell types expressed more ion channel genes than glial cells, and several ion channel genes showed enriched transcription in certain types of neurons. Moreover, both ionotropic and metabotropic Glu and GABA receptor genes showed a widespread expression in all neuronal cell types, while the receptors of other neurotransmitters, having mainly modulatory role, were transcribed at lower levels in particular cell types. The correspondence of transcriptomic profile with intrinsic and synaptic physiological properties in certain layer 1 neurons (most likely neurogliaform cells) was demonstrated by patch clamp recordings in brain slices of transgenic mice. Our previous findings are quite similar to these observations regarding the expression pattern of neurotransmitter and neuropeptide receptors in electrophysiologically defined cell types (59 pyramidal cells and 25 interneurons) of the mouse prefrontal cortex [62]. Since we were looking for cell-type-specific (i.e., differentially expressed) transcripts coding cell surface proteins, we identified many ion channel and receptor genes showing enriched expression in GABAergic FS interneurons (e.g., *Kcnc2*, *Kcnk2*, *Agtr2*, and *Prlr*) or glutamatergic pyramidal cells (e.g., *Cacna2d1*, *Kcnk4*, *Adora1*, and *Cckbr*). A comprehensive patch-seq study provided a detailed electrophysiological, morphological, and transcriptomic characterization of 38 pyramidal cells and 45 CCK interneurons harvested from mouse somatosensory cortical slices [38]. The authors identified dozens of transcripts encoding transporter, ion channel, and cell surface receptor subunits; many of them could be linked to specific electrophysiological and synaptic properties of the particular neuron type. For example, in layer 1 interneurons, the quantitative expression of Na^+/K^+-ATPase subunits (e.g., *Atp1a1*, *Atp1a3*) showed a positive correlation with the resting membrane potential, and the high level of *Kcnc1* transcripts (encoding Kv3.1) found in these cells can also correspond to certain physiological features defining the FS phenotype. Similarly, the lack of *Hcn1* expression could be coupled to the lack of hyperpolarization-activated sag depolarization, an electrophysiological effect of HCN1 activity. Another patch-seq study, focusing on layer 1 interneurons, also confirmed the correlation of whole-transcriptome profile and electrophysiological properties of single cortical neurons [39]. Földy et al. examined the correlation between the intrinsic excitability and transcriptomic profile of 27 interneurons and 14 pyramidal cells harvested after patch clamp recordings from the CA1 region of murine hippocampal slices [40]. The authors found that the typical characteristics of electrical activity recorded from interneurons (e.g., the high values of firing rate, threshold potential, and afterhyperpolarization) were consistent with the expression pattern of genes encoding voltage-gated ion channels (e.g., the enrichment of *Kcnc1* and *Kcnc2* transcripts) in the same neurons. Interestingly, the examination of the distribution of 140 ion channel transcripts revealed that the number of expressed genes encoding potassium and sodium channels was lower in pyramidal cells than in the smaller interneurons. Based on these coherent findings, it can be strongly assumed that there is a robust and causal relationship between the ion channel transcriptome and electrical properties of neurons. It suggests that the backbone of cellular electrogenesis and its regulation are mostly determined at the transcriptomic level, while protein-level modulatory processes contribute to the fine-tuning of neuronal activity. This principle seems to be confirmed by deep bulk RNA-sequencing of homogeneous populations of FACS-sorted mouse somatosensory neurons. Zheng et al. analyzed the transcriptome profiles of eight, electrophysiologically defined dorsal root ganglion neuron subtypes and found that distinct sets of voltage-gated ion channel transcripts have predictive value for unique membrane and electrical properties [70]. For example, the high expression of genes encoding Kv1.1 and Kv1.2 subunits can be related to strongly adapting firing patterns, and the high level of Kv4.3 subunit transcripts measurably correlates with the delayed firing properties of certain somatosensory neurons. Finally, it should be mentioned that the neuroinformatic analysis of transcriptomic and electrophysiological datasets identified more than 400 genes showing correlated expression with certain physiological parameters; some of them (e.g., *Scn1a*, *Gabrd*) have been mechanistically linked to specific functional phenotypes [103].

6 Considerations for Future Studies

Single-cell transcriptomics of CNS neurons mainly focused on the description of novel neuronal subtypes and revealed several molecularly distinct neuron populations on the basis of transcriptomic differences. Taxonomic studies, examining a large number of automatically separated individual neurons, identified several molecular subtypes, which are sometimes challenging to recognize in functional studies focusing on the intact brain tissue. Because of the extreme high costs, the large number of cells (in the order of thousands) does not allow ultra-deep sequencing; thus, phenotyping studies are commonly based on a few thousands of transcripts (usually under 3000). However, deep sequencing studies of different cells, not only neurons, revealed that the average number of transcripts in one cell often exceeds 20,000 [87]. Therefore, phenotyping studies probably see only a small fraction of the whole transcriptome. On the other hand, ultra-deep sequencing allows the recovery of transcripts with two copies. Therefore, there will be a lot of zero transcription values in the single-cell dataset, as genes are transcribed in bursts; thus, genes can be in "on" or "off" transcriptional state at a certain time point. Because of these technical difficulties, in many single-cell transcriptomic datasets, only a fraction of ion channel subunit genes is represented. Using deep or ultra-deep sequencing, a larger fraction of the transcriptome can be revealed (close to the 20,000 transcripts) but only in a small number of neurons. The investigation of CNS neurons by the patch-seq technique is slow and labor-intensive; however, it is the most appropriate way to get information of transcriptome-function correlations. It must be acknowledged that patch-seq method has several difficulties. The harvesting of somatic mRNA can be incomplete, and the amount of aspirated cytoplasm can vary from cell to cell. The normalization of the raw sequencing data is not properly standardized because the large number of zero expression values proposes difficulties. However, compromise is necessary since studying the electrical activity and the transcriptomic profile of the same neuron greatly facilitates future studies. Also, it has been revealed in single-cell electrophysiological studies that the neuronal activity in a brain slice and the in vivo activity of the same cell type in the intact brain may differ significantly [44, 45]. Single-cell sequencing data from neurons harvested from the intact mammalian brain [104], even from a freely moving and behaving animal, will certainly support the better understanding of how the brain and functional neuronal networks are working. Because the patch clamp recording of cortical neurons in behaving animals is already feasible using microdrives placed on the skull [105], we propose to conduct more functional studies to elucidate cellular activity-transcriptome interplay to help reveal novel drug targets and biomarkers in brain diseases.

References

1. Park H, Kaang B-K (2019) Balanced actions of protein synthesis and degradation in memory formation. Learn Mem 26:299–306
2. Grant SGN (2003) Synapse signaling complexes and networks: machines underlying cognition. Bioessays 25:1229–1235
3. Hawkins J, Ahmad S (2016) Why neurons have thousands of synapses, a theory of sequence memory in neocortex. Front Neural Circuits 10:23
4. Kole MHP, Ilschner SU, Kampa BM et al (2008) Action potential generation requires a high sodium channel density in the axon initial segment. Nat Neurosci 11:178–186
5. Milojkovic BA, Wuskell JP, Loew LM et al (2005) Initiation of sodium spikelets in basal dendrites of neocortical pyramidal neurons. J Membr Biol 208:155–169
6. Eberwine J, Sul J-Y, Bartfai T et al (2014) The promise of single-cell sequencing. Nat Methods 11:25–27
7. Payne SH (2015) The utility of protein and mRNA correlation. Trends Biochem Sci 40: 1–3
8. Hodgkin AL, Huxley AF (1939) Action potentials recorded from inside a nerve fibre. Nature 144:710–711
9. Bean BP (2007) The action potential in mammalian central neurons. Nat Rev Neurosci 8: 451–465
10. Azarfar A, Calcini N, Huang C et al (2018) Neural coding: a single neuron's perspective. Neurosci Biobehav Rev 94:238–247
11. Hasenstaub A, Otte S, Callaway E et al (2010) Metabolic cost as a unifying principle governing neuronal biophysics. Proc Natl Acad Sci U S A 107:12329–12334
12. Pissadaki EK, Bolam JP (2013) The energy cost of action potential propagation in dopamine neurons: clues to susceptibility in Parkinson's disease. Front Comput Neurosci 7: 13
13. Gadsby DC (2009) Ion channels versus ion pumps: the principal difference, in principle. Nat Rev Mol Cell Biol 10:344–352
14. Lai HC, Jan LY (2006) The distribution and targeting of neuronal voltage-gated ion channels. Nat Rev Neurosci 7:548–562
15. Wickman K, Clapham DE (1995) Ion channel regulation by G proteins. Physiol Rev 75: 865–885
16. Kann O, Kovács R (2007) Mitochondria and neuronal activity. Am J Physiol Cell Physiol 292:C641–C657
17. Schulz DJ, Temporal S, Barry DM et al (2008) Mechanisms of voltage-gated ion channel regulation: from gene expression to localization. Cell Mol Life Sci 65:2215–2231
18. Bongiorno D, Schuetz F, Poronnik P et al (2011) Regulation of voltage-gated ion

channels in excitable cells by the ubiquitin ligases Nedd4 and Nedd4-2. Channels 5: 79–88

19. Kaczmarek LK (2006) Non-conducting functions of voltage-gated ion channels. Nat Rev Neurosci 7:761–771
20. Ratcliffe CF, Qu Y, McCormick KA et al (2000) A sodium channel signaling complex: modulation by associated receptor protein tyrosine phosphatase beta. Nat Neurosci 3: 437–444
21. Stühmer W, Ruppersberg JP, Schröter KH et al (1989) Molecular basis of functional diversity of voltage-gated potassium channels in mammalian brain. EMBO J 8:3235–3244
22. Dolmetsch RE, Pajvani U, Fife K et al (2001) Signaling to the nucleus by an L-type calcium channel-calmodulin complex through the MAP kinase pathway. Science 294:333–339
23. Takezawa R, Schmitz C, Demeuse P (2004) Receptor-mediated regulation of the TRPM7 channel through its endogenous protein kinase domain. Proc Natl Acad Sci U S A 101:6009–6014
24. Isom LL (2002) The role of sodium channels in cell adhesion. Front Biosci 7:12–23
25. Black JA, Waxman SG (2013) Noncanonical roles of voltage-gated sodium channels. Neuron 80:280–291
26. Sontheimer H (1994) Voltage-dependent ion channels in glial cells. Glia 11:156–172
27. Bay V, Butt AM (2012) Relationship between glial potassium regulation and axon excitability: a role for glial Kir4.1 channels. Glia 60: 651–660
28. Bates E (2015) Ion channels in development and cancer. Annu Rev Cell Dev Biol 31: 231–247
29. Catterall WA (1995) Structure and function of voltage-gated ion channels. Annu Rev Biochem 64:493–531
30. Dai S, Hall DD, Hell JW (2009) Supramolecular assemblies and localized regulation of voltage-gated ion channels. Physiol Rev 89: 411–452
31. Vacher H, Trimmer JS (2011) Diverse roles for auxiliary subunits in phosphorylation-dependent regulation of mammalian brain voltage-gated potassium channels. Pflugers Arch 462:631–643
32. Catterall WA (2000) From ionic currents to molecular mechanisms: the structure and function of voltage-gated sodium channels. Neuron 26:13–25
33. Simms BA, Zamponi GW (2014) Neuronal voltage-gated calcium channels: structure, function, and dysfunction. Neuron 82:24–45
34. Kuang Q, Purhonen P, Hebert H (2015) Structure of potassium channels. Cell Mol Life Sci 72:3677–3693
35. Wang J, Ou SW, Wang YJ (2017) Distribution and function of voltage-gated sodium channels in the nervous system. Channels 11: 534–554
36. Gutman GA, Chandy KG, Grissmer S et al (2005) International Union of Pharmacology. LIII. Nomenclature and molecular relationships of voltage-gated potassium channels. Pharmacol Rev 57:473–508
37. Shah MM (2014) Cortical HCN channels: function, trafficking and plasticity. J Physiol 592:2711–2719
38. Fuzik J, Zeisel A, Máté Z et al (2016) Integration of electrophysiological recordings with single-cell RNA-seq data identifies neuronal subtypes. Nat Biotechnol 34:175–183
39. Cadwell CR, Palasantza A, Jiang X et al (2016) Electrophysiological, transcriptomic and morphologic profiling of single neurons using Patch-seq. Nat Biotechnol 34:199–203
40. Földy C, Darmanis S, Aoto J et al (2016) Single-cell RNAseq reveals cell adhesion molecule profiles in electrophysiologically defined neurons. Proc Natl Acad Sci U S A 113: E5222–E5231
41. Contreras D (2004) Electrophysiological classes of neocortical neurons. Neural Netw 17: 633–646
42. Petilla Interneuron Nomenclature Group, Ascoli G.A., Alonso-Nanclares L. et al (2008) Petilla terminology: nomenclature of features of GABAergic interneurons of the cerebral cortex. Nat Rev Neurosci 9:557–568
43. McCormick DA, Connors BW, Lighthall JW et al (1985) Comparative electrophysiology of pyramidal and sparsely spiny stellate neurons of the neocortex. J Neurophysiol 54:782–806
44. Overton PG, Greenfield SA (1995) Determinants of neuronal firing pattern in the guinea-pig subthalamic nucleus: an in vivo and in vitro comparison. J Neural Transm Park Dis Dement Sect 10:41–54
45. Fernandez FR, Rahsepar B, White JA (2018) Differences in the electrophysiological properties of mouse somatosensory layer 2/3 neurons in vivo and slice stem from intrinsic sources rather than a network-generated high conductance state. eNeuro 5:2

46. Sanchez-Vives MV, Nowak LG, McCormick DA (2000) Cellular mechanisms of long-lasting adaptation in visual cortical neurons in vitro. J Neurosci 20:4286–4299
47. Naundorf B, Wolf F, Volgushev M (2006) Unique features of action potential initiation in cortical neurons. Nature 440:1060–1063
48. Pennartz CMA, De Jeu MTG, Geurtsen AMS et al (1998) Electrophysiological and morphological heterogeneity of neurons in slices of rat suprachiasmatic nucleus. J Physiol 506: 775–793
49. Halabisky B, Shen F, Huguenard JR et al (2006) Electrophysiological classification of somatostatin-positive interneurons in mouse sensorimotor cortex. J Neurophysiol 96: 834–845
50. Gouwens NW, Sorensen SA, Berg J et al (2019) Classification of electrophysiological and morphological neuron types in the mouse visual cortex. Nat Neurosci 22: 1182–1195
51. Yang CR, Seamans JK, Gorelova N (1996) Electrophysiological and morphological properties of layers V-VI principal pyramidal cells in rat prefrontal cortex in vitro. J Neurosci 16:1904–1921
52. Zaitsev AV, Povysheva NV, Gonzalez-Burgos G et al (2012) Electrophysiological classes of layer 2/3 pyramidal cells in monkey prefrontal cortex. J Neurophysiol 108:595–609
53. Chu Z, Galarreta M, Hestrin S (2003) Synaptic interactions of late-spiking neocortical neurons in layer 1. J Neurosci 23:96–102
54. Apicella AJ, Wickersham IR, Seung HS et al (2012) Laminarly orthogonal excitation of fast-spiking and low-threshold-spiking interneurons in mouse motor cortex. J Neurosci 32:7021–7033
55. Hedrick T, Waters J (2011) Spiking patterns of neocortical L5 pyramidal neurons in vitro change with temperature. Front Cell Neurosci 5:1
56. Okaty BW, Miller MN, Sugino K et al (2009) Transcriptional and electrophysiological maturation of neocortical fast-spiking GABAergic interneurons. J Neurosci 29:7040–7052
57. Hu W, Tian C, Li T et al (2009) Distinct contributions of Na(v)1.6 and Na(v)1.2 in action potential initiation and backpropagation. Nat Neurosci 12:996–1002
58. Duflocq A, Le Bras B, Bullier E et al (2008) Nav1.1 is predominantly expressed in nodes of Ranvier and axon initial segments. Mol Cell Neurosci 39:180–192
59. Magee JC, Carruth M (1999) Dendritic voltage-gated ion channels regulate the action potential firing mode of hippocampal CA1 pyramidal neurons. J Neurophysiol 82: 1895–1901
60. Irie T, Trussell LO (2017) Double-nanodomain coupling of calcium channels, ryanodine receptors, and BK channels controls the generation of burst firing. Neuron 96:856–870
61. Martina M, Schultz JH, Ehmke H (1998) Functional and molecular differences between voltage-gated K+ channels of fast-spiking interneurons and pyramidal neurons of rat hippocampus. J Neurosc 18:8111–8125
62. Ravasz L, Kékesi KA, Mittli D (2020) Cell surface protein mRNAs show differential transcription in pyramidal and fast-spiking cells as revealed by single-cell sequencing. Cereb Cortex 31(2):731–745. https://doi.org/10.1093/cercor/bhaa195
63. Rudy B, McBain CJ (2001) Kv3 channels: voltage-gated K+ channels designed for high-frequency repetitive firing. Trends Neurosci 24:517–526
64. Erisir A, Lau D, Rudy B et al (1999) Function of specific K+ channels in sustained high-frequency firing of fast-spiking neocortical interneurons. J Neurophysiol 82:2476–2489
65. Goldberg EM, Watanabe S, Chang SY et al (2005) Specific functions of synaptically localized potassium channels in synaptic transmission at the neocortical GABAergic fast-spiking cell synapse. J Neurosci 25:5230–5235
66. Sacco T, De Luca A, Tempia F (2006) Properties and expression of Kv3 channels in cerebellar Purkinje cells. Mol Cell Neurosci 33: 170–179
67. Soares D, Goldrick I, Lemon RN et al (2017) Expression of Kv3.1b potassium channel is widespread in macaque motor cortex pyramidal cells: a histological comparison between rat and macaque. J Comp Neurol 525: 2164–2174
68. Vigneswaran G, Kraskov A, Lemon RN (2011) Large identified pyramidal cells in macaque motor and premotor cortex exhibit "thin spikes": implications for cell type classification. J Neurosci 31:14235–14242
69. Tasic B, Menon V, Nguyen TN et al (2016) Adult mouse cortical cell taxonomy revealed by single cell transcriptomics. Nat Neurosci 19:335–346
70. Zheng Y, Liu P, Bai L et al (2019) Deep sequencing of somatosensory neurons reveals molecular determinants of intrinsic physiological properties. Neuron 103:598–616
71. Yuste R, Hawrylycz M, Aalling N (2020) A community-based transcriptomics

classification and nomenclature of neocortical cell types. Nat Neurosci 23:1456–1468
72. Traynelis SF, Wollmuth LP, McBain CJ et al (2010) Glutamate receptor ion channels: structure, regulation, and function. Pharmacol Rev 62:405–496
73. Watanabe M, Maemura K, Kanbara K et al (2002) GABA and GABA receptors in the central nervous system and other organs. Int Rev Cytol 213:1–47
74. Willard SS, Koochekpour S (2013) Glutamate, glutamate receptors, and downstream signaling pathways. Int J Biol Sci 9:948–959
75. Karakas E, Regan MC, Furukawa H (2015) Emerging structural insights into the function of ionotropic glutamate receptors. Trends Biochem Sci 40:328–337
76. Lüscher C, Malenka RC (2012) NMDA receptor-dependent long-term potentiation and long-term depression (LTP/LTD). Cold Spring Harb Perspect Biol 4:a005710
77. Wu C, Sun D (2015) GABA receptors in brain development, function, and injury. Metab Brain Dis 30:367–379
78. Tomita S (2019) Molecular constituents and localization of the ionotropic GABA receptor complex in vivo. Curr Opin Neurobiol 57: 81–86
79. Ferguson BR, Gao WJ (2018) PV interneurons: critical regulators of E/I balance for prefrontal cortex-dependent behavior and psychiatric disorders. Front Neural Circuits 12:37
80. O'Leary T, Williams AH, Caplan JS et al (2013) Correlations in ion channel expression emerge from homeostatic tuning rules. Proc Natl Acad Sci U S A 110:E2645–E2654
81. O'Leary T, Williams AH, Franci A et al (2014) Cell types, network homeostasis, and pathological compensation from a biologically plausible ion channel expression model. Neuron 82:809–821
82. Temporal S, Lett KM, Schulz DJ (2014) Activity-dependent feedback regulates correlated ion channel mRNA levels in single identified motor neurons. Curr Biol 24: 1899–1904
83. Schulz DJ, Goaillard JM, Marder EE (2007) Quantitative expression profiling of identified neurons reveals cell-specific constraints on highly variable levels of gene expression. Proc Natl Acad Sci U S A 104:13187–13191
84. Parrish JZ, Kim CC, Tang L et al (2014) Krüppel mediates the selective rebalancing of ion channel expression. Neuron 82:537–544
85. Wolfram V, Southall TD, Günay C et al (2014) The transcription factors islet and Lim3 combinatorially regulate ion channel gene expression. J Neurosci 34:2538–2543
86. Schulz DJ, Goaillard JM, Marder E (2006) Variable channel expression in identified single and electrically coupled neurons in different animals. Nat Neurosci 9:356–362
87. Dueck H, Khaladkar M, Kim TK et al (2015) Deep sequencing reveals cell-type-specific patterns of single-cell transcriptome variation. Genome Biol 16:1–17
88. Klein JP, Tendi EA, Dib-Hajj SD et al (2003) Patterned electrical activity modulates sodium channel expression in sensory neurons. J Neurosci Res 74:192–198
89. Khazen G, Hill SL, Schürmann F et al (2012) Combinatorial expression rules of ion channel genes in juvenile rat (Rattus norvegicus) neocortical neurons. PLoS One 7:e34786
90. Hixson KM, Cogswell M, Brooks-Kayal AR et al (2019) Evidence for a non-canonical JAK/STAT signaling pathway in the synthesis of the brain's major ion channels and neurotransmitter receptors. BMC Genomics 20: 677
91. Nair SM, Werkman TR, Craig J (1998) Corticosteroid regulation of ion channel conductances and mRNA levels in individual hippocampal CA1 neurons. J Neurosci 18: 2685–2696
92. Zhou C, Qi C, Zhao J et al (2011) Interleukin-1β inhibits voltage-gated sodium currents in a time-and dose-dependent manner in cortical neurons. Neurochem Res 36:1116–1123
93. Kim TK, Hemberg M, Gray JM et al (2010) Widespread transcription at neuronal activity-regulated enhancers. Nature 465:182–187
94. Tyssowski KM, DeStefino NR, Cho JH et al (2018) Different neuronal activity patterns induce different gene expression programs. Neuron 98:530–546
95. Dascal N (2001) Ion-channel regulation by G proteins. Trends Endocrinol Metab 12: 391–398
96. Brown DA (2018) Regulation of neural ion channels by muscarinic receptors. Neuropharmacology 136:383–400
97. Pochynyuk O, Stockand JD, Staruschenko A (2007) Ion channel regulation by Ras, Rho, and Rab small GTPases. Exp Biol Med (Maywood) 232:1258–1265
98. Misonou H, Menegola M, Mohapatra DP et al (2006) Bidirectional activity-dependent regulation of neuronal ion channel phosphorylation. J Neurosci 26:13505–13514
99. Shipston MJ (2011) Ion channel regulation by protein palmitoylation. J Biol Chem 286: 8709–8716

100. Sucher NJ, Deitcher DL (1995) PCR and patch-clamp analysis of single neurons. Neuron 14:1095–1100
101. Toledo-Rodriguez M, Blumenfeld B, Wu C et al (2004) Correlation maps allow neuronal electrical properties to be predicted from single-cell gene expression profiles in rat neocortex. Cereb Cortex 14:1310–1327
102. Subkhankulova T, Yano K, Robinson H et al (2010) Grouping and classifying electrophysiologically-defined classes of neocortical neurons by single cell, whole-genome expression profiling. Front Mol Neurosci 3: 10
103. Tripathy SJ, Toker L, Li B et al (2017) Transcriptomic correlates of neuron electrophysiological diversity. PLoS Comput Biol 13: e1005814
104. Liu J, Wang M, Sun L (2020) Integrative analysis of in vivo recording with single-cell RNA-seq data reveals molecular properties of light-sensitive neurons in mouse V1. Protein Cell 11:417–432
105. Valero M, English DF (2019) Head-mounted approaches for targeting single-cells in freely moving animals. J Neurosci Methods 326: 108397

Chapter 11

Genetically Encoded Biosensors Reveal Spatiotemporal Dynamics and Cellular Heterogeneity of Neuronal Cells

Jeremiah Keyes, Sohum Mehta, and Jin Zhang

Abstract

Neuronal networks are among the most fascinating, yet complex, systems within biology. From neurotransmitter and neuromodulator signaling to intracellular changes in second messenger levels and enzymatic activities, the biochemical architecture of neuronal cells is in a constant state of flux. Biochemical dynamics are at the root of such complex processes such as neuronal development, learning and memory, and behavior. Noninvasively measuring and quantifying such dynamic biochemical states, however, remains a major challenge. In this review, we discuss the use of genetically encodable biosensors to address this challenge. Such biosensors are noninvasive, are readily modifiable to measure a vast array of biochemical processes, are able to provide real-time, quantitative measurements in single living cells, and link cellular biochemistry to physiological function. We provide a brief overview of genetically encodable biosensors and discuss several studies in which biosensors were used to answer specific questions in neuroscience, highlighting the vital role of genetically encodable biosensors in studying neuronal processes in a variety of contexts.

Key words Single-cell imaging, Fluorescence, Signaling, FRET, Biosensor

1 Introduction

The complex molecular mixture in cells is in constant flux. From nutrient metabolism to Ca^{2+} dynamics to protein posttranslational modifications (PTMs), cells are constantly undergoing dynamic changes to their biochemical makeup in response to both internal and external cues. However, these biochemical changes are not always uniform from cell to cell. Indeed, slight variations in the biochemical dynamics within cells are a major contributor to cellular heterogeneity and can lead to drastic differences in cellular behavior. For example, cell-to-cell differences in enzyme activity profiles or even organelle dynamics have been shown to impact cell-fate decisions such as proliferation, differentiation, or entry into senescence or quiescence [1–3]. Thus, the ability to probe the spatiotemporal dynamics of biochemical processes within

Jonathan V. Sweedler et al. (eds.), *Single Cell 'Omics of Neuronal Cells*,
Neuromethods, vol. 184, https://doi.org/10.1007/978-1-0716-2525-5_11,

individual cells is imperative for elucidating how variations in the biochemical architecture of a cell influence cell function and behavior. However, traditional methods of qualitatively identifying or quantifying molecules or events, such as ATP concentration or protein phosphorylation, typically require trade-offs between spatial and temporal resolution. Indeed, even techniques such as immunofluorescence, which has the added bonus over immunoblotting in that one can visualize differences between cells and tissues, are limited to providing single snapshots of fixed samples. Over the past couple of decades, however, the creation and use of genetically encoded biosensors has fundamentally transformed our ability to track the spatiotemporal dynamics of biochemical activities [4, 5], ushering in a new era of single-cell functional phenotyping.

Genetically encodable fluorescent biosensors are able to visualize the spatiotemporal dynamics of a specific biochemical event in its endogenous environment in a sensitive, rapid, and noninvasive way. Through the use of biosensors, researchers are therefore not only able to visualize endogenous biochemical processes in real time, but they can do so in individual cells both in vitro and in vivo, allowing the ability to determine cell heterogeneity and link this heterogeneity to cell or organismal function. Thus, questions pertaining to protein-protein or protein-DNA interactions, enzymatic activities, the activation dynamics of signaling nodes, and the spatiotemporal behavior of small molecules, among others, are now able to be delineated to better determine the functional phenotype of individual cells. In this chapter, we briefly describe the history of and provide a general overview of genetically encodable biosensors, after which we discuss several recent examples of how biosensors are being used to tackle specific questions in neuroscience.

2 Overview of Biosensors

2.1 Historical Development

Before the cloning and development of green fluorescent protein (GFP) and subsequent generations of fluorescent proteins (FPs) [6] and their incorporation into genetically encodable biosensors, the early days of the live-cell imaging field were dominated by the use of small-molecule fluorescent dyes [7], such as the Ca^{2+} indicator Fura2 [8] or the reactive oxygen species (ROS) sensor *dichlorofluorescein diacetate* [9, 10]. However, although these chemical dyes have proven to be powerful molecular tools that are still widely used, they suffer from limitations in subcellular targeting, and their designs are not easily generalized to monitor a broad range of biochemical activities. In contrast, genetically encoded biosensors based on FP technology allow researchers to overcome the limitations of small-molecule dyes through the relative ease of including subcellular targeting sequences to target the biosensor to specific subcellular domains, while their modular nature has proven extremely versatile for monitoring diverse biochemical processes. Indeed, the design of one of the first genetically encoded sensors, a Ca^{2+} indicator known as "cameleon" [11, 12], has been adapted to develop biosensors for numerous biochemical events, ranging from the dynamics of ion and small-molecule concentrations to activity of specific enzymes. Today, the literature contains hundreds of different genetically encodable biosensors, which we recently reviewed in great detail [5] and have also compiled into a searchable online database (http://biosensordb.ucsd.edu).

2.2 Types of Biosensors

As the basic purpose of a genetically encodable biosensor is to convert a specific molecular event into a measurable optical signal, the two essential elements of any biosensor are the sensing unit and reporting unit (Fig. 1a). While the sensing unit can sometimes be incorporated directly into the reporting unit where the cellular analyte directly affects the chromophore, more commonly the sensing unit takes the form of a molecular switch derived from endogenous peptides or proteins that are intrinsically sensitive to the analyte of interest. Meanwhile, the reporting unit consists of one or more FPs, configured such that the sensing unit, when responding to the analyte of interest, modulates the fluorescence behavior of the reporting unit. These behaviors can include changes in fluorescence intensity or excitation and emission ratios, as well as fluorescence resonance energy transfer (FRET), in which excited-state energy from a donor FP is non-radiatively transferred to a compatible acceptor FP depending on their relative distance and orientation [13, 14].

Not every biosensor is perfect for every experiment, and many factors contribute to selecting the best biosensor to use in a given system. A comprehensive discussion of these various considerations can be found elsewhere [5], but for the purposes of this review, we will briefly summarize some of the main factors below. First and foremost, is the biosensor sensitive enough to measure the desired biochemical changes? For example, when one is seeking to measure small changes in analyte concentrations, it is important to have a sensing unit with a low K_D so that low levels of the analyte bind efficiently enough to produce a signal. Similarly, one must consider whether the biosensor also has sufficient dynamic range (i.e., the difference between maximal and minimal fluorescent signal) to report these biochemical changes, which is affected by both sensitivity and saturation of the sensor (Fig. 1c). Some thought should also be given to whether the sensor is bright enough to use in the desired experimental system, given that different FPs can vary greatly in terms of brightness, and brighter FPs will provide increased signal-to-noise (S/N) ratios. Another key feature is reversibility, which is essential for observing decay dynamics in

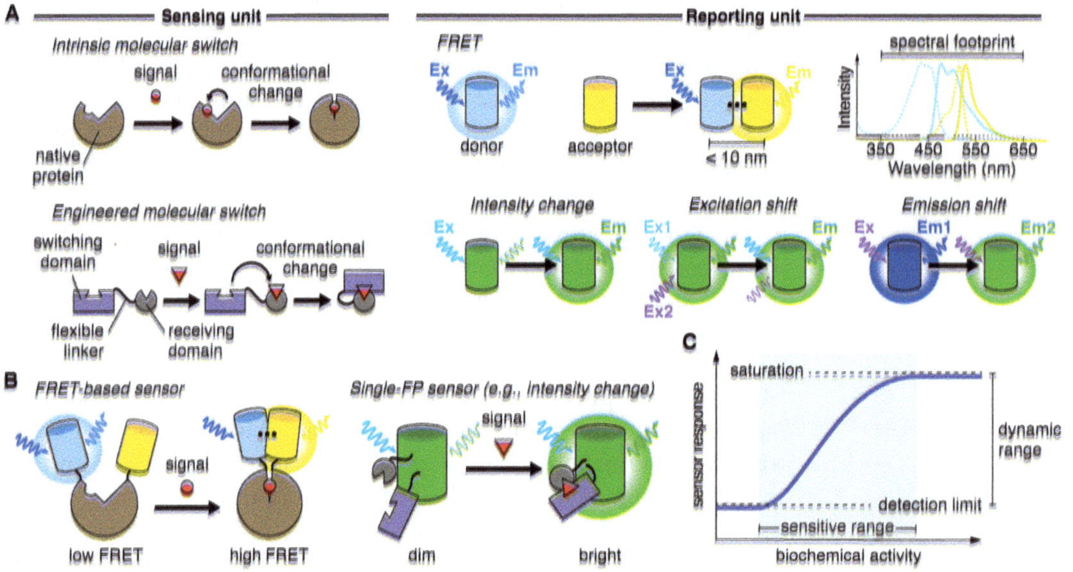

Fig. 1 Basic features of genetically encodable biosensors. (**a**) All genetically encodable biosensors feature a sensing unit that responds to a specific biochemical input and a reporting unit that provides the optical readout. (Left) The sensing unit typically comprises a molecular switch that undergoes a conformational change in response to a particular signal, through either (upper) an intrinsic conformational change within an intact native protein or (lower) an engineered molecular switch that is assembled from native protein or peptide fragments tethered by a flexible linker. (Right) Meanwhile, the reporting unit provides a quantifiable optical readout in the form of changes in the fluorescence behavior of one or more fluorescent proteins (FPs), such as fluorescence resonance energy transfer (FRET) between a donor and acceptor FP pair or the intensity, excitation maximum, or emission maximum of a single FP. Different readout methods vary in terms of their spectral footprint, which is a function of their excitation (dashed lines) and emission (solid lines) spectra and determines their suitability for multiplexing with other biosensor or optical tools. (**b**) Examples of how a sensing unit and reporting unit can be combined to yield a FRET-based (left) or single-FP (right) biosensor. (**c**) Schematic diagram illustrating the general properties of biosensor performance

biochemical processes. For the biosensor signal to reverse when an analyte is removed, the sensing unit must not bind too tightly, while for PTMs such as phosphorylation, the sensing unit must be accessible to erasing enzymes such as phosphatases. Finally, is the biosensor compatible with other tools in use, such as photostimulation tools? Each FP has a spectral footprint that determines its compatibility with other light-based tools (Fig. 1a). Thus, if one needs to image biosensors together or in combination with photostimulation tools, it is necessary to verify that there is sufficient spectral separation between the tools.

The sensor readout method is another key consideration that should be taken into account, as different reporting unit designs offer specific advantages and disadvantages. For example, the signal from FRET-based biosensors, which remain the most commonly used sensor design, is typically reported as the ratio between the acceptor and donor FP intensities (i.e., acceptor-to-donor emission ratio). This type of ratiometric readout has the advantage of being able to cancel out experimental variations such as differences in cell thickness, probe concentration, or sample illumination that might affect fluorescence intensity, and thus allowing for absolute, quantitative measurements [8, 15]. Yet because FRET-based biosensors always require two spectrally distinct FPs (Fig. 1a, b), they take up a significant amount of spectral space, making it challenging to use FRET-based biosensors in multiplexing experiments, although recent studies have reported significant progress in this area [16, 17]. Alternatively, single-FP biosensors occupy much less spectral space and provide greater opportunities for multiplexing with other biosensors or light-based tools compared with FRET-based biosensors [18–21] and also tend to have much higher dynamic ranges. However, the intensiometric readout of most single-FP biosensors may be susceptible to the aforementioned experimental variations and is therefore less quantitative compared to ratiometric (e.g., FRET-based) biosensors. While adding a second FP that is not affected by the sensing unit can achieve a ratiometric readout, there are also examples of intrinsically ratiometric single-FP biosensors. In some cases, these sensors shift their fluorescence emission peaks, such as a Ca^{2+} indicator that shifts from blue (450 nm) to green (510 nm) upon 400 nm excitation [22]. Other single-FP sensors change their excitation peak, such as our recently developed excitation-ratiometric kinase activity reporters (ExRai-KARs), which shift their maximum excitation wavelength from ~400 nm to ~480 nm in response to phosphorylation [19, 23]. Such intrinsically ratiometric single-FP sensors offer the benefit of enabling quantitative measurements while still exhibiting relatively small spectral footprints. Nevertheless, single-FP biosensors are not yet available for many targets, and more development is needed to broaden their utility.

3 Biosensors in Action

We now turn our attention to the use of fluorescent biosensors for single-cell functional phenotyping, with a specific focus on applications in neuroscience. From *C. elegans*'s 300 neurons to the complexity of the human brain, rapid intra- and intercellular communication is at the heart of neuronal function. Thus, the biochemical activity architecture of individual neurons, and how this architecture is regulated in space and time, is highly fertile ground for genetically encodable biosensors. Below, we discuss several recent studies in which fluorescent biosensors were used to answer specific questions in neuroscience while also highlighting the different biosensors used in these studies.

3.1 Activity Phenotyping in Cultured Neurons

Neuronal cells grown in culture are a powerful model for elucidating key biochemical phenomena such as the dynamics of signaling pathways and the subcellular compartmentalization of diverse molecular species. As we discuss below, in vitro analyses of cultured neurons afford us the ability to examine enzymatic activities and molecular concentrations with remarkable spatial and temporal precision, providing crucial subcellular insights and dramatically advancing our understanding of basic neuronal physiology and function.

3.1.1 Investigating Spatiotemporal Control of Neuronal Development

The universal messenger cyclic adenosine monophosphate (cAMP), which is generated by adenylyl cyclases (ACs) and degraded by phosphodiesterases (PDEs), is a key signaling node in a variety of neuronal contexts. In neuronal development specifically, tight control over the spatiotemporal dynamics of cAMP formation has been shown to directly influence the formation of the highly polarized neuronal morphology, characterized by a single axon and multiple branching dendrites [24–26]. However, until recently, the mechanisms by which cAMP compartmentalization was achieved in developing neurons were not well understood.

Recently, Gorshkov et al. [27] set out to determine the role of cAMP compartmentalization in developing neurons and to investigate how this compartmentalization occurs. To do so, Gorshkov and colleagues used a variant of the FRET-based cAMP biosensor ICUE, which contains a fragment of the cAMP-binding protein exchange factor directly activated by cAMP 1 (Epac1) sandwiched between cyan- and yellow-emitting FPs [28, 29], to compare cAMP spatiotemporal dynamics in cultured hippocampal neurons at different developmental stages. Upon binding to cAMP, ICUE undergoes a hinge-like conformational change that leads to a decrease in FRET (Fig. 2). Interestingly, when they expressed this sensor in mature neurons cultured for 5 days in vitro (DIV5), the authors observed a dramatic gradient of ICUE FRET responses (i.e., cAMP accumulation) in response to stimulation, with the highest cAMP levels in the distal axon, whereas younger DIV3 neurons displayed weaker, more uniform cAMP accumulation (Fig. 2). In seeking to understand the differences observed between these developmental stages, the authors found that inhibiting PDE activity in unstimulated cells produced modest cAMP gradients in both DIV5 and DIV3 cells, suggesting that PDEs play a major role in shaping developmental-specific cAMP gradients. Furthermore, they were also able to observe cAMP gradient formation in DIV3 cells treated with peptides that disrupt interactions between the cAMP-dependent protein kinase (PKA), a major downstream effector of cAMP, and A-kinase anchoring proteins (AKAPs), which play a key role in compartmentalizing cAMP/PKA signaling [30]. Their results ultimately revealed that PKA anchoring via AKAPs is a

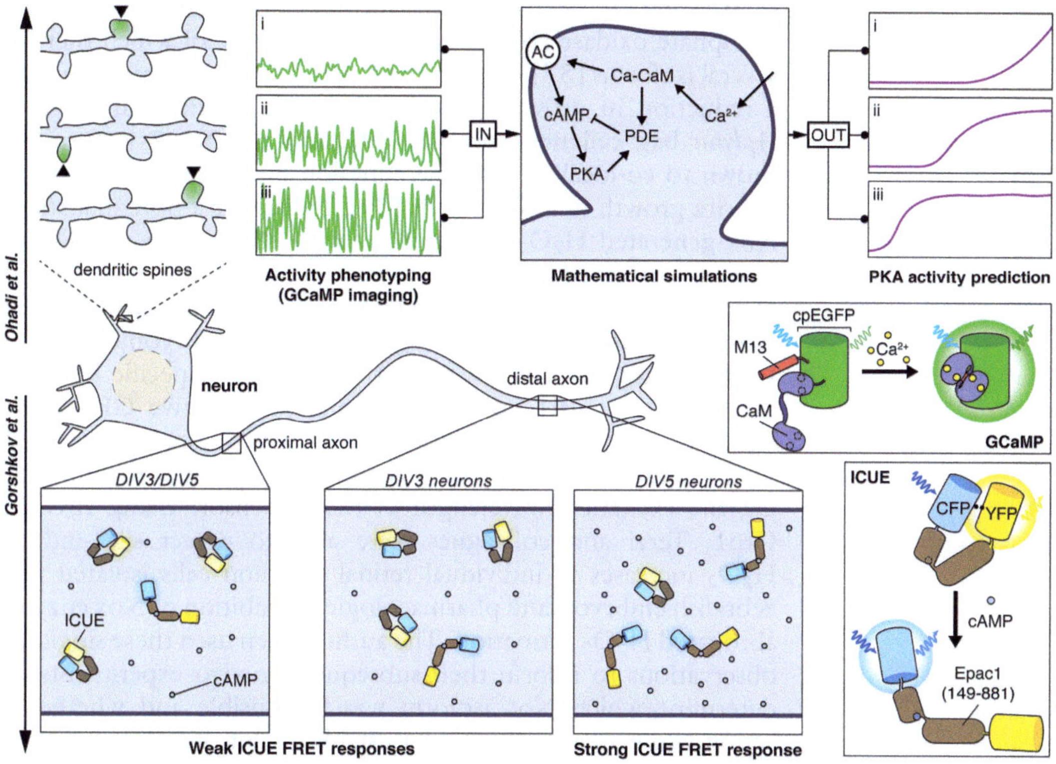

Fig. 2 Illuminating biochemical diversity in cultured neurons. (Upper) In the study by Ohadi et al., the authors utilized the single-FP Ca^{2+} sensor GCaMP to visualize spontaneous Ca^{2+} oscillations within the dendritic spines of cultured rat hippocampal neurons. GCaMP consists of calmodulin (CaM) and the CaM-binding M13 peptide inserted into circularly permuted EGFP (cpEGFP), such that the binding of Ca^{2+} to CaM induces a conformational change that increases fluorescence intensity. The authors observed distinct GCaMP responses of varying amplitudes and frequencies in individual spines, and by using these dynamic Ca^{2+} recordings as input values for mathematical simulations, they were then able to predict how the cAMP/PKA signaling pathway functions to integrate synaptic inputs. (Lower) Gorshkov et al. similarly used cultured hippocampal neurons to study the compartmentalization of cAMP during axon development by employing the FRET-based sensor ICUE, in which a fragment of the cAMP-binding protein Epac1 is sandwiched between cyan fluorescent protein (CFP) and yellow fluorescent protein (YFP). The binding of cAMP induces a conformational change and alters the ICUE FRET signal. Using this sensor, the authors observed weak ICUE responses (i.e., cAMP accumulation) throughout hippocampal neurons cultured for 3 days in vitro (DIV3) versus increasingly strong ICUE responses along the developing axon in DIV5 neurons, revealing the existence of a developmentally timed cAMP gradient

pivotal component of a negative-feedback mechanism that suppresses cAMP gradient formation in DIV3 neurons to control the timing of neuronal development [27].

In another recent study of neuronal development, Terzi et al. investigated the role of reactive oxygen species (ROS), specifically H_2O_2, in axon guidance [31]. ROS such as H_2O_2 are now well established as second messengers in a variety of contexts, including nervous system development and function [32–36]. A major source of ROS and H_2O_2 production is nicotinamide adenine dinucleotide phosphate oxidase (NADPH oxidase, or Nox), of which there are several isoforms [37]. In a number of studies, Nox inhibition led to a reduction in growth cone motility and neurite outgrowth in *Aplysia* bag cell neurons [38, 39], and the Nox2 isoform was shown to co-localize with the adhesion molecule apCAM during neurite growth in *Aplysia* cells [39], indicating a potential role of Nox-generated H_2O_2 in neurite growth. To further establish the role of Nox and H_2O_2 in an in vivo model of neuronal development, Terzi et al. investigated connection establishment between retinal ganglion cells and the optic tectum in embryonic zebrafish. In this study, the authors turned to the H_2O_2-specific biosensor roGFP2-Orp1 [40], composed of a redox-sensitive GFP variant linked to the yeast peroxiredoxin Orp1 which, after reaction with H_2O_2, undergoes a disulfide exchange with roGFP2, thereby altering the excitation wavelength of the biosensor. Using roGFP2-Orp1, Terzi and colleagues were able to detect slit2-induced H_2O_2 increases in individual retinal ganglion cells isolated from zebrafish embryos, and pharmacological inhibition of Nox enzymes abrogated H_2O_2 formation. The authors then used these single-cell observations to inform their subsequent in vivo experiments. To determine which Nox isoform was responsible and whether its action was necessary for proper establishment of retinotectal connections between retinal ganglion cells and the optic tectum, Terzi and colleagues developed specific $Nox^{-/-}$ KO zebrafish and found that Nox2 is necessary for chemotropic growth cone guidance responses.

3.1.2 Exploring the Mechanisms of Synaptic Plasticity

Dendritic spines, typically composed of a bulbous head with a neck connected to dendrites, are a major focal point of signaling-dependent synaptic plasticity, learning, and memory [41–43]. PKA and cAMP are also well known to play important roles in long-term potentiation (LTP), memory formation, consolidation, and retrieval [44–47]. As described above, PKA compartmentalization and gradients are a key feature in regulating cAMP/PKA to induce proper responses [82]. Despite their known importance, it has been difficult to measure spatiotemporal signaling dynamics in spines due to their relatively small size. Recently, Tang et al. developed modified biosensors for PKA and extracellular signal-regulated kinase (ERK), both known to play critical roles in plasticity [44–49], targeted to dendritic spines [50]. In this report, the authors successfully measured ERK and PKA activities in spines and dendrites in response to glutamate uncaging near individual spines and observed differences in the onset and duration of these two kinases, along with increases in dendritic spine size [50].

In another recent study underscoring how biosensors are a vital tool in cell dynamic modeling, Ohadi et al. sought to determine how Ca^{2+} oscillations, which are necessary for specific physiological roles [51–54], affected cAMP/PKA dynamics in dendritic spines. To investigate this relationship, Ohadi and colleagues performed cellular modeling in conjunction with Ca^{2+} imaging using GCaMP6f, an optimized member of the GCaMP sensor family [55], expressed in the spines of cultured rat hippocampal neurons [56]. GCaMPs contain the Ca^{2+}-effector calmodulin (CaM) and the CaM-binding M13 peptide inserted into circularly permuted GFP (cpGFP), where Ca^{2+} binding greatly increases GFP fluorescence intensity (Fig. 2). Interestingly, Ohadi et al. found that spines from different neurons exhibited a range in Ca^{2+} concentrations and spiking behaviors. Using their measurements of Ca^{2+} oscillations to inform their simulations, the authors were able to predict the cAMP and PKA dynamics induced by these heterogeneous Ca^{2+} patterns (Fig. 2). They found the frequency and amplitude of Ca^{2+} oscillations in the spines led to drastic differences in the onset of cAMP and PKA activation. Specifically, low Ca^{2+} concentrations and oscillations led to a drastically delayed onset of cAMP formation and PKA activity that never reached a steady state after several minutes. However, moderate Ca^{2+} concentrations and oscillations resulted in much faster cAMP/PKA onset, which reached a steady state within 2 min, while rapid oscillations with high Ca^{2+} concentrations resulted in rapid cAMP/PKA activation that reached steady state within the first minute (Fig. 2). Thus, their study shows not only that cAMP/PKA is a leaky integrator of Ca^{2+} dynamics in dendritic spines but that subcellular domains, such as dendritic spines, play a role in cellular heterogeneity [56].

Another central kinase implicated in synaptic plasticity is protein kinase C (PKC) [57, 58]. PKC has a relatively large number of isozymes [59], and it has been historically difficult to dissect which enzymes are activated in dendritic spines. Furthermore, while PKC has been implicated in integrating multiple signals across time and space, it has not been possible to accurately measure such dynamics. To address this issue, Colgan et al. developed and used a set of isozyme-specific PKC biosensors to directly visualize PKC activation and investigate how PKC integrates signals for the refinement of neuronal development and learning [60]. These FRET-based sensors contain a full-length PKC isozyme fused to GFP, which is paired with the red-emitting FP mCherry either tethered to the plasma membrane or fused to a pseudosubstrate peptide. By using these isozyme-specific biosensors, Colgan and colleagues were able to track PKC activation in individual spines and demonstrate, both in primary cells and in hippocampal slices, that synaptic plasticity was reliant only on PKCα. In another example of how cell-based biosensor imaging can be used to directly inform in vivo experiments, the authors then used PKCα KO mice to show that these mice have behavioral phenotypes associated with impairment in spatial learning and memory.

3.2 Activity Phenotyping Inspiepr146 Vivo

Even though cell-based studies using biosensors have provided many insights into neuronal function, numerous key processes simply cannot be mimicked in vitro, such as the recruitment and interaction of different neuronal circuits during behavior. Thus, the ultimate goal of neuronal imaging is to perform single-cell phenotyping and elucidate the dynamics of neuronal signaling directly in live, behaving animals. However, the use of genetically encodable biosensors in vivo is challenging. One of the major hurdles to in vivo imaging relates to tissue penetration of visible light. The excitation and emission wavelengths of most commonly used FPs are readily scattered and absorbed by living tissue, thus making the relatively modest fluorescence signals of most biosensors even harder to detect over noise. Furthermore, many of the biochemical changes associated with physiological processes are very subtle and can happen on extremely fast timescales. In vivo imaging thus requires powerful biosensors with very high dynamic ranges and S/N ratios to overcome these challenges. Recent advances over the past few years have enabled considerable progress in achieving in vivo imaging using biosensors, and we discuss select examples below.

3.2.1 Neuromodulator Imaging

In the brain, billions of neurons communicate through trillions of synapses, where neurotransmitters are released to excite or inhibit specific neurons and neuromodulators are secreted throughout the brain to modify neuronal activity. Understanding how neuronal firing and neurotransmitter release are coordinated in space and time to influence brain pathways and govern complex behavior remains a fundamental goal in neuroscience to elucidate the relationships that underlie brain function and behaviors. Addressing this challenge requires the ability to capture the in vivo dynamics of neurotransmitters with high spatiotemporal resolution, sensitivity, and selectivity. Recently, there has been immense progress in the development of genetically encodable FP-based biosensors for neurotransmitters. In contrast to analytical chemistry techniques, these biosensors offer high temporal precision and single-cell resolution. For example, Marvin et al. [61] developed iGluSnFR, a single-FP sensor for glutamate (Glu), which contains of a portion of the ABC transporter complex for glutamate and aspartate (GltI) inserted into cpGFP, that exhibits a high affinity and dynamic range for Glu binding to allow rapid visualization of Glu release at frequencies up to ~10 Hz (10 times per second). To resolve Glu dynamics at a higher frequency, Helassa et al. [62] generated variants of iGluSnFR that have reduced Glu affinity by mutating residues in the GluBP binding pocket. The resulting fast (iGluf) and ultrafast (iGluu) variants have comparable brightness to iGluSnFR, and their decreased Glu affinity enables more rapid glutamate dissociation for rapid and sensitive imaging of synaptic Glu release events at 100 Hz in individual boutons. The generalized design for iGluSnFR has

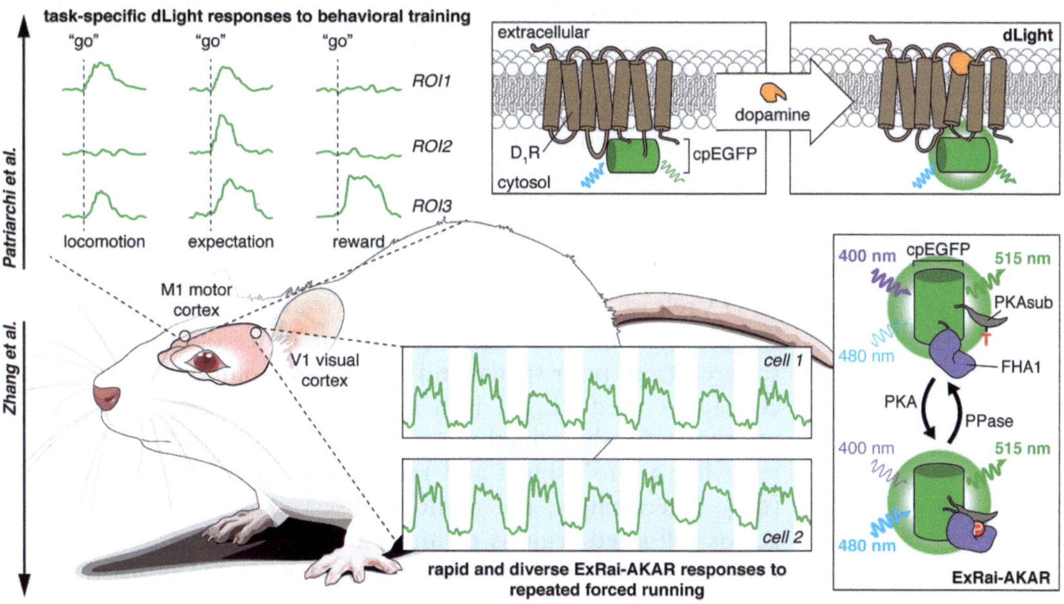

Fig. 3 Genetically encodable biosensors reveal biochemical heterogeneity in vivo. (Upper) In Patriarchi et al., the authors developed a genetically encoded dopamine sensor, dLight, in which circularly permuted EGFP (cpEGFP) is inserted into the third intracellular loop of the G protein-coupled dopamine 1 receptor (D1R), such that the dopamine-induced conformational change in the receptor will lead to increase GFP fluorescence intensity. As part of their study, the authors utilized dLight to visualize dopamine release dynamics within the primary motor cortex (M1) of mice that were trained to run in response to an auditory "go" signal in order to receive a reward. In vivo dLight imaging reveals functionally heterogeneous responses from cell-sized regions of interest (ROIs) that were distinctly activated by running ("locomotion"), reward expectation ("expectation"), or reward receipt ("reward"). (Lower) Recently, Zhang et al. were able to perform rapid and sensitive imaging of protein kinase A (PKA) activity in the brains of awake mice using a new, ultrasensitive, excitation-ratiometric PKA sensor (ExRai-AKAR). This sensor contains a PKA-specific substrate peptide and FHA1 phosphoamino acid-binding domain inserted into cpEGFP. The phosphorylation-triggered interaction of the substate and FHA1 domains induces a conformational change that causes the maximum excitation wavelength of cpEGFP to shift from ~400 nm to ~480 nm, while the emission wavelength remains ~515 nm. The high sensitivity and S/N ratio of ExRai-AKAR allowed the authors to detect subtle PKA activity changes in the primary visual cortex (V1) of mice subjected to repeated periods of mild force running, revealing heterogeneous responses between individual neurons and among running episodes in the same neuron

also been adapted for a variety of other neurotransmitters and analytes, including GABA [63], ATP [64], nicotine [65], and most recently serotonin [66].

In contrast to the above sensors, which are based on various bacterial periplasmic binding proteins, the other major class of neurotransmitter and neuromodulator biosensors uses G protein-coupled receptors (GPCRs) as the sensing unit, wherein the binding of neuromodulators such as dopamine (DA) [20, 21, 67, 68] or acetylcholine (ACh) [69, 70] induces a conformation change in the cognate GPCR (Fig. 3). DA in particular plays an important role in motivation, movement, cognition, and reward-driven learning, and two series of DA biosensors have been developed to date: dLight [67] and GPCR activation-based DA ($GRAB_{DA}$) sensors [68]. Patriarchi et al. generated and optimized variants of the dLight probe family (dLight1.1–1.5) which exhibit low nanomolar to micromolar affinity for DA. Using dLight, Patriarchi et al. were able to observe DA release changes with in vitro pharmacological manipulation and electrophysiological and optogenetic stimulation and to co-image DA release with Ca^{2+} imaging. Of particular interest, however, is how dLight enabled the rapid in vivo imaging of DA release dynamics in mice. Previous tools, such as microdialysis, electrochemical microsensors, or other optical approaches, suffered from poor spatial and temporal resolution. Patriarchi and colleagues used dLight in a stimulus-behavior-reward model in which previously trained mice would run in a fixed position upon receiving a "go" signal and receive a drop of water as a reward [67]. Interestingly, the authors observed brief increases in DA across cell-sized regions of interest (ROIs) in their field of view, indicating spatially distinct and heterogeneous DA transients at the cellular level (Fig. 3). Furthermore, they were able to distinguish reward responses between layer 1 and layer 2/3 of the M1 region of the motor cortex, with layer 2/3 showing more active cell-sized ROIs during the water reward than layer 1, indicating DA projections are spatially heterogeneous.

Meanwhile, Sun et al. developed $GRAB_{DA1m}$ and $GRAB_{DA1h}$, which have moderate (~130 nM) and high (~10 nM) DA binding affinity, respectively. The authors then sought to observe DA release during Pavlovian conditioning by expressing these $GRAB_{DA}$ sensors in the nucleus accumbens (NAc) in naïve mice, which were then trained with either a reward (water drop) or punishment (brief puff of air to the face) in response to an auditory cue. Sun and colleagues observed no DA release in response to the auditory cues in naïve mice, but a strong positive response to water and a moderate negative response to the puff of air. However, once the mice learned to correlate the cue with the reward or punishment, the positive or negative DA response increased for the auditory cue while the DA response for the actual reward diminished. These consistent responses led Sun and colleagues to investigate DA release in naturally rewarding social behaviors. The authors introduced sexually receptive female mice into the home cages of NAc $GRAB_{DA1h}$-expressing male mice. They observed small increases in $GRAB_{DA1h}$ fluorescence in some males during investigation of the female and consistent fluorescence increases upon mounting and intromission, with the peak occurring during ejaculation [68]. This ability to directly observe spatiotemporally precise DA dynamics in live animals during naturally rewarding social behaviors greatly expands the field's ability to quantitatively measure neurotransmitter release in vivo.

These DA biosensors have subsequently been used to examine how DA supports brain function. In a recent study, Mohebi et al. [71] shed light on how DA release from dopaminergic neurons that project from the ventral tegmental area (VTA) to the nucleus accumbens (NAc), one of the four major dopaminergic pathways, is regulated in the brains of rats performing a decision-making task. In this pathway, the long axons of the VTA dopaminergic neurons extend to other parts of the brain, including the NAc, the dorsal striatum, and the prefrontal cortex, where the activation of these VTA neurons generates electrical signals (neuronal firing) that cause local DA release at these axon terminals. To dissect how DA release is regulated to achieve two distinct functions (i.e., learning and motivation), the authors measured DA release in the VTA using single-cell optrode recording together with measurements of DA release in the NAc core using multiple approaches, including dLight imaging, during the same task in freely moving rats [71]. In this test, Mohebi and colleagues observed elevated DA release only in the NAc and regions of the prefrontal cortex, and not in the dorsal striatum. Furthermore, the authors observed dLight fluorescence, and thus DA release, to ramp up when the rate of reward increases, which motivated rats to participate in the task. However, neuronal activity in the VTA of rats was not affected, indicating that motivation-coding DA release is independent of the firing activity of VTA DA neurons. By contrast, both VTA DA neuron spiking activity and DA release were linked to reward availability and the reward itself, suggesting that DA release associated with DA neuron firing encodes reward prediction errors (RPEs) to promote learning. Because the dLight probe enables the detection of highly dynamic fluctuations in DA release on subsecond timescales, which correlates with both state values and RPEs, it suggests that DA release coupled with corresponding changes in the firing rates of DA neurons serves as a global "broadcast" to facilitate learning, whereas local DA release without DA neuron firing drives motivation [71].

3.2.2 Enzymatic Activities

While neurotransmitters and neuromodulators are a core component of neuronal signaling, they do not act alone to transduce signals into a cellular response. Necessary components of these pathways include enzymes such as kinases and phosphatases, which relay critical information throughout the cell, transducing the signal into a response. As demonstrated earlier, understanding how these enzymes are regulated in space and time is key to elucidating neuronal signaling and function; however, measuring enzymatic activities in vivo has proven difficult; the majority of kinase biosensors are FRET-based sensors with too low dynamic range or sensitivity for in vivo imaging. Although some technical advances have enabling in vivo FRET-based imaging [72–75], these are hampered by limitations in temporal resolution or reliance on spatial averaging that masks single-cell responses. However, the recent expansion of single-FP sensors shows great promise for adapting this design to generate enzyme sensors with high dynamic range and sensitivity for robust in vivo enzyme activity imaging.

As we mentioned briefly above (*see* Subheading 2.2), we previously succeeded in developing a family of single-FP, excitation-ratiometric biosensors for monitoring the activity of kinases such as PKA [19]. These ExRai-KARs are constructed by inserting a kinase-specific substrate peptide and phosphoamino acid-binding domain such as FHA1 [76] into cpGFP (Fig. 3). More recently, we were able to engineer an enhanced, second-generation excitation-ratiometric A-kinase activity reporter (ExRai-AKAR2) that enabled us to visualize dynamic PKA responses in individual L2/L3 neurons in the primary visual cortex (V1) in living mice undergoing mild forced running [23], which induces noradrenergic and cholinergic activation [77–81]. Using the high S/N ratio and sensitivity of ExRai-AKAR2, we were able to acquire PKA activity measurements at 5.83 Hz, >600-fold faster than the existing state-of-the-art approaches, while maintaining single-cell resolution, revealing rapid and diverse PKA dynamics induced by neuromodulatory signaling [23] (Fig. 3). The availability of powerful, single-FP biosensors to visualize kinase activities in individual neurons on a millisecond timescale in living mice opens the door for future studies of heterogeneous, cell-specific enzyme activity dynamics among diverse neuronal populations to determine the molecular mechanisms of neuronal signaling in vivo.

Given their high dynamic ranges and optimized performance, the newly developed single-FP-based biosensors for PKA, DA, Glu, ACh, and GABA are increasingly enabling the quantification of neuronal signaling dynamics in the brains of freely behaving animals, a crucial step toward understanding the relationship between neuronal signaling pathways and brain functions.

4 Concluding Remarks

Genetically encodable biosensors are a key part of our molecular toolkit for determining the biochemical activity architecture of individual cells, as they permit dynamic measurements of endogenous processes such as signaling activities and nutrient availabilities in live cells in real time, thus linking the spatiotemporal characteristics of these processes to cellular functions. Although the use of biosensors is not without caveats, such as sufficient dynamic range, sensitivity, or S/N ratio, biosensor development is constantly improving to address these issues. However, improvements in biosensor performance (e.g., sensitivity and specificity), expanding the biosensor toolkit to simultaneously detect multiple signaling activities [20], and advanced microscopy techniques for in vivo functional brain imaging will further enable us to delineate how the heterogeneity of neuronal signaling in individual cells correlates with complex behavioral outputs.

In this review, we have covered ways in which genetically encodable biosensors can be used for single-cell phenotyping and provided a number of examples of both intracellular and synaptic signaling events that have been monitored using such biosensors. As genetically encodable biosensors improve, our ability to connect the biochemical patterns of individual cells to cellular outcomes and even organismal behaviors will dramatically increase.

Acknowledgment

Work in the Zhang laboratory is supported by the National Institutes of Health (R35 CA197622, R01 DK073368, and R01 DE030497 and R01 HL162302).

References

1. Ryu H, Chung M, Dobrzyński M et al (2015) Frequency modulation of ERK activation dynamics rewires cell fate. Mol Syst Biol 11: 838. https://doi.org/10.15252/msb.20156458
2. Laporte D, Gouleme L, Jimenez L et al (2018) Mitochondria reorganization upon proliferation arrest predicts individual yeast cell fate. elife 7:e35685. https://doi.org/10.7554/elife.35685
3. Wood NE, Kositangool P, Hariri H et al (2020) Nutrient signaling, stress response, and inter-organelle communication are non-canonical determinants of cell fate. Cell Rep 33: 108446. https://doi.org/10.1016/j.celrep.2020.108446
4. Mehta S, Zhang J (2017) Illuminating the cell's biochemical activity architecture. Biochemistry 56:5210–5213. https://doi.org/10.1021/acs.biochem.7b00561
5. Greenwald EC, Mehta S, Zhang J (2018) Genetically encoded fluorescent biosensors illuminate the spatiotemporal regulation of signaling networks. Chem Rev 118: 11707–11794. https://doi.org/10.1021/acs.chemrev.8b00333
6. Rodriguez EA, Campbell RE, Lin JY et al (2017) The growing and glowing toolbox of fluorescent and photoactive proteins. Trends Biochem Sci 42:111–129. https://doi.org/10.1016/j.tibs.2016.09.010
7. Tsien RY (1989) Fluorescent probes of cell signaling. Annu Rev Neurosci 12:227–253. https://doi.org/10.1146/annurev.ne.12.030189.001303
8. Grynkiewicz G, Poenie M, Tsien RY (1985) A new generation of Ca2+ indicators with greatly improved fluorescence properties. J Biol Chem 260:3440–3450
9. Hempel SL, Buettner GR, O'Malley YQ et al (1999) Dihydrofluorescein diacetate is superior for detecting intracellular oxidants: comparison with 2′,7′-dichlorodihydrofluorescein diacetate, 5(and 6)-carboxy-2′,7-′-dichlorodihydrofluorescein diacetate, and dihydrorhodamine 123. Free Radical Bio Med 27:146–159. https://doi.org/10.1016/s0891-5849(99)00061-1
10. Miller EW, Tulyathan O, Tulyanthan O et al (2007) Molecular imaging of hydrogen peroxide produced for cell signaling. Nat Chem Biol 3:263–267. https://doi.org/10.1038/nchembio871
11. Miyawaki A, Llopis J, Heim R et al (1997) Fluorescent indicators for Ca2+ based on green fluorescent proteins and calmodulin. Nature 388:882–887. https://doi.org/10.1038/42264
12. Miyawaki A, Griesbeck O, Heim R, Tsien RY (1999) Dynamic and quantitative Ca2+ measurements using improved cameleons. Proc

Natl Acad Sci 96:2135–2140. https://doi.org/10.1073/pnas.96.5.2135

13. Jares-Erijman EA, Jovin TM (2003) FRET imaging. Nat Biotechnol 21:1387–1395. https://doi.org/10.1038/nbt896

14. Cardullo RA (2013) Theoretical principles and practical considerations for fluorescence resonance energy transfer microscopy. Methods Cell Biol 114:441–456. https://doi.org/10.1016/b978-0-12-407761-4.00019-1

15. Hanson GT, McAnaney TB, Park ES et al (2002) Green fluorescent protein variants as Ratiometric dual emission pH sensors. 1. Structural characterization and preliminary application †. Biochemistry 41:15477–15488. https://doi.org/10.1021/bi026609p

16. Shcherbakova DM, Cammer NC, Huisman TM et al (2018) Direct multiplex imaging and optogenetics of rho GTPases enabled by near-infrared FRET. Nat Chem Biol 14:591–600. https://doi.org/10.1038/s41589-018-0044-1

17. Mo GCH, Posner C, Rodriguez EA et al (2020) A rationally enhanced red fluorescent protein expands the utility of FRET biosensors. Nat Commun 11:1848. https://doi.org/10.1038/s41467-020-15687-x

18. Akerboom J, Calderón NC, Tian L et al (2013) Genetically encoded calcium indicators for multi-color neural activity imaging and combination with optogenetics. Front Mol Neurosci 6:2. https://doi.org/10.3389/fnmol.2013.00002

19. Mehta S, Zhang Y, Roth RH et al (2018) Single-fluorophore biosensors for sensitive and multiplexed detection of signalling activities. Nat Cell Biol 20:1215–1225. https://doi.org/10.1038/s41556-018-0200-6

20. Patriarchi T, Mohebi A, Sun J et al (2020) An expanded palette of dopamine sensors for multiplex imaging in vivo. Nat Methods 17:1147–1155. https://doi.org/10.1038/s41592-020-0936-3

21. Sun F, Zhou J, Dai B et al (2020) Next-generation GRAB sensors for monitoring dopaminergic activity in vivo. Nat Methods 17:1156–1166. https://doi.org/10.1038/s41592-020-00981-9

22. Zhao Y, Araki S, Wu J et al (2011) An expanded palette of genetically encoded Ca2+ indicators. Science 333:1888–1891. https://doi.org/10.1126/science.1208592

23. Zhang J-F, Liu B, Hong I et al (2021) An ultrasensitive biosensor for high-resolution kinase activity imaging in awake mice. Nat Chem Biol 17:39–46. https://doi.org/10.1038/s41589-020-00660-y

24. Barnes AP, Polleux F (2009) Establishment of axon-dendrite polarity in developing neurons. Annu Rev Neurosci 32:347–381. https://doi.org/10.1146/annurev.neuro.31.060407.125536

25. Shelly M, Lim BK, Cancedda L et al (2010) Local and long-range reciprocal regulation of cAMP and cGMP in axon/dendrite formation. Science 327:547–552. https://doi.org/10.1126/science.1179735

26. Pasterkamp RJ (2012) Getting neural circuits into shape with semaphorins. Nat Rev Neurosci 13:605–618. https://doi.org/10.1038/nrn3302

27. Gorshkov K, Mehta S, Ramamurthy S et al (2017) AKAP-mediated feedback control of cAMP gradients in developing hippocampal neurons. Nat Chem Biol 13:425–431. https://doi.org/10.1038/nchembio.2298

28. Violin JD, DiPilato LM, Yildirim N et al (2008) β2-adrenergic receptor signaling and desensitization elucidated by quantitative modeling of real time cAMP dynamics. J Biol Chem 283:2949–2961. https://doi.org/10.1074/jbc.m707009200

29. DiPilato LM, Zhang J (2009) The role of membrane microdomains in shaping β2-adrenergic receptor -mediated cAMP dynamics. Mol BioSyst 5:832–837. https://doi.org/10.1039/b823243a

30. Wong W, Scott JD (2004) AKAP signalling complexes: focal points in space and time. Nat Rev Mol Cell Bio 5:959–970. https://doi.org/10.1038/nrm1527

31. Terzi A, Roeder H, Weaver CJ, Suter DM (2020) Neuronal NADPH oxidase 2 regulates growth cone guidance downstream of slit2/robo2. Dev Neurobiol. https://doi.org/10.1002/dneu.22791

32. Halliwell B (1992) Reactive oxygen species and the central nervous system. J Neurochem 59:1609–1623. https://doi.org/10.1111/j.1471-4159.1992.tb10990.x

33. Acker T, Acker H (2004) Cellular oxygen sensing need in CNS function: physiological and pathological implications. J Exp Biol 207:3171–3188. https://doi.org/10.1242/jeb.01075

34. Finkel T (2011) Signal transduction by reactive oxygen species. J Cell Biol 194:7–15. https://doi.org/10.1083/jcb.201102095

35. Reczek CR, Chandel NS (2015) ROS-dependent signal transduction. Curr Opin Cell Biol 33:8–13. https://doi.org/10.1016/j.ceb.2014.09.010

36. Bórquez DA, Urrutia PJ, Wilson C et al (2016) Dissecting the role of redox signaling in

neuronal development. J Neurochem 137: 506–517. https://doi.org/10.1111/jnc.13581

37. Bedard K, Krause K-H (2007) The NOX Family of ROS-generating NADPH oxidases: physiology and pathophysiology. Physiol Rev 87: 245–313. https://doi.org/10.1152/physrev.00044.2005
38. Munnamalai V, Suter DM (2009) Reactive oxygen species regulate F-actin dynamics in neuronal growth cones and neurite outgrowth. J Neurochem 108:644–661. https://doi.org/10.1111/j.1471-4159.2008.05787.x
39. Munnamalai V, Weaver CJ, Weisheit CE et al (2014) Bidirectional interactions between NOX2-type NADPH oxidase and the F-actin cytoskeleton in neuronal growth cones. J Neurochem 130:526–540. https://doi.org/10.1111/jnc.12734
40. Gutscher M, Sobotta MC, Wabnitz GH et al (2009) Proximity-based protein thiol oxidation by H2O2-scavenging peroxidases. J Biol Chem 284:31532–31540. https://doi.org/10.1074/jbc.m109.059246
41. Arellano JI, Benavides-Piccione R, DeFelipe J, Yuste R (2007) Ultrastructure of dendritic spines: correlation between synaptic and spine morphologies. Front Neurosci 1:131–143. https://doi.org/10.3389/neuro.01.1.1.010.2007
42. Bourne JN, Harris KM (2008) Balancing structure and function at hippocampal dendritic spines. Annu Rev Neurosci 31:47–67. https://doi.org/10.1146/annurev.neuro.31.060407.125646
43. Nägerl UV, Willig KI, Hein B et al (2008) Live-cell imaging of dendritic spines by STED microscopy. Proc Natl Acad Sci 105: 18982–18987. https://doi.org/10.1073/pnas.0810028105
44. Yagishita S, Hayashi-Takagi A, Ellis-Davies GCR et al (2014) A critical time window for dopamine actions on the structural plasticity of dendritic spines. Science 345:1616–1620. https://doi.org/10.1126/science.1255514
45. Diering GH, Gustina AS, Huganir RL (2014) PKA-GluA1 coupling via AKAP5 controls AMPA receptor phosphorylation and cell-surface targeting during bidirectional homeostatic plasticity. Neuron 84:790–805. https://doi.org/10.1016/j.neuron.2014.09.024
46. Li S, Jin M, Zhang D et al (2013) Environmental novelty activates β2-adrenergic signaling to prevent the impairment of hippocampal LTP by Aβ oligomers. Neuron 77:929–941. https://doi.org/10.1016/j.neuron.2012.12.040
47. Hell JW (2016) How Ca^{2+}-permeable AMPA receptors, the kinase PKA, and the phosphatase PP2B are intertwined in synaptic LTP and LTD. Sci Signal 9:pe2–pe2. https://doi.org/10.1126/scisignal.aaf7067
48. Kelleher RJ, Govindarajan A, Jung H-Y et al (2004) Translational control by MAPK signaling in long-term synaptic plasticity and memory. Cell 116:467–479. https://doi.org/10.1016/s0092-8674(04)00115-1
49. Thomas GM, Huganir RL (2004) MAPK cascade signalling and synaptic plasticity. Nat Rev Neurosci 5:173–183. https://doi.org/10.1038/nrn1346
50. Tang S, Yasuda R (2017) Imaging ERK and PKA activation in single dendritic spines during structural plasticity. Neuron 93:1315–1324.e3. https://doi.org/10.1016/j.neuron.2017.02.032
51. Uhlén P, Fritz N (2010) Biochemistry of calcium oscillations. Biochem Bioph Res Co 396: 28–32. https://doi.org/10.1016/j.bbrc.2010.02.117
52. Dupont G, Combettes L, Bird GS, Putney JW (2011) Calcium oscillations. Csh Perspect Biol 3:a004226. https://doi.org/10.1101/cshperspect.a004226
53. Smedler E, Uhlén P (2014) Frequency decoding of calcium oscillations. Biochimica Et Biophysica Acta Bba – Gen Subj 1840:964–969. https://doi.org/10.1016/j.bbagen.2013.11.015
54. Sneyd J, Han JM, Wang L et al (2017) On the dynamical structure of calcium oscillations. Proc Natl Acad Sci 114:1456–1461. https://doi.org/10.1073/pnas.1614613114
55. Chen T-W, Wardill TJ, Sun Y et al (2013) Ultrasensitive fluorescent proteins for imaging neuronal activity. Nature 499:295–300. https://doi.org/10.1038/nature12354
56. Ohadi D, Schmitt DL, Calabrese B et al (2019) Computational modeling reveals frequency modulation of calcium-cAMP/PKA pathway in dendritic spines. Biophys J 117: 1963–1980. https://doi.org/10.1016/j.bpj.2019.10.003
57. Hu G-Y, Hvalby Ø, Walaas SI et al (1987) Protein kinase C injection into hippocampal pyramidal cells elicits features of long term potentiation. Nature 328:426–429. https://doi.org/10.1038/328426a0
58. Hongpaisan J, Alkon DL (2007) A structural basis for enhancement of long-term associative memory in single dendritic spines regulated by PKC. Proc Natl Acad Sci 104:19571–19576. https://doi.org/10.1073/pnas.0709311104

59. Dempsey EC, Newton AC, Mochly-Rosen D et al (2000) Protein kinase C isozymes and the regulation of diverse cell responses. Am J Physiol Lung Cell Mol Physiol 279: L429–L438. https://doi.org/10.1152/ajplung.2000.279.3.l429
60. Colgan LA, Hu M, Misler JA et al (2018) PKCα integrates spatiotemporally distinct Ca2+ and autocrine BDNF signaling to facilitate synaptic plasticity. Nat Neurosci 21: 1027–1037. https://doi.org/10.1038/s41593-018-0184-3
61. Marvin JS, Borghuis BG, Tian L et al (2013) An optimized fluorescent probe for visualizing glutamate neurotransmission. Nat Methods 10:162–170. https://doi.org/10.1038/nmeth.2333
62. Helassa N, Dürst CD, Coates C et al (2018) Ultrafast glutamate sensors resolve high-frequency release at Schaffer collateral synapses. Proc Natl Acad Sci 115:5594–5599. https://doi.org/10.1073/pnas.1720648115
63. Marvin JS, Shimoda Y, Magloire V et al (2019) A genetically encoded fluorescent sensor for in vivo imaging of GABA. Nat Methods 16: 763–770. https://doi.org/10.1038/s41592-019-0471-2
64. Lobas MA, Tao R, Nagai J et al (2019) A genetically encoded single-wavelength sensor for imaging cytosolic and cell surface ATP. Nat Commun 10:711. https://doi.org/10.1038/s41467-019-08441-5
65. Shivange AV, Borden PM, Muthusamy AK et al (2019) Determining the pharmacokinetics of nicotinic drugs in the endoplasmic reticulum using biosensors. J Gen Physiol 151:738–757. https://doi.org/10.1085/jgp.201812201
66. Unger EK, Keller JP, Altermatt M et al (2020) Directed evolution of a selective and sensitive serotonin sensor via machine learning. Cell 183:1986–2002.e26. https://doi.org/10.1016/j.cell.2020.11.040
67. Patriarchi T, Cho JR, Merten K et al (2018) Ultrafast neuronal imaging of dopamine dynamics with designed genetically encoded sensors. Science 360:eaat4422. https://doi.org/10.1126/science.aat4422
68. Sun F, Zeng J, Jing M et al (2018) A genetically encoded fluorescent sensor enables rapid and specific detection of dopamine in flies, fish, and mice. Cell 174:481–496.e19. https://doi.org/10.1016/j.cell.2018.06.042
69. Jing M, Zhang P, Wang G et al (2018) A genetically encoded fluorescent acetylcholine indicator for in vitro and in vivo studies. Nat Biotechnol 36:726–737. https://doi.org/10.1038/nbt.4184
70. Jing M, Li Y, Zeng J et al (2020) An optimized acetylcholine sensor for monitoring in vivo cholinergic activity. Nat Methods 17: 1139–1146. https://doi.org/10.1038/s41592-020-0953-2
71. Mohebi A, Pettibone JR, Hamid AA et al (2019) Dissociable dopamine dynamics for learning and motivation. Nature 570:65–70. https://doi.org/10.1038/s41586-019-1235-y
72. Chen Y, Saulnier JL, Yellen G, Sabatini BL (2014) A PKA activity sensor for quantitative analysis of endogenous GPCR signaling via 2-photon FRET-FLIM imaging. Front Pharmacol 5:56. https://doi.org/10.3389/fphar.2014.00056
73. Ma L, Jongbloets BC, Xiong W-H et al (2018) A highly sensitive A-kinase activity reporter for imaging Neuromodulatory events in awake mice. Neuron 99:665–679.e5. https://doi.org/10.1016/j.neuron.2018.07.020
74. Lee SJ, Chen Y, Lodder B, Sabatini BL (2019) Monitoring behaviorally induced biochemical changes using fluorescence lifetime photometry. Front Neurosci 13:766. https://doi.org/10.3389/fnins.2019.00766
75. Lee SJ, Lodder B, Chen Y et al (2020) Cell-type-specific asynchronous modulation of PKA by dopamine in learning. Nature 590: 451–456. https://doi.org/10.1038/s41586-020-03050-5
76. Liao H, Yuan C, Su M-I et al (2000) Structure of the FHA1 domain of yeast Rad53 and identification of binding sites for both FHA1 and its target protein Rad9. J Mol Biol 304:941–951. https://doi.org/10.1006/jmbi.2000.4291
77. Polack P-O, Friedman J, Golshani P (2013) Cellular mechanisms of brain state–dependent gain modulation in visual cortex. Nat Neurosci 16:1331–1339. https://doi.org/10.1038/nn.3464
78. Lee AM, Hoy JL, Bonci A et al (2014) Identification of a brainstem circuit regulating visual cortical state in parallel with locomotion. Neuron 83:455–466. https://doi.org/10.1016/j.neuron.2014.06.031
79. Fu Y, Tucciarone JM, Espinosa JS et al (2014) A cortical circuit for gain control by behavioral state. Cell 156:1139–1152. https://doi.org/10.1016/j.cell.2014.01.050
80. Paukert M, Agarwal A, Cha J et al (2014) Norepinephrine controls Astroglial responsiveness to local circuit activity. Neuron 82:

1263–1270. https://doi.org/10.1016/j.neuron.2014.04.038

81. Reimer J, McGinley MJ, Liu Y et al (2016) Pupil fluctuations track rapid changes in adrenergic and cholinergic activity in cortex. Nat Commun 7:13289. https://doi.org/10.1038/ncomms13289

82. Omar, Torres-Quesada Johanna E., Mayrhofer Eduard, Stefan (2017) The many faces of compartmentalized PKA signalosomes. Cellular Signalling 371–311. https://doi.org/10.1016/j.cellsig.2017.05.012

INDEX

Jonathan V. Sweedler et al. (eds.), *Single Cell 'Omics of Neuronal Cells*,
Neuromethods, vol. 184, https://doi.org/10.1007/978-1-0716-2525-5,

MIX
Papier aus verantwortungsvollen Quellen
Paper from responsible sources
FSC® C105338

If you have any concerns about our products,
you can contact us on
ProductSafety@springernature.com

In case Publisher is established outside the EU,
the EU authorized representative is:
Springer Nature Customer Service Center GmbH
Europaplatz 3, 69115 Heidelberg, Germany

Printed by Libri Plureos GmbH
in Hamburg, Germany